STUDY GUIDE

JONES/CHILDERS

CONTEMPORARY COLLEGE
PHYSICS
2ND EDITION

STUDY GUIDE

JONES/CHILDERS

CONTEMPORARY COLLEGE
PHYSICS
2ND EDITION

JOHN SAFKO

THE UNIVERSITY OF SOUTH CAROLINA

Addison-Wesley Publishing Company

Reading, Massachusetts • Menlo Park, California • New York
Don Mills, Ontario • Wokingham, England • Amsterdam • Bonn
Sydney • Singapore • Tokyo • Madrid • San Juan • Milan • Paris

Reproduced by Addison-Wesley from camera-ready copy supplied by the author.

ISBN 0-201-62961-5

1 2 3 4 5 6 7 8 9 10-BA-95949392

Preface

This study guide is designed to accompany Edwin R. Jones and Richard L. Childers *Contemporary College Physics*, *2nd Ed*. The purpose of any study guide is to aid students in understanding physical concepts and problem solving techniques. Specifically, aid is provided through exercises that illustrate the concepts and worked examples of the application of these concepts. Physics teachers have found from experience that practice in solving problems is the only way that most students can truly understand the basic physical concepts.

Ideally each student should have a personal tutor and should have worked examples of all possible problems and all possible ways of posing problems. Because neither of these tactics is possible, a study guide is provided to aid students. The first edition of this study guide was developed during a two-year period while physics was offered via a television course. Draft copies of both the text and the study guide were then used by students. Revisions were made to both books, based on student response and success. The video lectures used were *The Mechanical Universe*© and *Beyond the Mechanical Universe*©. These tapes did not provide many examples of problem solving, and, as a result, the students relied on the study guide and the text provided as major sources of worked examples.

The structure of the study guide is as follows:

1. **Important Terms**

   The text gives a list of Important Terms at the beginning of each chapter. An understanding of the vocabulary is necessary for a discussion of the material included in this chapter. In the long run, you will find it a great help to use the exercises in this portion of the study guide. A lack of facility with the Important Terms used in each chapter often causes half the difficulty a student has with problem solving. The practice consists of three parts:

   a) Sentences with blank space(s) are provided for students to enter the Important Term that properly completes the sentence.

   b) A list of Important Terms requires students to enter the definition of each expression. The corresponding chapter and section number are also included to direct the student to "look up" the term in the textbook, if necessary.

   c) The answers to the sentence completion section are given.

2. **Sample Questions**

   These are samples of the thought questions at the end of each chapter. They can be answered with little or no calculation. Suggested answers are provided.

3. Problems with Sample Solutions

The Problems with Sample Solutions are the core of this study guide. They consist of fairly detailed solutions to problems that are similar to those at the end of each chapter. An attempt has been made to explain fully the symbols and laws used, to draw any relevant and useful figure, and to clarify each step as completely as possible. The highest section needed from the text material is indicated in parentheses at the end of the statement of each example question.

Most of the samples involve several steps and are designed to help the student to develop a fair understanding of the concepts needed for solving problems. The emphasis is on the quality of the examples given, rather than on the quantity. Thus these Sample Solutions are not just "putting the numbers into an equation." As a consequence of this philosophy, many of the Sample Solutions are for the complicated problems. It is hoped that providing these complicated problems with their solutions will enable the student to deal with other complicated problems.

If the student will take the effort to follow each step in these samples and to ask why each was taken, he or she will find problem solving less difficult. As with any study aid, students will get out of this material only what they are willing to put into it.

Several templates or alternative methods of solving the problems are often presented. These methods include:

a) Before starting any problem, always state a strategy that you intend to use.

b) Always make a sketch or a motion diagram of the posed problem.

c) Clearly identify the symbols used. State relevant physical laws and show substitutions.

d) Give the final result with units and appropriate significant figures.

4. Comments

Comment sections are included in only some chapters for two reasons:

a) There may be comments that expand upon ideas or sections of the text. Comments are included only if they seem to be necessary in solving some types of problems. These comments often represent the material that would be covered during normal classroom lectures. They are included as help for the nontraditional student and as reinforcement for the traditional student.

b) Comments are included to help clarify the real-world models that are implicit in the text discussion. Scientists simplify reality by building models that have symbols to represent objects in the real world. The ideas expressed in the Comments can be found more fully explained in articles by D. Hestenes and others. See, for example, *American Journal of Physics*, **55**, 440 (1987) and references therein.

Students are strongly urged to study the comments before attempting to solve any problems.

5. Suggested Readings

This is a short list of articles or books that consider topics from the chapter in more detail. Many of these readings consider material that the authors can only touch upon in the text. Many of these readings relate to interactions between the physics discussed and society.

I wish to thank those students who, in the two years of television courses, have provided input and suggestions, and who suffered from errors of omission as well as commission in the original material. Special thanks must also be extended to William M. Lee, Jr. for his efforts to eliminate errors from the final manuscript of the first edition. Any remaining errors are my responsibility. Students and instructors who have corrections or suggestions for improvements are urged to submit them to me.

John L. Safko
Columbia, SC
1992

STUDY GUIDE
for
CONTEMPORARY COLLEGE PHYSICS, 2nd Ed.
Text by E. R. Jones, Jr. and R. L. Childers

Study Guide by John L. Safko

Contents

Chapter 1
Measurement and Analysis

IMPORTANT TERMS

Fill in the blank with the appropriate word or words.

1. _____ law of planetary motion says that the orbits of the planets are ellipses with the sun at one focus.

2. The number of _____ tells us how well we know a given quantity.

3. The ratio of the arc length of a sector of a circle to the radius of that sector is a measure of the included angle of the sector in _____.

4. "An imaginary line from the sun to a moving planet will sweep out equal areas in equal times" is a statement of _____ law.

5. The meter is an example of a base or _____ unit.

6. Kepler's _____ law states that "the ratio of the square of a planet's period to the cube of its average distance from the sun is a constant."

7. All physical quantities have _____.

8. The second of time is an example of a(n) _____ or fundamental unit.

9. The orbits of the planets are _____.

10. The process of putting approximate numbers into relations to get rough estimates is called _____ calculation.

11. _____ notation is designed to show explicitly the number of significant figures known for a physical quantity.

Write the definitions of the following:

12. Kepler's third law (1.2):

13. significant figures (1.4):

14. base or fundamental unit (1.5):

15. dimension of a physical quantity (1.5):

16. the radian (1.1):

17. ellipse (1.2):

18. order of magnitude (1.7):

19. Kepler's first law (1.2):

20. scientific notation (1.2):

21. Kepler's second law (1.2):

ANSWERS TO 1–11, IMPORTANT TERMS

1. Kepler's first; 2. significant figures; 3. radians; 4. Kepler's second; 5. fundamental; 6. third; 7. dimension; 8. base; 9. ellipses; 10. order of magnitude; 11. Scientific.

A WORD ON SOLVING PHYSICS PROBLEMS

During this course, you will have to solve problems such as those you find in the text at the end of each chapter. Simply writing an equation, putting in numbers, and getting an answer is not how to solve a physics problem. It will be necessary for you to identify your model and explain the physical meaning of the variables used. You must also interpret the results for the posed physical problem. The sample solutions in each chapter of this study guide should provide a template for this method.

COMMENTS ON MAKING SKETCHES

As a general procedure, you will be well advised to make a sketch of each problem you have to solve. This sketch is a visual model of the problem; it does not need to be drawn to scale or even to be well drawn to be useful. Of course, the better the sketch is drawn, the more helpful it may be to you.

On the sketch you can note which directions are considered positive, what symbols are used for the directions, and the objects that are involved in the problem. If motion is involved, draw a path. When you solve the model and the path length turns out be be negative, that would mean that the path is in the direction opposite from what you assumed when you made the sketch. The sketch and numbers tell you this, so you can properly interpret the results and redraw the sketch if desired.

CONVERSION OF UNITS (1.5)

It will often be necessary to convert units when solving problems. This may be a conversion of units within a system of units or a conversion between two different systems. Examples of the former are millimeters to meters (SI) or inches to feet (English). An example of the latter is inches to meters (English to SI).

As a general policy, it is best to put all your initial information into the **base** or **fundamental units** of one consistent system. Usually the SI system is the better choice; although, in a few cases, problems may be posed so that the English system is preferable. This means, for example, that all distances should be expressed in meters; do not express some distances in meters, some in cm, and others in mm.

If you adopt the policy of consistent units, you have another check on your calculations; that is, the result will come out in an expected combination of units. If you are calculating a quantity that is a length, the answer will come out to be meters, not m/s or some other unit. You should always carry the units along with your calculations and work them out separately.

PRACTICE WITH UNIT CONVERSIONS (1.6)

Following the procedures outlined in Section 1.6 of the text, work out the following conversions. The answers are given below as a check. Some helpful factors are:

$$1 = \frac{2.54 \text{ cm}}{1 \text{ in.}}; \qquad 1 = \frac{3600 \text{ s}}{1 \text{ hr}} = \frac{60 \text{ s}}{1 \text{ min}} \times \frac{60 \text{ min}}{1 \text{ hr}}$$

All of these conversions are exact definitions; i.e., 1 inch is exactly equal to 2.54 cm, etc.

A. 1 yard = _____ m

B. 30 days = _____ s

C. 50 feet = _____ m

D. 3.5×10^{-3} mm = _____ m

E. $(3.0 \text{ inches})^3$ = _____ m^3

F. 1.0 ft^2 = _____ cm^2

Solutions:

A. $\quad 1 \text{ yard} \times \frac{3 \text{ ft}}{1 \text{ yard}} \times \frac{12 \text{ in}}{1 \text{ ft}} \times \frac{2.54 \text{ cm}}{1 \text{ in.}} \times \frac{1 \text{ m}}{100 \text{ cm}} = 0.9144 \text{ m}$

B. $\quad 30 \text{ days} \times \frac{24 \text{ hr}}{1 \text{ day}} \times \frac{3600 \text{ s}}{1 \text{ hr}} = 2,592,000 \text{ s} \approx 2.6 \times 10^6 \text{ s}$

C. $\quad 50 \text{ ft} \times \frac{12 \text{ in.}}{1 \text{ ft}} \times \frac{2.54 \text{ cm}}{1 \text{ in.}} \times \frac{1 \text{ m}}{100 \text{ cm}} = 15.24 \text{ m}$

D. $\quad 3.5 \times 10^{-3} \text{ mm} \times \frac{1 \text{ m}}{1000 \text{ mm}} = 3.5 \times 10^{-6} \text{ m}$

E. $\quad \left(3.0 \text{ in.} \times \frac{2.54 \text{ cm}}{1 \text{ in.}} \times \frac{1 \text{ m}}{100 \text{ cm}} \right)^3 = (7.6 \times 10^{-2} \text{ m})^3 = (7.6)^3 \times (10^{-2})^3 \times m^3$

$$= (4.4 \times 10^2) \times 10^{-6} \text{ m}^3 = 4.4 \times 10^{-4} \text{ m}^3$$

or

$$(3.0 \text{ in.})^3 = 27 \left(\text{in.} \times \frac{2.54 \text{ cm}}{1 \text{ in.}} \times \frac{1 \text{ m}}{100 \text{ cm}} \right)^3 = 4.4 \text{ m}^3$$

F. $\quad 1.0 \left(\text{ft} \times \frac{12 \text{ in.}}{1 \text{ ft}} \times \frac{2.54 \text{ cm}}{1 \text{ in.}} \right)^2 = 1.0 \times (30.48 \text{ cm})^2 = 9.3 \times 10^2 \text{ cm}^2$

Note that for questions A through C the answers are exact, so all the figures are significant; however, for D through F there are only 2 significant figures since the numbers 3.5, 3.0, and 1.0 appear, respectively.

SAMPLE QUESTIONS

QUESTION 1.1: According to Kepler, what must all the orbital planes of the planets have in common?

Answer:

Since the sun is at one focus of each planet's orbit, Kepler would expect the sun to be a common point in the orbital planes of all the planets.

QUESTION 1.2: A piece of paper is 4.5 cm wide and 11.8 cm long. How many significant figures can you give for the answer?

Answer:

The paper's width is given to 2 significant figures and the length to 3 significant figures, so the area is known to only 2 significant figures.

SAMPLE SOLUTIONS

The method of solving these problems is the same as described in Section 1.8 of the text. We will also point out what modeling assumptions have gone into the solution.

PROBLEM 1.1: What is the diagonal of a rectangular appearing object with measured sides of 37 m and 153 m? Make an order of magnitude calculation also. (1.4, 1.7)

Solution:

> **Strategy:** Consider the diagonal as forming a right triangle with two of the sides. Since we know one side to 2 significant figures and know the other side to 3 significant figures, we can have only 2 significant figures in the answer.

We will model the physical object with a mathematical rectangle. This mathematical rectangle has sides denoted by a and b; for our problem $a = 37$ m and $b = 153$ m.

Figure 1.1 shows the relation between a and b and the diagonal.

Fig. 1.1:   Rectangle of sides a and b and diagonal c.

The diagonal, c, of a rectangle is given by $c = \sqrt{a^2 + b^2}$.

In terms of our model the variables are
    a — the short side of the ideal rectangle
    b — the long side of the ideal rectangle
    c — the diagonal of the ideal rectangle

An order of magnitude calculation gives

$$c \approx \sqrt{(50 \text{ m})^2 + (150)^2} = \sqrt{(2500 + 22500) \text{ m}^2} = \sqrt{25000} \text{ m}$$

$$\approx \sqrt{4 \times 10^4} \text{ m} = 2 \times 10^2 \text{ m},$$

choosing to replace 2.5 by 4 since the square root of 4 is known without need of a calculator.

To get the exact solution to the mathematical equation, the numbers with units can be substituted.

$$c = \sqrt{(37 \text{ m})^2 + (153 \text{ m})^2} = \sqrt{1369 \text{ m}^2 + 23409 \text{ m}^2}$$

$$= \sqrt{24778 \text{ m}^2} = \sqrt{24778} \sqrt{\text{m}^2} = 157.4102919 \text{ m}$$

We must now interpret this model diagonal to be the diagonal of the physical object with which we started. When making the interpretation, we must consider the significant figures. The quantity represented by a had 2 significant figures, and the side represented by b had 3 significant figures. Since 2 is the smaller number, our final answer has only 2 significant figures. With two figures, we round 157.4. . . to 160, so $c = 160$ m. However, this implies that we know values to within ± 1 m, which is 3 significant figures; the correct form is $c = 1.6 \times 10^2$ m, which shows there are 2 significant figures.

The result is in agreement with the order of magnitude calculation.

PROBLEM 1.2: Calculate the volume of the rectangular board with a height 8.7 cm, width 21.2 cm, and length 121 cm. Also make an order of magnitude calculation. (1.4, 1.7)

Solution:

> **Strategy**: Use the formula for the volume of a rectangular box. Keep track of significant figures.

The volume, V, of a box of dimensions h, w, and l as shown in Fig. 1.2 is

$$V = h \times w \times l.$$

Fig. 1.2:   Box of h, w, and l.

First, we do the order of magnitude calculation:

$$V_{approx} = 10 \text{ cm} \times 20 \text{ cm} \times 120 \text{ cm} = 24{,}000 \text{ cm}^3.$$

We also note that there are 2 significant figures given for h, 3 significant figures for w, and 3 significant figures for l. There should thus be 2 significant figures in the volume, V.

$$V = 8.7 \text{ cm} \times 21.2 \text{ cm} \times 121 \text{ cm}$$

$$= 22{,}317.24 \text{ cm}^3$$

$$= 2.2 \times 10^4 \text{ cm}^3$$

where in the final step we considered the significant figures. The order of magnitude calculation agrees with the careful calculation as expected.

PROBLEM 1.3: For the preceding problem (1.2), what is an estimate of the error in the volume if the height, h, and the width, w, are uncertain by ± 0.1 cm and the length, l, is uncertain by ± 1 cm? (1.4)

Solution:

> **Strategy:** Apply the significant figure rule.

Since the quoted uncertainties are $\pm$ in the last figures, our standard rule for the uncertainty in a product applies. The uncertainty is $\pm 0.1 \times 10^4$ cm^3. So we give the volume as
$$V = (2.2 \pm 0.1) \times 10^4 \text{ cm}^3.$$

PROBLEM 1.4: As seen from the earth, the moon repeats its phases every 29.5 days (time between full moons). This means, when seen from the earth, the moon appears to rotate about the earth once each 29.5 days. This rate corresponds to "how many radians/s?" Note: Contrary to most units, the radian unit is often omitted if no confusion will result. Thus if no unit is given for an angle, the unit of radian is understood. Angular motions in radians/s (rad/s) appear as "/s" or as "s^{-1}." We will explicitly put the radian in for the first few chapters. (1.6)

Solution:

> **Strategy:** A complete cycle of the moon will be 2π radians. Since we know the time a complete revolution takes, we should be able to determine the average rate of revolution.

We replace the moon with an idealized point body revolving about the earth on an ideal circle. That is, we model the moon's orbit as a circle and the motion of the moon as uniform on this circle. We will let the symbol ω be the variable representing the radians/s for the moon's motion. (This quantity is called the angular velocity of the moon.) We let the symbol P be the variable representing the period of the moon's motion about the earth.

Fig. 1.3 shows the model we are using:

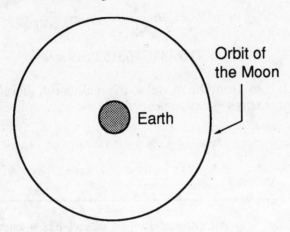

Fig. 1.3: Orbit of moon about the earth.

Since there are 2π radians in a circle, we expect that the angular change in radians/(time interval) times the period of rotation is the angular distance traveled — i.e., 2π radians. As an equation, this statement is

$$\omega P = 2\pi, \quad \text{so} \quad \omega = \frac{2\pi}{P}.$$

An order of magnitude calculation is

$$\omega \approx \frac{10 \text{ rad}}{30 \text{ day}} \approx 0.3 \text{ rad/day},$$

and

$$1 \text{ day} \approx 20 \times 50 \times 50 \text{ s} = 5 \times 10^4 \text{ s},$$

giving

$$\omega \approx 6 \times 10^{-6} \text{ rad/s}$$

as an order of magnitude estimate.

We can now enter the numbers appropriate for the model; that is,

$$P = 29.5 \text{ days},$$

$$\omega = \frac{2 \times 3.14159 \text{ radians}}{29.5 \text{ days}} = 0.21299 \text{ radians/day}.$$

We must convert radians per day to radians per second. There are 60 seconds in a minute and 60 minutes in an hour and 24 hours in a day, so

$$1 \text{ day} = 24 \text{ hr} \times 60 \frac{\text{min}}{\text{hr}} \times 60 \frac{\text{s}}{\text{min}}$$

$$= (24 \times 60 \times 60) \left(\frac{\text{hr} \times \text{min} \times \text{s}}{\text{hr} \times \text{min}} \right) = (86400)\,(\text{s}).$$

Thus

$$\frac{\text{day}}{86400\ \text{s}} = 1,$$

so

$$\omega = 0.21299 \frac{\text{radians}}{\text{day}} \times \frac{\text{day}}{86400\ \text{s}}$$

$$= 0.000002465162\ \text{radians/s}.$$

When we pull the result back from the model to the real world, there are only 3 significant figures (π is known to as many figures as needed), so the answer is

$$\omega = 2.47 \times 10^{-6}\ \text{rad/s}.$$

PROBLEM 1.5: A small conical-shaped object has a base diameter of 5.0 inches and a height of 12.0 inches. What is its volume in cubic centimeters (cm^3)? Make an order of magnitude calculation as a check. (1.6, 1.7)

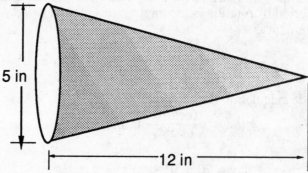

5 in

12 in

Fig. 1.4:   Cone of base 5 inches and height 12 inches.

Solution (Fill in the Missing Steps):

Strategy: There are two factors to consider: units and significant figures. First, convert the units to the requested centimeters, and then use the formula for the volume of a cone. Finally, consider the significant figures.

The conical object is replaced by an idealized cone whose volume is given by

$$V = \frac{\pi}{3} \times r^2 \times h,$$

where r is the radius of the cone and h is its height. Label r and h on Fig. 1.4.

For this problem:

$$r = \underline{\hspace{1cm}} \text{ inches} \quad \text{and} \quad h = \underline{\hspace{1cm}} \text{ inches.}$$

For an order of magnitude calculation take $V \approx r^2 h$, $r \approx 10$ cm, and $h \approx 30$ cm, giving

$$V \approx 9000 \text{ cm}^3 \approx 10^4 \text{ cm}^3,$$

as an order of magnitude calculation.

It is easiest to **convert units first** if more than one unit needs be converted. The conversion is

$$1 \text{ inch } = \underline{\hspace{1cm}} \text{ cm.}$$

So

$$1 = \frac{2.54 \text{ cm}}{\text{inch}} = \frac{2.54}{1} \frac{\text{cm}}{\text{inch}} = 2.54 \frac{\text{cm}}{\text{inch}} \; ;$$

hence

$$r = 2.5 \text{ inches} \times 2.54 \times \frac{\text{cm}}{\text{inch}} = \underline{\hspace{1cm}} \text{ cm}$$

and

$$h = \underline{\hspace{1cm}} \text{ cm.}$$

Thus the ideal volume, V, is

$$V = 1287.036997 \text{ cm}^3.$$

For the physical object, there are 2 significant figures for the radius and 3 significant figures for the height, so there are only 2 significant figures in the answer. The volume must be given as

$$V = 1.3 \times 10^3 \text{ cm}^3.$$

This agrees with our order of magnitude calculation.

PROBLEM 1.6: If an automobile is traveling at 55 miles/hr, what is its speed in m/s? Again do an order of magnitude calculation. (1.6, 1.7)

Solution:

> **Strategy:** Convert each unit to its metric equivalent.

The conversions that are needed are:

$$1 \text{ mile} = 5{,}280 \text{ ft}; \quad 1 \text{ ft} = 12 \text{ inches}; \quad 1 \text{ inch} = 2.54 \text{ cm};$$

$$1 \text{ hr} = 60 \text{ min} = 60 \text{ min} \times 60 \text{ s/min} = 3600 \text{ s.}$$

We will present the order of magnitude calculation first for this problem. As an order of magnitude expression,

$$1 \text{ mile} \approx 6{,}000 \times 10 \times 5 \text{ cm} = 3 \times 10^5 \text{ cm} = 3 \times 10^3 \text{ m},$$

$$1 \text{ hour} \approx 4000 \text{ s.}$$

So, as an order of magnitude,

$$60 \text{ miles/hr} \times \frac{3 \times 10^3 \text{ m}}{\text{mile}} \times \frac{1 \text{ hr}}{4000 \text{ s}} \approx \frac{180000 \text{ m}}{4000 \text{ s}} \approx 50 \text{ m/s}.$$

The calculation is

$$55 \text{ miles/hr} \times \frac{5280 \text{ ft}}{\text{mile}} \times \frac{12 \text{ inch}}{\text{ft}} \times \frac{2.54 \text{ cm}}{\text{inch}} \times \frac{1 \text{ m}}{100 \text{ cm}} \times \frac{1 \text{ hr}}{3600 \text{ s}} = 24.58720 \text{ m/s}.$$

The only number with relevant significant figures is the 55 mi/hr, which has 2 significant figures. Thus

$$55 \text{ miles/hr} = 25 \text{ m/s}.$$

This number agrees with the order of magnitude calculation.

PROBLEM 1.7: How far in feet will light go in a vacuum in 1 nanosecond? (Speed of light $= 3.0 \times 10^8$ m/s.) (1.6)

Solution:

Strategy: Convert the speed of light to feet/s and multiply by 1 nanosecond.

Since the speed of light is given to 2 significant figures, we will have only 2 significant figures in the answer:

$$1 \text{ m} = 100 \text{ cm} \times \frac{1 \text{ inch}}{2.54 \text{ cm}} \times \frac{1 \text{ foot}}{12 \text{ inches}} = 3.28 \text{ feet},$$

so

$$\text{speed of light} = 9.8 \times 10^8 \text{ feet/s}.$$

$$1 \text{ nanosecond} = 10^{-9} \text{ s},$$

so

$$\text{speed of light} = 0.98 \text{ feet/ns}.$$

Light in a vacuum travels about 1 foot in one nanosecond.

GENERAL COMMENTS

This chapter is our introduction to physics. Scientific knowledge can be separated into two types:

Factual knowledge — theories, models, and empirical data interpreted with the models and the theories. The material in the Important Terms practice provides some examples of this factual knowledge.

Procedural knowledge — the collection of methods, tactics, and techniques for using and checking the factual knowledge. Procedural knowledge involves the skills needed to solve the problems that will be posed in the text and on your exams. This knowledge is sometimes referred to as the scientific method.

You must be careful not to let the need for the procedural knowledge obscure the need for the factual knowledge, nor to believe that the factual knowledge is, by itself, physics. In this guide, we will refer to these two aspects of physics again as further examples are studied.

The word *model* was used in the previous paragraphs. In physics we build models to represent the real world. These models are conceptual representations of the real world. They often are simplifications of the real world. Kepler, for example, in his orbital models ignored the other planets when studying the orbit of Mars. We now recognize that this is an oversimplification of the real world. One of the skills you will need to develop is the ability to create the correct model to use to restate problems and to pose solutions. The models of physics are mathematical; that is, physical properties are represented by quantitative variables. As we will develop during this course, mathematical models have four components:

Names — names are assigned to the objects and the agents that interact with the objects. Kepler, for example, modeled the planets and the sun with ideal objects he called planets and sun.

Descriptive variables — these represent the properties of the object. Kepler, for example, used the positions of the planets and sun with respect to the then supposedly fixed stars as some of his variables. He also introduced the concept of the period of the planets. Since the earth was one of these planets, this period had to be a period measured with respect to the fixed stars as seen from the sun.

Equations of the model — these describe the structure and time evolution of the model. Kepler's third law provides an example of this. To Kepler, the third law was an equation of his model. When we study Newtonian mechanics, we will have a model in which the laws of Kepler are deductions rather than equations of the model.

Interpretation — the interpretation relates the descriptive variables of the model to some properties of the object that the model is representing. Often this interpretation is simple; for example, the planetary positions of Kepler are interpreted as telling us where the planet would appear in the sky. Until we make an interpretation, the equations of a model represent nothing — they would simply be abstract relations among mathematical variables. An ellipse and its equation are abstract mathematical relations and concepts; however, the ellipse has meaning when we interpret it as the path through space traveled by a planet.

Each of these four components of a mathematical model are important. As you proceed through this course, reference will be made to them many times.

COMMENTS ON SYMBOLS AND EQUATIONS

We will have a number of equations in our models. The equations are a mathematical shorthand way of writing sentences that relate the descriptive variables of the model. The equations, in and of themselves, have no meaning unless we know what the symbols are to represent — that is, an interpretation is required. We write Kepler's third law (Equation 1.2) as

$$\frac{T^2}{R^3} = k.$$

This equation has meaning only when we know the interpretation of the letters T, R, and k. We could just as well have written

$$\frac{P^2}{A^3} = h,$$

with P being the time for a complete orbit, A the average distance of the planet from the sun, and h a constant. Both of these equations have the same physics; only the names of the variables and their interpretation have changed.

Each model usually has a set of symbols that are commonly used to represent the variables and the interpretation of those variables, such as F for force; but the common convention need not be followed. To be safe, it is always best to give the interpretation of the variables used. To repeat — unless you know what the symbols (variables) represent, you have no physics, only mathematics.

SUGGESTED READINGS

Barnes, G., "Food, Eating, and Mathematical Scaling." *The Physics Teacher,* December 1990, p. 614.

Boorstin, D. J., *The Discoverers.* New York: Random House, 1983. A broad-ranging and interesting account of the origins of clocks and calendars as well as of the telescope.

Bronowski, J., *The Ascent of Man.* Boston: Little, Brown, 1973. The chapters titled "The Starry Messenger" and "The Majestic Clockwork" trace the development of our views of the universe, space, and time.

Coel, M., "Keeping Time by the Atom." *American Heritage of Invention & Technology,* Winter 1988, p. 42.

Heilbron, J. L., "The Politics of the Meter Stick." *American Journal of Physics,* November 1989, p. 988.

Mathieson, D. L., "Why Round the Answer?" *The Physics Teacher,* October 1990, p. 471.

Morrison, P., P. Morrison, and Office of Charles and Ray Eames, *Powers of Ten.* New York: Scientific American Library, 1982. A book about the relative size of things in the universe.

Rosenberg, N., and L. E. Birdzell, Jr., "Science, Technology and the Western Miracle." *Scientific American,* November 1990, p. 42.

von Baeyer, H. C., "How Fermi Would Have Fixed It." *The Sciences,* September 1988, p. 2. A discussion of Fermi's style of doing physics and of making order-of-magnitude calculations.

Wilson, C. "How Did Kepler Discover His First Two Laws?" *Scientific American,* March 1972, p. 92.

Chapter 2
Motion in One Dimension

IMPORTANT TERMS

Fill in the blanks with the appropriate word or words.

1. The study of motion is called _____.

2. The quantity that describes both direction and speed is _____.

3. The _____ of a body is the rate at which the velocity of the body is changing.

4. The horizontal axis of a graph is called the _____.

5. The study of the causes of motion is called _____.

6. The _____ speed of a body is its displacement divided by the elapsed time.

7. The _____ acceleration is the change in velocity divided by the change in time.

8. The text value quoted for the _____ , 9.8 m/s^2, is the approximate value at the earth's surface.

9. If we neglect air resistance, a body falling towards the earth is said to be in _____.

10. The difference in the initial and final positions of a body is termed the body's _____.

11. Draw a curve of position versus time. The tangent to this curve at any time is called the _____ velocity.

12. The vertical axis of a graph is called the _____.

13. The slope of the velocity versus displacement curve for a body is that body's _____ acceleration.

14. The _____ of a line is the ratio of the change in ordinate to the change in abscissa.

15. The branch of physics that deals with kinematics and dynamics is called _____.

16. On a two-dimensional plot the axis running left to right is called the _____ .

17. On a two-dimensional plot the axis running bottom to top is called the _____.

Write the definitions of the following:

18. velocity (2.2):

19. distinguish between distance and displacement:

20. average acceleration (2.5):

21. kinematics (2.0):

22. dynamics (2.0):

23. slope of a line (2.3):

24. mechanics (2.0):

25. instantaneous acceleration (2.5):

26. acceleration of gravity (2.7):

27. freely falling body (2.7):

28. ordinate (2.3):

29. abscissa (2.3):

30. average velocity (2.2):

31. instantaneous velocity (2.4):

32. mechanics (2.0):

33. displacement (2.2):

34. constant speed (2.1):

35. average speed (2.1):

ANSWERS TO 1–17, IMPORTANT TERMS

1. kinematics; 2. velocity; 3. acceleration; 4. abscissa; 5. dynamics; 6. average; 7. average; 8. acceleration of gravity; 9. free fall; 10. displacement; 11. instantaneous; 12. ordinate; 13. instantaneous; 14. slope; 15. mechanics; 16. abscissa; 17. ordinate.

PRACTICE WITH WRITING EQUATIONS

Let the following symbols have the interpretation given:

t	the clock time
x	position at time t
x_0	position at time t_0
v	velocity at time t
v_0	velocity at time t_0
a	acceleration — assumed constant

A bar over any symbol means to take the average value of the quantity represented by the symbol.

The six most frequently used kinematic equations were given in the summary. Write them here, stopping to think carefully of the interpretation of each variable.

$\bar{v} =$ $\bar{a} =$

$x =$ $v =$

$\bar{x} =$ $v^2 =$

In this space write what each of the symbols with a subscript means (x_0, etc.).

COMMENTS ON USING EQUATIONS

When you answered the preceding, you should have realized that the equations were written such that x_0 was the position at $t = 0$ and v_0 was the velocity at $t = 0$. We often have problems where we have position or velocity given for $t \neq 0$. It is then necessary to replace t on the right hand side of the equations by $(t - t_0)$ and to interpret x_0 as the position at t_0 and v_0 as the velocity at t_0. Eq. 2.7 of the text then becomes

$$x(t) = x_0 + v_0(t - t_0) + \frac{1}{2} a (t - t_0)^2,$$

where x_0 is the position at t_0 and v_0 is the velocity at t_0.

MOTION PLOTS AND TIME PLOTS

There are two types of drawings we can make to help understand what is happening in a given case. In Fig. 2.1 of the text, we show the first of these. This is called a **motion plot.** A motion plot shows the various positions that a body has. It shows where the body has moved, not when it moved.

The other type of plot has two axes: the horizontal axis is usually time (the independent variable), and the vertical axis can be position, velocity, or acceleration. We will call such plots **time plots.** Plots such as these allow the definition of instantaneous velocity and acceleration as a limit of the average velocity as the time intervals approach zero (see text Eq. 2.3).

Both of these plots can be useful, but a motion plot should always be drawn at some point in the process of solving a problem. A comparison of the motion plot with the equations used can often help pick the approach to the solution or point out errors in your approach.

Either plot will help you keep the distinction between distance and displacement clear. The displacement is the difference between the initial and final positions without regard to the path traveled between them. The distance traveled will depend on the path traveled and may have to be calculated by summing up the displacements on various parts of the path.

PRACTICE READING TIME PLOTS

Time plots are used in Sections 2.4, 2.5, and 2.6 of the text to derive the idea of instantaneous velocity and acceleration. Consider the position versus time plot shown in Fig. 2.1.

We have calculated the average velocity by the slope method for a few points. Fill in the remaining positions and/or average velocities in Table 2.1. Two example calculations are given beside the table.

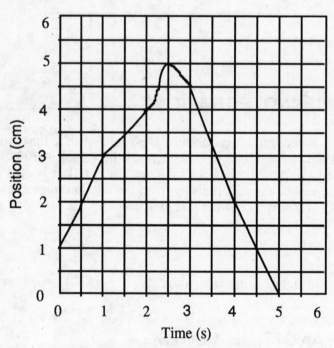

Fig. 2.1:   A position versus time plot for a moving body.

Use the equation

$$\bar{v} = (x - x_0)/(t - t_0).$$

From the figure, we find

for $t = 1.0$ s $x = 3.0$ cm
for $t_0 = 0.5$ s $x_0 = 1.9$ cm.

Table 2.1:
Time in seconds, position in cm, and average velocities in cm/s for Fig. 2.1

$\bar{v} = (3.0 - 1.9)\text{cm}/(1.0 - 0.5)\text{s}$

$= (1.1/0.5) \times (\text{cm/s})$

$= 2.2$ cm/s

for $t = 3.0$ s, $x = 4.5$ cm
for $t_0 = 1.0$ s, $x_0 = 3.0$ s

t	t_0	x	x_0	$\bar{v}$
1.0	0.5	3.0	1.9	2.2
3.0	1.0	4.5	3.0	0.75
5.0	4.0			−2.0
2.0	1.0			
4.0	3.0			
3.0	2.0			

$\bar{v} = \dfrac{(4.5 - 3.0)\text{cm}}{(3.0 - 1.0)\text{s}}$

$= 0.75$ cm/s

All numbers have been rounded to 2 significant figures.

The instantaneous speed can be calculated from a time plot by the method of tangent lines. Figure 2.2 shows the same position versus time curve as in Fig. 2.1. The tangent lines at $t = 1$ s and at $t = 3$ s are shown and the slope of these lines is calculated. Examine this process and then repeat for the indicated values.

Fig. 2.2:   A position versus time plot for a moving body.

To draw the tangent line, draw a line that just touches the position versus time curve as shown in the examples. A length of the tangent line is drawn for your convenience. Pick two points on the line, and get the x and t values. Then calculate the slope by the rule:

$$\text{slope} = \Delta x / \Delta t.$$

This slope will be the instantaneous velocity.

For $t = 1$ s, the tangent line drawn includes the points

$$(x, t); (3.0, 1.0) \text{ and } (2.25, 0.5).$$

The slope is then

$$v(1 \text{ s}) = \text{slope at 1 s} = \frac{\Delta x}{\Delta t} = \frac{(3.0 - 2.25)\text{cm}}{(1.0 - 0.5)\text{s}}$$

$$= \frac{0.75}{0.5} \times \frac{\text{cm}}{\text{s}} = 1.5 \text{ cm/s}.$$

We could have used

$$v(1 \text{ s}) = \frac{(2.25 - 3.0)\text{cm}}{(0.5 - 1.0)\text{s}} = 1.5 \text{ cm/s}$$

with no change in the result.

Either the first x and t can be subtracted from the second x and t, or the second x and t can be subtracted from the first x and t; but do the same for both x and t to get Δx and Δt .

For example, if you have the pair (x_1, t_1) and (x_2, t_2), then

$$\Delta x = (x_1 - x_2) \text{ and } \Delta t = (t_1 - t_2)$$

or

$$\Delta x = (x_2 - x_1) \text{ and } \Delta t = (t_2 - t_1)$$

but not

$$\Delta x = (x_1 - x_2) \text{ and } \Delta t = (t_2 - t_1).$$
[*THIS CHOICE GIVES THE WRONG SIGN*]

This velocity (the instantaneous velocity) is not the same as the average velocity we calculated before using the same t values. This is to be expected whenever the curve representing the motion is not a straight line.

For $t = 3$ s, the tangent line drawn includes the points (3.0, 4.5) and (2.5, 5.5). The slope then gives

$$v(3 \text{ s}) = \frac{\Delta x}{\Delta t} = \frac{(4.5 - 5.5)\text{cm}}{(3.0 - 2.5)\text{s}} = -2.0 \text{ cm/s}.$$

The minus sign (–) tells us that the value of the position is decreasing (moving to smaller values), which is what we see on the plot.

Exercises:
Show that the velocity at 2.5 s is zero, and find the velocity at 2.0 s and at 4.0 s. You may get slightly different values, since graphical analysis involves the judgment as to exactly where to put the tangent line.

A usual difficulty for beginning students is confusing displacement, velocity, and acceleration when examining time plots. There are several steps that must be taken when you are confronted with a time plot.

1. What is being plotted against what? Is the plot a position versus time (a displacement plot) or a velocity versus time or an acceleration versus time plot?
2. What scales are used for the plot axes? Look at the numbers and the legend — if position, is it in cm, m, feet, km? If velocity, is it in m/s, radians/s, etc.? Is the lower left corner of the plot zero or some other number?
3. What can you say about the motion from the plot? If it is a position versus time plot, does the velocity change magnitude or sign? If it is a velocity versus time plot, does the acceleration change magnitude or sign?

In Fig. 2.3a the position versus time plots for two different objects are shown—one we call red, the other blue. First examine only Fig. 2.3a.

The red object starts at 0 cm and at the end of 10 s is at 50 cm. The blue object starts at 20 cm and at the end of 10 s is at 40 cm. Consider the following questions (write your answers and check them before reading further):

Exercise:
    a) At time equals 9 s, which has the larger velocity, red or blue?
    b) At time equals 3 s, which has the larger velocity, red or blue?
    c) Do the red and blue objects ever have the same speed in the interval 0–10 s?

There are several common and easy-to-make errors when you first start to read graphs: confusing the height of a line with its slope, saying the blue object's speed is larger at 3 s since the blue curve is higher at that time, and saying the speeds are the same at 7 s since the curves cross at that point.

In Fig. 2.3a, the height of the line gives us the displacement, while the slope gives the velocity. The slopes of the two curves are given in Fig. 2.3b. Clearly the red object always has the larger velocity.

Fig. 2.3:   Motion of two objects: (a) Position versus time plot. (b) Velocity versus time plot.

Another skill you need in time plot reading is the ability to construct the motion that is associated with a given time plot. Conversely, given a moving body or a description of the motion of a body, you should be able to construct a time plot description of that motion.

Figure 2.4a shows the position as a function of time for a moving body. When the curve is rising, the body's velocity is positive; when the curve is falling, the body's velocity is negative. Thus at point A, we see that the slope of the position curve is zero, so the velocity is zero. Before the time A, the velocity is positive; just after A, it is negative as we can see in Fig. 2.4b. The velocity changes from negative to positive between points C and D.

Looking at Fig. 2.4a, what is the velocity at B? At E? Use Fig. 2.4b for help if needed.

Fig. 2.4:   The motion of a body: (a) Position versus time. (b) Velocity versus time.

Finally, a motion plot for Fig. 2.4a would look like Fig. 2.5.

Position
= 0 cm

Position
= 6 cm

Fig. 2.5:   Rough motion plot for the time plot Fig. 2.4a. The vertical displacement is just so we can see the true displacement given by the horizontal axis.

ANSWERS TO TIME PLOT QUESTIONS

Table 2.1: 3rd line 0.0, 2.0; 4th line 4.0, 3.0, 1.0; 5th line 2.0, 4.5, – 2.5.
Fig. 2.2: (answers may differ slightly) v(2 s) = 1.6 cm/s; v(4 s) = – 2.4 cm/s.
Fig. 2.3: red; red; never.
Fig. 2.4: negative; positive.

SAMPLE QUESTIONS

QUESTION 2.1: Distinguish between speed and velocity.

Answer:

Speed has only a magnitude, while velocity has both a magnitude and a direction. The speed of your auto could be given as 55 miles per hour, while your velocity would give the direction, for example, southeast.

QUESTION 2.2: A car's speedometer is calibrated for tires of a specific size. To save money, you put on smaller, cheaper tires when the first ones wear out. What effect, if any, will this have on your speedometer reading?

Answer:

The speedometer on your car counts the revolutions of the drive shaft and converts this to a speed. If your tires are smaller than assumed, you will go a shorter distance with each revolution of the drive shaft. So, smaller tires will cause your speedometer to read a higher number (speed) than is correct.

SAMPLE SOLUTIONS

PROBLEM 2.1: What are the average speed and the average velocity for a trip that is made by traveling slowly for two hours at a constant speed of 10 km/hr and then traveling for three hours at a constant speed of 30 km/hr in the opposite direction? (2.2)

Solution:

> **Strategy:** Describe the model and draw a motion plot. The problem asks for average values; the average speed is the total distance traveled divided by the elapsed time and the average velocity is the net displacement divided by the elapsed time.

We first model this problem by letting t represent the time elapsed since the trip started, x the distance traveled, and v the velocity. We will assume the initial motion is to the $+x$ direction (this is an arbitrary choice that will affect the sign of the answer).

An important aspect of our model is that we will ignore the acceleration that must occur when the velocity changes from 10 km/hr to the right to 30 km/hr to the left.

A **motion plot** will make the discussion simpler. The motion plot first drawn is a preliminary version, which does not really tell if we end up to the right or left of the starting point.

Fig. 2.6: Motion plot, preliminary version; $x = 0$ is the starting point and positive distances are to the right.

We need to decide what we know and what we want to find.

GIVEN:

Time (t)	Velocity (v)
0 to 2 hr	10 km/hr
2 hr to 2 hr + 3 hr = 5 hr	− 30 km/hr

The minus sign for $t = 2$ hr to $t = 5$ hr is needed since the motion is now in the negative x direction.

FIND:

the average speed $\quad \bar{s} = \dfrac{\text{distance traveled}}{t_f - t_i}$

the average velocity $\quad \bar{v} = \dfrac{x_f - x_i}{t_f - t_i},$

where t_f denotes $t = 5$ hr and t_i denotes $t = 0$ hr.

We see that in order to get either average we will need to find the final displacement at $t = 5$ hr. We can do this if we calculate the displacement while traveling 10 km/hr to the right and 30 km/hr to the left. The final displacement is the sum of these two when we keep in mind the directions.

We will use the relation

$$x = x_0 + v_0(t - t_0) + \frac{1}{2} a (t - t_0)^2$$

where $a = 0$.

For

$$0 \le t \le 2 \text{ hr}, x_0 = 0 \text{ and } v_0 = 10 \text{ km/hr},$$

then

$$x(2 \text{ hr}) = 0 + (10 \text{ km/hr}) \times (2 \text{ hr}) = 20 \text{ km}.$$

Then

$$3 \text{ hr} \le t \le 5 \text{ hr}, t_0 = 2 \text{ hr}, x_0 = 20 \text{ km, and } v_0 = -30 \text{ km/hr}.$$

Thus

$$x(5 \text{ hr}) = 20 \text{ km} + (-30 \text{ km/hr}) \times (5 \text{ hr} - 2 \text{ hr})$$

$$= 20 \text{ km} + (-30 \times 3) \times \frac{\text{km} \times \text{hr}}{\text{hr}} = 20 \text{ km} - 90 \text{ km} = -70 \text{ km}.$$

We can now draw our final motion diagram as shown in Fig. 2.7.

Fig. 2.7: Motion diagram for Problem 2.1.

The average speed is given by

$$\bar{s} = \frac{\text{distance traveled}}{t_f - t_i} = \frac{20 \text{ km} + 90 \text{ km}}{5 \text{ hrs}} = 22 \frac{\text{km}}{\text{hr}} \, ,$$

while the average velocity is given by

$$\bar{v} = \frac{x_f - x_i}{t_f - t_i} = \frac{-70 \text{ km} - 0 \text{ km}}{5 \text{ hr} - 0 \text{ hr}} = -14 \frac{\text{km}}{\text{hr}} \, .$$

The average speed is always positive, while the negative sign of the average velocity depends on our choice of the initial direction being called the $+x$ direction. You should repeat this calculation with the initial velocity to the left (towards $-x$) and show that you obtain 14 km/hr for the average velocity.

Note: This problem clearly shows the difference between distance traveled and displacement. It also shows that average velocity may be misleading since the true velocity was never the average value except at the turn-around.

PROBLEM 2.2: A boat is traveling on a river whose current flows at 5.0 miles/hr. The boat is capable of traveling 12.0 miles/hr in still water. The boat travels downstream for 1.2 hr. It then turns quickly around and returns to its starting position. What is the total duration of the trip? (2.3)

Fig. 2.8: a) Sketch from the frame of the river. b) Sketch from the riverbank.

Solution:

> **Strategy:** Since this problem is posed in the English system and an answer with the units of hours will be acceptable, we will not convert the initial data to the SI system.
> **Method 1: Shore as the frame** — Add or subtract the river velocity.
> **Method 2: River as the frame** — Choose the river as the rest frame.

Method 1: Shore as the Frame
We first make a sketch of the problem as shown in Fig. 2.8b. The velocity of the boat with respect to the water is 12 mi/hr, so if we let v_d be the downstream velocity and v_u be the upstream velocity,

$$v_d = 12.0 \text{ mi/hr} + 5.0 \text{ mi/hr} = 17 \text{ mi/hr}$$

and

$$v_u = -12.0 \text{ mi/hr} + 5.0 \text{ mi/hr} = -7.0 \text{ mi/hr}$$

where we have chosen the downstream direction as positive.

We take the initial position of the boat to be $x = 0$ at $t = 0$. Then for

$$0 < t < 1.2 \text{ hr}, \qquad x(t) = v_d t$$

and for

$$1.2 \text{ hr} < t < t_f \qquad x(t) = x_0 + v_u (t - t_0),$$

where

$$x_0 = v_d \times 1.2 \text{ hr} = 20 \text{ mi} \quad \text{and} \quad t_0 = 1.2 \text{ hr}.$$

During the upstream portion of the trip,

$$x(t) = 20 \text{ mi} - 7.0 \text{ mi/hr} \times (t - 1.2 \text{ hr}).$$

The boat returns to the starting position when $t = t_f$ and $x(t_f) = 0$.

This gives

$$x(t) = 0 = 20 \text{ mi} - 7.0 \text{ mi/hr} \times (t_f - 1.2 \text{ hr}), \quad \text{or} \quad t_f = 1.2 \text{ hr} + \frac{20 \text{ mi}}{7.0 \text{ mi/hr}} = 4.1 \text{ hr}.$$

The trip takes 4.1 hr and the boat travels 20 mi downstream with respect to the shore before returning to its starting position.

Method 2: River as the Frame

We first make a sketch of the problem as shown in Fig. 2.8a. The velocity of the boat is 12 mi/hr in both directions. From the frame of the river, the starting position is moving 5.0 mi/hr to the left. We choose right to be positive. The position of the boat relative to its starting position with respect to the river is given by

$$x_{boat} = \begin{cases} 12 \text{ mi/hr} \times t & \text{for } t < 1.2 \text{ hr} \\ 14.4 \text{ mi} - 12 \text{ mi/hr} \times (t - 1.2 \text{ hr}) & \text{for } t > 1.2 \text{ hr} \end{cases}$$

while

$$x_{start(shore)} = -5.0 \text{ mi/hr} \times t.$$

The boat returns to the starting location on the shore when $x_{boat} = x_{start(shore)}$, or when

$$-5.0 \text{ mi/hr} \times t = 14.4 \text{ mi} - 12 \text{ mi/hr} \times (t - 1.2 \text{ hr}),$$

which gives

$$t = \frac{28.8 \text{ mi}}{7.0 \text{ mi/hr}} = 4.1 \text{ hr}.$$

A proper choice makes the problem simple.

PROBLEM 2.3: A ball is thrown straight upward at an initial speed of 10 m/s. How long does it take for the ball to return to its starting position? At what time and how far away from the starting point are its velocity zero and the acceleration nonzero? (2.6)

Solution:

Strategy: This is a constant acceleration problem, so use the relation for constant acceleration. Make a motion plot.

We model the ball with a point and assume the motion is one-dimensional. The problem is stated such as to assume the motion occurs in the earth's gravitational field. Choose upward as a positive direction, z. Take $z = 0$ to be the initial starting point. Figures 2.9 and 2.10 show the motion diagrams.

Fig. 2.9: Initial motion diagram for Problem 2.3.

Fig. 2.10: Final motion diagram for Problem 2.3.

Examine the items known for the problem

GIVEN:

initial position	z_0	=	0.0 m
initial velocity	v_0	=	10.0 m/s
acceleration	a	=	-9.8 m/s^2

Since we want a time,

$$z(t) = z_0 + v_0 t + \frac{1}{2} a\, t^2 . \qquad \text{(Eq. 2.7 of text)}$$

Since we can take $t = 0$ at the start of the motion, we may use the unmodified form of the text equation.

This gives

$$z(t) = 10 \text{ m/s} \times t - 4.9 \text{ m/s}^2 \times t^2.$$

For $z(t) = 0$ we have

$$z(t) = 10 \text{ m/s} \times t - 4.9 \text{ m/s}^2 \times t^2$$

$$= 0 = t(10 \text{ m/s} - 4.9 \text{ m/s}^2 \times t).$$

This has solutions

$$t = 0 \text{ s}$$

and

$$t = \frac{10 \text{ m/s}}{4.9 \text{ m/s}^2} = 2.041 \text{ s}.$$

The first time ($t = 0$ s) is the starting time, and the second time ($t = 2.0$ s) would be the return time. Note that we have now rounded to 2 significant figures.

To answer the second part of the question, we note that the acceleration is constant so the acceleration is never zero. From the motion diagram, we see that the velocity is zero at the top of the motion since the direction changes. We can get the z value for this from

$$v^2 = v_0^2 + 2a(x - x_0). \text{ (Eq. 2.7 of the text)}$$

For $v = 0$,

$$v^2 = 0 = (10 \text{ m/s})^2 + 2 \times (-9.8 \text{ m/s}^2) \times (z - 0).$$

This gives

$$z = 5.1 \text{ m}.$$

So the ball would rise with decreasing velocity to a height of 5.1 m, stop, and then fall down. It would pass its starting point 2 s after it started.

To calculate the time at which the ball is at rest (but accelerated), use Eq. 2.6 again. We find that t must satisfy

$$5.1 \text{ m} = (10 \text{ m/s}) \times t - (4.9 \text{ m/s}^2) \times t^2,$$

which can be rewritten as

$$t^2 - 2.04 \text{ s} \times t - 1.04 \text{ s}^2 = 0.$$

This can be solved using the binomial theorem. For example, the equation

$$a x^2 + b x + c = 0$$

has the solution

$$x = \frac{-b \pm \sqrt{b^2 - 4 ac}}{2a},$$

so

$$t = (1.02 \pm 0.00) \text{ s} = 1.02 \text{ s} = 1.0 \text{ s}.$$

The ball will be at rest 1 s after leaving the ground at a height of 5.1 m.

We can also use symmetry to calculate this last time. Since the entire motion takes 2.0 s, we expect the turn-around to occur halfway through the motion.

PROBLEM 2.4: A person starts from rest and moves with a constant acceleration of -2 m/s^2 for 11 seconds and then moves with an acceleration of 2 m/s for another 11 seconds: (a) What is the maximum speed attained? (b) What is the average speed? (c) How far does the person go during the whole trip? (2.6)

Solution:

> **Strategy:** Make a motion plot that has the essential features of the description. Choose an arbitrary direction for the initial motion. The person initially has no velocity, and the acceleration is opposite to the direction chosen as positive. After a period of time (11 s), the acceleration changes direction and magnitude. For each portion of the motion the acceleration is constant. Treat each constant acceleration portion separately so you can use the rules for constant acceleration.

Model the motion with the variable x, which measures the displacement from the starting position. x is positive to the right, and the acceleration is assumed to change instantaneously from -2 m/s to $+2$ m/s. A preliminary motion diagram is shown in Fig. 2.11. A final motion diagram is shown in Fig. 2.12.

Fig. 2.11: Preliminary motion diagram for Problem 2.3.

Fig. 2.12: Final motion diagram for Problem 2.3.

What do we know?

GIVEN:

Initial acceleration	a_i	-2 m/s, lasting for 11 seconds
Final constant acceleration	a_f	$+2$ m/s, lasting for 11 seconds.

FIND:

maximum speed	v_{max}
average speed	$v_{average}$
final displacement	x_{final}

The person starts from rest at $t = 0$ s, decreases her velocity (that is, the velocity has a larger negative value) until $t = 11$ s, then increases her velocity for the next 11 s. Using the relation

$$v = v_0 + at,$$

we have

$v = -2 \ (\text{m/s}^2) \times t$	for	$0 < t < 11$ s
$v = -22$ m/s	for	$t = 11$ s
$v = -22 \text{ m/s} + 2(\text{m/s}^2) \times (t - 11 \text{ s})$	for	$11 \text{ s} < t < 22$
$v = 0$ m/s	for	$t \geq 22$ s

From this we see that the answer to (a), the maximum speed (speed is the magnitude of velocity), is 22 m/s.

To answer (b), we need to find the final displacement. Use Eq. 2.7 of the text for

$$x = x_0 + v_0 t + \frac{1}{2} a\, t^2.$$

Since the acceleration changes at 11 seconds, we will have to break the motion into two parts:

$x = \frac{1}{2} \times (-2 \text{ m/s}^2) \times t^2$ for $0 < t < 11$ s

$x = -121$ m   for $t = 11$ s

$x = -121 \text{ m} - 22 \text{ m/s} \times (t - 11 \text{ s})$

        $+ \frac{1}{2} \times (2 \text{ m/s}) \times (t - 11 \text{ s})^2$ for $11 < t < 22$ s

where the -22 m/s at 11 s was found in part (a), and

$x = -121 \text{ m} - 242 \text{ m} + 121 \text{ m} = -242 \text{ m}$ for $t \geq 22$ s.

Note that even though the person ends up at rest, there is a net displacement.

The symmetry and equal times of the accelerations will not bring us back to the initial starting position. Assuming that she would return to the initial starting position is a common and easily made mistake.

We can now draw our final motion diagram, as we previously did in Fig. 2.9, and calculate the average velocity using Eq. 2.2:

$$v_{average} = \frac{-242 \text{ m} - 0 \text{ m}}{22 \text{ s} - 0 \text{ s}} = -11 \text{ m/s}.$$

Part (c) can now be answered. Since the velocity is always negative, the total distance traveled is 242 m and is equal to the displacement. Had the person changed directions, the total distance traveled would not be the displacement. We would have to add (neglecting signs) the individual parts of her trip to get the total distance traveled.

PROBLEM 2.5: A stone is thrown downward at 5.0 m/s from a height of 30 meters. At the same time, from directly below, a stone is thrown upward from the ground with a speed of 20 m/s. At what height do their paths intersect? (2.6)

Solution:

> **Strategy:** Draw a motion diagram. To avoid confusion, use a different variable for each stone, but measure both variables from the same starting position and in the same direction. The first stone has its initial acceleration and velocity in the same direction, while the second stone has these quantities in opposite directions.

We model the two stones as points and measure all distances positively upward from the ground. We will use the symbol t to represent time measured from zero when both stones start out on their paths. The position of the stone starting downward will be denoted by $x(t)$ and the position of the stone starting upward by $y(t)$. This is shown in Fig. 2.13. In the figure we have assumed that the stones meet before the upward moving stone starts to fall back. The calculations are independent of this assumption.

Fig. 2.13: The two stones and their motion. The downward thrown stone's position is given by $x(t)$ and the upward thrown stone's position by $y(t)$. The motions are shown as curved paths. We assume the paths cross.

Thus $x(t)$ is the height above the ground of the downward moving stone and $y(t)$ is the height above the ground of the upward moving stone. We are asked to find the height when their paths intersect. That is, we want to find the x or the y for which $x(t) = y(t)$. Since this is a constant acceleration problem, we expect Eq. 2.6 of the text to be the needed relation. What are we given?

> ### GIVEN:
> Stone 1
>> position at $t = 0$ is 30 m $\qquad\qquad\qquad\qquad\qquad\quad$ $x_0 = 30$ m
>> velocity at $t = 0$ is -5.0 m/s (minus since up is $+$) $\qquad$ $v_{ox} = -5$ m/s
>> acceleration is -9.8 m/s^2
>
> Stone 2
>> position at $t = 0$ is 0 m $\qquad\qquad\qquad\qquad\qquad\qquad\quad$ $y_0 = 0$ m
>> velocity at $t = 0$ is $+20$ m/s $\qquad\qquad\qquad\qquad\qquad$ $v_{oy} = +20$ m/s
>> acceleration is -9.8 m/s^2

So

$$x(t) = 30 \text{ m} - (5 \text{ m/s}) \times t - (4.9 \text{ m/s}^2) \times t^2,$$

and

$$y(t) = (20 \text{ m/s}) \times t - (4.9 \text{ m/s}^2) \times t^2.$$

If we set these two equal [$x(t) = y(t)$], we can solve for t, the time that the paths cross, and then use either equation to find the height.

Setting $x(t) = y(t)$ gives

$$30 \text{ m} - (5 \text{ m/s}) \times t - (4.9 \text{ m/s}^2) \times t^2 = (20 \text{ m/s}) \times t - (4.9 \text{ m/s}^2) \times t^2$$

or

$$30 \text{ m} - (5 \text{ m/s}) \times t = (20 \text{ m/s}) \times t.$$

This gives

$$t = \frac{30 \text{ m}}{25 \text{ m/s}} = 1.2 \text{ s}.$$

So

$$x(1.2 \text{ s}) = y(1.2 \text{ s}) = 24 \text{ m} - 7.1 \text{ m} = 17 \text{ m}.$$

Hence the two stones would pass 1.2 s after we started and would be at a height of 17 m.

Note that it is often necessary to find information, in this case t, that is not requested in order to be able to find the desired quantities.

Has the initially upward moving body reached its maximum height? To find this use the expression for velocity, $v(t) = v_0 + a\,t$, to find when $v(t) = 0$. For the initially upward moving body,

$$v(t) = 0 = 20 \text{ m/s} - 9.8 \text{ m/s} \times t,$$

whose solution is

$$t = 2.0 \text{ s}.$$

Figure 2.13 was drawn correctly.

PROBLEM 2.6: Figure 2.14 shows a position versus time plot for the motion of a block. The line from p_1 to p_2 is a straight line, the line from p_2 to p_3 is curved, and the line from p_3 to p_4 is straight but with a steeper slope than the portion from p_1 to p_2. Describe the velocity both in words and with a sketch. Do the same for the acceleration.

Solution:

Strategy: Separately analyze the motion of the block for each of the three portions of the described displacement. Each portion has different velocities and accelerations.

We will denote the time that the block is at position p_1 by t_1, the time at position p_2 by t_2, and so on as shown in Fig. 2.14.

Since the line between p_1 and p_2 is straight, the velocity must be constant. That velocity is nonzero and positive since the line has a positive slope.

The same comments apply to the line between p_3 and p_4. The velocity is constant, nonzero, and positive. The velocity between p_3 and p_4 is also greater than the velocity between p_1 and p_2.

Fig. 2.14: Position versus time.

The curve between p_2 and p_3 is not a straight line, so there must always be acceleration during this interval, and the velocity is not constant. The curve is concave upward between p_2 and p_3, so the velocity increases throughout this range. This analysis allows us to draw the velocity versus time curve in Fig 2.15a.

Since the final velocity is greater than the initial velocity, the net acceleration must be positive. Also, since the curve is always concave upward between p_2 and p_3, we know that the acceleration is always positive. We do not have enough information to tell if the acceleration is constant. With this in mind, we can draw the acceleration versus time curve in Fig. 2.15b, realizing that, although we know the acceleration is positive, we don't know its exact shape.

Fig. 2.15:   (a) Velocity versus time.   (b) Acceleration versus time.

PROBLEM 2.7: A bug crawls into an open drain pipe. It enters the pipe at a time we'll call t = 0.0 s, moving at a velocity of 2.0 cm/s. 5.0 s after it enters the pipe, it starts slowing down with a constant acceleration, stopping 3.0 s later (8.0 s after starting into the pipe). After a 1.0 s pause, the bug turns around and accelerates for 4.0 s at the same rate as before toward the end of the pipe it first entered. The bug then moves at constant velocity until it leaves the pipe. (2.6)

(a)   Make a sketch of the velocity versus time, the position versus time, and the acceleration versus time.

(b)   What are the acceleration and the final velocity of the bug and at what time does it leave the pipe?

Solution:

Strategy: The first part of this problem is a practice in translating a word description into a motion plot. Since we are given the time periods of acceleration, we will have to start with an acceleration plot, total it to get a velocity plot, and then construct the position plot.

To analyze this problem it is helpful to make a table of known and unknown quantities. The columns will be the time intervals, and the rows the position, velocity, and acceleration.

Table 2.2

time (s)	position (cm)	velocity(cm/s)	acceleration (cm/s^2)
0.0	0.0	2.0	0
$0.0 < t < 5.0$	$x_1(t)$	2.0	0
$5.0 < t < 8.0$	$x_2(t)$	$v_2(t)$	a
$8.0 < t < 9.0$	x_3	0.0	0
$9.0 < t < 13.0$	$x_4(t)$	$v_4(t)$	a
$13.0 < t < t_f$	$x_5(t)$	v_5	0

Examine Table 2.2. It contains all the information you were given about the bug's motion. For the first 5.0 s, the velocity is a constant 2.0 cm/s. For the next three seconds, the acceleration is a constant, which we call a. The motion stops by $t = 8.0$ s, and the position remains fixed for 1.0 s. Then the same acceleration acts for 4.0 s, and the velocity is constant until the bug leaves the pipe. This information alone is sufficient to allow us to sketch the acceleration, the velocity, and the position as functions of time.

Let us take the initial direction of motion as positive. Then the acceleration is a negative value between 5.0 and 8.0 seconds and between 9.0 and 13.0 seconds and is zero for all other times. This is shown in Fig. 2.16(a).

The velocity starts at +2.0 cm/s and remains constant until $t = 5.0$ s; then, since the acceleration is constant, it decreases linearly until $t = 8.0$ s when the velocity is zero. The velocity remains zero until $t = 9.0$ s. It then decreases linearly (with the same slope as for $5 s < t < 8 s$) until $t = 13.0$ s. After that time the velocity is constant. The velocity is negative after $t = 9.0$ s, and the slopes for the two intervals of acceleration must be the same since the accelerations are equal. This is shown in Fig. 2.16(b).

The position must start at 0.0 for $t = 0.0$ s and move upward with a constant slope (slope = velocity) until $t = 5$ s. The position curve then bends over to a fixed value of position for 8.0 s $< t < 9.0$ s.

The position curve must then bend down until $t = 13.0$ s after which the slope is constant. The slope of the final straight line portion is negative while the initial slope is positive. Since the acceleration time on the return path is longer than during the first part of the motion, the final slope does not have the same magnitude as the initial slope. This is shown in Fig. 2.16(c).

Study these figures and verify that they contain the initial information:

Fig. 2.16: Plots for Problem 2.7: (a) Acceleration versus time. (b) Velocity versus time. (c) Position versus time.

To answer part (b), which is quantitative, we put the given information into the relevant kinematic equation:

$$x(t) = x_0 + v_0(t - t_0) + \frac{1}{2} a(t - t_0)^2$$

and

$$v(t) = v_0 + a(t - t_0).$$

During any time interval, the values of x_0 and v_0 are the final values of x and v of the preceding time period.

For $\qquad$ $0.0 \text{ s} < t < 5.0 \text{ s}$,
we have

$$t_0 = 0, v_0 = 2.0 \text{ cm/s}, x_0 = 0, \text{ and } a = 0.$$

This gives

$$x_1(t) = v_0 t = 2.0 \text{ cm/s} \times t;$$

hence,

$$x_1(5.0 \text{ s}) = 10 \text{ cm}.$$

For $\qquad$ $5.0 \text{ s} < t < 8.0 \text{ s}$,
we have

$$t_0 = 5.0 \text{ s}, x_0 = 10 \text{ cm}, v_0 = 2.0 \text{ cm/s},$$

so

$$x_2(t) = 10 \text{ cm} + (2.0 \text{ cm/s}) \times (t - 5.0 \text{ s}) + \tfrac{1}{2} a (t - 5.0 \text{ s})^2$$

and

$$v_2(t) = 2.0 \text{ cm/s} + a(t - 5.0 \text{ s}).$$

We know that v_2 is zero at

$$t = 8.0 \text{ s},$$

so

$$v_2(8 \text{ s}) = 2.0 \text{ cm/s} + a \times 3.0 \text{ s} = 0.$$

Hence,

$$a = -0.67 \text{ cm/s}^2$$

and

$$x_3 = x_2(8 \text{ s}) = 10 \text{ cm} + 6.0 \text{ cm} - 1.0 \text{ cm} = 15 \text{ cm}.$$

Thus, for $\qquad$ $9.0 \text{ s} < t < 13.0 \text{ s}$,
we have

$$x_0 = 15 \text{ cm}, v_0 = 0 \text{ and } a = -0.67 \text{ cm/s}^2.$$

So

$$v_4(t) = a(t - 9.0 \text{ s}) = -0.67 \text{ cm/s}^2 (t - 9.0 \text{ s})$$

and

$$x_4(t) = 15 \text{ cm} - 0.33 \text{ cm/s}^2 (t - 9.0 \text{ s})^2.$$

When $\qquad$ $t = 13.0 \text{ s}$,

these give

$$v_4(13.0 \text{ s}) = -2.7 \text{ cm/s} \text{ and } x_4(13.0 \text{ s}) = 4.7 \text{ cm}.$$

Hence,

$$v_5 = v_4(13.0 \text{ s}) = -2.7 \text{ cm/s},$$

and

$$x_5(t) = 4.7 \text{ cm} - 2.7 \text{ cm/s} (t - 13.0 \text{ s}).$$

The bug exits the pipe when $x_5(t) = 0$. This occurs when

$$t = 13.0 \text{ s} + \frac{4.7 \text{ cm}}{2.7 \text{ cm/s}} = 14.7 \text{ s} = 15 \text{ s}.$$

So the acceleration is -0.67 cm/s^2, the final velocity is -2.7 cm/s, and the bug leaves the tube 15 s after entering.

PROBLEM 2.8: An automobile starts from rest and reaches a speed of 100 km/hr in 10.0 seconds. Find the average acceleration. What distance would be covered if the acceleration is constant? (2.6)

Solution:

> **Strategy:** Use the definition of average acceleration. This average acceleration can be used with the displacement law to find the distance traveled.

If we let a_{avg} be the average acceleration, Δv the change in velocity, and Δt the elapsed time,

$$v_{avg} = \frac{\Delta v}{\Delta t}$$

$$= \frac{(100 - 0) \text{ km/hr}}{10 \text{ s}}$$

$$= \frac{100 \text{ km/hr} \times 10^3 \text{ m/km} \times 3600 \text{ s/hr}}{10 \text{ s}} = 2.78 \text{ m/s}^2.$$

If this average acceleration is the true constant acceleration, we may write the position $x(t)$ as

$$x(t) = x_0 + v_0 t + \frac{1}{2} a t^2,$$

where x_0 is the initial position and v_0 is the initial velocity (both assumed zero). Using the average acceleration, we find that when $t = 10$ s, displacement $= 139$ m.

The last step may also be calculated from

$$v^2 = v_0^2 + 2ax.$$

Solving this for x gives, since $v_0 = 0$,

$$x = \frac{v^2}{a} = 139 \text{ m}.$$

The average acceleration is 2.78 m/s^2, and the displacement after 10 s is 139 m.

GENERAL COMMENTS

In this introduction to one-dimensional kinematics we meet the first and the simplest of our many models. Real bodies are idealized as point particles that we call objects (**name** for the object) that have the **variables** position, velocity, and acceleration. In the real world, we have a reference point from which we measure position. The usual convention is that positive positions are to the right of this point and negative positions are to the left. Velocity or speed is introduced, with positive velocity tending to increase the numerical value of position and negative velocity decreasing the position value (remember –5 cm is a smaller value of position than –2 cm). Likewise, accelerations change the value of velocity. Positive acceleration increases velocity, while negative acceleration decreases velocity.

Usually the mathematical symbols x, v, a, and t are used with the **interpretation** that x is position, v is instantaneous velocity, a is instantaneous acceleration, and t is the elapsed time. This interpretation pulls the mathematical formulation back to the real world.

Even though we consider only constant acceleration in this chapter, we still call acceleration a variable. In this chapter acceleration is independent of time, but the acceleration can have different values depending on the problem.

The factual knowledge is the definition of position, velocity, acceleration and the relationships that hold among them, and time. The procedural knowledge is how to solve problems using these relations with prescribed accelerations.

SOLVING QUADRATIC EQUATIONS

In this chapter and in many of the following chapters, you will need to solve equations of the form:

$$a\,x^2 + b\,x + c = 0,$$

where a, b, and c are known numbers and x is the unknown quantity to be determined. This is a general quadratic equation, which has two solutions:

$$x_1 = \frac{-b + \sqrt{b^2 - 4ac}}{2a}$$

and

$$x_2 = \frac{-b - \sqrt{b^2 - 4ac}}{2a}.$$

The form of these solutions is a result of the binomial theorem. The nature of these solutions, x_1 and x_2, depends on $b^2 - 4ac$. In order for real solutions to exist

$$b^2 - 4ac \geq 0.$$

If this is not true, there are no real numbers that satisfy the equation. For a physics problem, this means one of two things:

1. The physical behavior is different than assumed.

2. You have made a mistake, either mathematically or in the physical quantities you used.

In general, for a physics problem, you will find two real solutions (or only one real solution if $b^2 - 4ac = 0$). It is possible that only one of the solutions is physically acceptable. There is no general principle to tell you which is the acceptable solution. Each problem must be analyzed to determine which solution is the one you want by asking if the solution makes physical sense. For example, if x must be greater than zero, either physically or because of initial conditions, a negative x should be discarded. Another possibility is that one of the solutions implies that two solid bodies must pass though each other. By asking such questions, you should be able to determine which solutions to keep and which to discard.

SOLVE THE FOLLOWING QUADRATIC EQUATIONS:

I.
$$3x^2 - 7x + 2 = 0$$

For this equation, $a = 3$, $b = -7$, and $c = 2$, so

$$x_1 = \frac{7 + \sqrt{49 - 24}}{2 \times 3} = 6 \quad \text{and} \quad x_2 = \frac{7 - \sqrt{49 - 24}}{6} = \frac{1}{3}.$$

II.
$$2x^2 + 2x - 3 = 0$$

Since $\sqrt{b^2 - 4ac} = \sqrt{28} = 5.3$, the solutions are $x_1 = 0.82$ and $x_2 = -1.8$.

III.
$$x^2 - 2x + 1 = 0$$

Since $\sqrt{b^2 - 4ac} = 0$, there are two identical roots: $x_1 = x_2 = 1$.

Verify each of the above answers.

SUGGESTED READINGS

Brancazio, P. J., *Sport Science*. New York: Simon and Schuster, 1984.

Casper, B. M., "Galileo and the Fall of Aristotle: A Case of Historical Injustice?" *American Journal of Physics*, Vol. 45, 1977, p. 325.

Crummett, B., "Measurements of Acceleration of Gravity." *The Physics Teacher*, May 1990, p. 291.

Drake, S., "The Role of Music in Galileo's Experiments." *Scientific American*, June 1975, p.98.

Schroeer, D., "Brecht's Galileo: A Revisionist View." *American Journal of Physics*, Vol. 48, 1980, p. 125.

Seeger, R. J., *Men of Physics: Galileo Galilei, His Life and His Works*. New York: Pergamon Press, 1966.

Winters, L. M., "High-Speed Flash Photography with Sound Triggers." *The Physics Teacher*, January 1990, p. 12.

Chapter 3
Motion in Two Dimensions

IMPORTANT TERMS

Fill in the blanks with the appropriate word or words.

1. A quantity having only magnitude is called a(n) _____.

2. The sum of two vectors is called the _____ vector.

3. The distance a projectile travels from its launch on the earth until it hits the earth is called the _____.

4. A(n) _____ has both magnitude and direction.

5. The _____ velocity of two moving bodies is just the vector difference of the individual velocities.

Write the definitions of the following:

6. scalar (3.0):

7. vector (3.0):

8. relative velocity (3.3):

9. resultant (3.1):

10. range of a projectile (3.6):

ANSWERS TO 1–5, IMPORTANT TERMS

1. scalar; 2. resultant; 3. range of a projectile; 4. vector; 5. relative.

ADDITION OF VECTORS

Vectors can be added in two ways: graphically and by adding components.

We will consider both ways in the following examples. These examples are also used to show you how we model reality with figures. Since our examples involve trips of several km, we could not possibly draw the figure to full scale, so we need a scaling factor. The first example will show how the scaling factor is used to convert to the real world the figure that we are modeling with the graph.

EXAMPLE 1

A student makes a trip which has two parts as shown in Fig. 3.1. First, he traveled to the north-northeast as shown and then to the east as shown. The scale is 1/2 inch = 2 km. The parts of the trip are labeled A and B.

Fig. 3.1:   Two part trip.

The parts of the trip can be defined either as a magnitude and a direction or as two components. With a ruler and a protractor, measure the lengths and directions. Measure lengths to the nearest 1/10 inch (or nearest 1/16 and round the fraction to the 1/10), and measure angles to the nearest 0.5 degree. For example, the length of vector A measures 2 and 3/16 inch; 3/16 = 0.18, so A is 2.2 inches.

Since 1/2 inch = 2 km, we can write

$$1 = \frac{2 \text{ km}}{1/2 \text{ inch}} = 4 \text{ km/inch}.$$

Multiply the measured distance in inches by this conversion factor (4 km/inch) to get the represented length in km.

Read the scales to get components. Also draw a vector **C**, which is the sum of **A** and **B**, and get its magnitude, direction, and components. Check your answers with the entries and verify that the components of **C** are the algebraic (signed) sums of the components of **A** and **B**.

| | Magnitude | | | Component | | | |
| | | | | Northern | | Eastern | |
Vector	Inch	Km	Direction	Inch	Km	Inch	Km
A	1.1	4.4	26.5°	1.0	4.0	0.5	2.0
B	2.0	8.0	90°	0.0	0.0	2.0	8.0
C	2.7	10.8	75°	1.0	4.0	2.5	10.0

EXAMPLE 2

Repeat the above procedure for the 3 vector sum shown in Fig. 3.2. Also draw the vector **D = A + B + C**. Calculate the scale factor and measure length and directions. All directions are measured from north again. Check your entries against the answers and reconcile any major differences. The scale is

$$\frac{1}{2} \text{ inch} = 1 \text{ km.}$$

Fig. 3.2: A vector sum of 3 vectors.

Scale factor : 1 = (____ km/ ____ inch) = ____ km/inch

| Vector | Magnitude | | | Component | | | |
| | | | | Northern | | Eastern | |
	Inch	Km	Direction	Inch	Km	Inch	Km
A				0.5	1.0		
B							
C							
D							

Answers to EXAMPLE 2 (scale factor 1 inch = 2 km

A: 0.70 in. = 1.4 km; 45°; 0.5 in. = 1 km N; 0.5 in. = 1 km E
B: 1.1 in. = 2.2 km; 60.5°; 0.5 in. = 1 km N; 1.0 in. = 2 km E
C: 0.70 in. = 1.4 km; 135°; –0.5 in. = –1 km N (– means S); 0.50 in. = 1 km E
D: 2.1 in. = 4.2 km; 76°; 0.5 in. = 1 km N; 2 in. = 4 km E

These examples provided practice with calculating vector sums using the two different methods (graphical analysis and components). Which method is best to use will depend on the particular problem under consideration.

SAMPLE QUESTIONS

QUESTION 3.1: A train is passing us, moving to our left. We observe a passenger on the train drop a ball. According to the passenger, the ball falls vertically downward. How do we see the ball fall?

Answer:

We see the ball as having the same velocity to the left as the train and passenger. The ball would appear to fall on a parabola, curving towards the left.

QUESTION 3.2: What effect does doubling the initial speed have on the range of a projectile?

Answer:

Since the range of a projectile is proportional to the square of the initial speed, doubling the initial speed will increase the range by a factor of four.

SAMPLE SOLUTIONS

PROBLEM 3.1: A boat capable of traveling 10 km/hr in still water is crossing a river near the ocean that is flowing at 5.0 km/hr. The river is 0.80 km wide at this point. The boat captain wishes to land at a point 0.90 km upstream. Assuming the captain steers on a shortest distance path, how long does the trip take? (3.2)

Solution:

> **Strategy:** The boat's velocity is given relative to the water, so we will have to consider that factor when calculating the distance the boat travels. The time of travel of the boat is found by considering both the velocity and the distance traveled.
> **Method 1:** Use the riverbank frame.
> **Method 2:** Use the frame of the water.

Method 1: Use the Riverbank as the Measuring Frame

A preliminary motion diagram is shown in Fig. 3.3a with the coordinates in the riverbank called x and y. Since the velocity of the boat is constant relative to the water, the velocity of the boat will be different in the two directions. In order to travel along a straight line from the start to the finish, the boat must be steered upstream at some angle so that the resulting velocities across and along the stream move the boat along the line from start to finish. We denote this angle by θ as shown in Fig. 3.3a, and let t be the time of travel from start to finish.

Fig. 3.3: Motion of boat across a river as seen by (a) an observer on the riverbank and (b) an observer at rest on the river.

We can write:

$$-0.90 \text{ km} = v_x t \qquad \text{and} \ +0.80 \text{ km} = v_y t$$

where v_x and v_y are the velocities of the boat in the x and y directions as measured relative to the riverbank. Thus

$$\frac{v_x}{v_y} = -1.1.$$

In terms of the angle θ the velocities are also expressed as

$$v_y = 10 \text{ km/hr} \times \cos \theta$$

and

$$v_x = -10 \text{ km/hr} \times \sin \theta + 5.0 \text{ km/hr}.$$

Combining these gives

$$v_x = -1.1 v_y = -11 \text{ km/hr} \times \cos \theta = -10 \text{ km/hr} \times \sin \theta + 5.0 \text{ km/hr}$$

or

$$\cos \theta = 0.91 \sin \theta - 0.45 = \sqrt{1 - \sin^2 \theta} ,$$

where the last identity is used to give us an equation only in terms of $\sin \theta$. This equation can be solved by squaring and rearranging the terms into the standard quadratic form:

$$1.8 \sin^2 \theta - 0.82 \sin \theta - 0.80 = 0.$$

The positive solution of this equation is $\sin \theta = 0.92$, and $\cos \theta = 0.39$.

This gives $v_y = 3.9$ km/hr, and $t = \dfrac{0.80 \text{ km}}{v_y} = 0.21$ hr.

Method 2: Use the River as the Reference Frame

We have sketched the motion in this frame in Fig. 3.3b. In this frame, it is obvious that the straight line path is the shortest and fastest path. We use x' and y' for our coordinates. The advantage of this frame is that the motion is along the path drawn. The disadvantage is that the landing point is moving relative to the river. We again let t be the time the trip takes. During this time, the boat travels at a speed of 10 km/hr along the dark arrow for a distance of 10 km/hr $\times t$, while the landing point moves 5 km/hr $\times t$ along the $-x'$ direction for a total displacement of 5 km/hr $\times t$ + 0.90 km.

We know all three sides of a right triangle, so the Pythagorean theorem gives

$$(0.80 \text{ km})^2 + (5 \text{ km/hr} \times t + 0.90 \text{ km})^2 = (10 \text{ km/hr} \times t)^2$$

or

$$75 \, (t/\text{hr})^2 - 9.0 \, (t/\text{hr}) - 1.5 = 0.$$

The positive solution is $t = 0.21$ hr.

Clearly, choosing the frame of the river makes the problem simple and easy to solve.

PROBLEM 3.2: A stone is thrown over a cliff with a velocity of 20 m/s and at an upwards angle of 45°. If the cliff is 25 m high, how far from the base of the cliff will the stone hit? (3.5)

Solution:

> **Strategy:** Draw a motion plot. The motion in the vertical and the horizontal directions is independent. First, find when the stone hits the ground; then determine the horizontal distance traveled during this time interval.

The usual first step in any problem with two or more dimensions is to draw the motion picture as in Fig. 3.4.

We will idealize the cliff as a vertical wall and assume the stone was thrown from the exact top and edge. We will let $x(t)$ represent the distance in the horizontal direction from the cliff's edge as shown, with $t = 0$ being the throwing of the stone. We will also let $y(t)$ be the height of the stone above the ground measured positively upward. The effect of any air resistance is neglected.

The approach is to use the independence of the two components of the position and velocity to write separate equations for each quantity.

The stone hits the ground when $y(t) = 0$. If we can find the time for which this is true, we can then put that value of t into x(t) to find the distance traveled in the x direction.

First we need to express the initial velocity in x and y components.
From the drawing we see that the initial x velocity is 20 m/s × cos(45°) = 14 m/s in the positive x direction. The initial y velocity is 20 m/s × sin(45°) = 14 m/s upward (positive y).

Fig. 3.4:   The thrown stone.

GIVEN:

Quantity	Symbol	Value
initial x position	x_0	0.0 m
initial y position	y_0	50.0 m
initial x velocity	v_{x0}	+14.0 m/s
initial y velocity	v_{y0}	+14.0 m/s
x acceleration	a_{x0}	0.0 m/s^2
y acceleration	a_{y0}	–9.8 m/s^2

The relevant model equations are then

$$x(t) = v_{x0}\, t = (14 \text{ m/s}) \times t$$

$$y(t) = y_0 + v_{y0}\, t + \frac{1}{2} a\, t^2$$

$$= 50 \text{ m} + (14 \text{ m/s}) \times t - (4.9 \text{ m/s}^2) \times t^2.$$

Now, using the binomial theorem for $y(t) = 0$ gives

$$t = \frac{-14 \text{ m/s} \pm \sqrt{(14 \text{ m/s})^2 - 4 \times (-4.9 \text{ m/s}^2) \times (50 \text{ m})}}{-9.8 \text{ m/s}^2}$$

$$= 4.9 \text{ s} \quad \text{or} \quad -2.1 \text{ s}.$$

Since the stone hits the ground at a time $t > 0$, $t = 4.9$ s.

Then using this value of t, we find

$$x(4.9 \text{ s}) = (14 \text{ m/s}) \times 4.9 \text{ s} = 69 \text{ m}.$$

So the stone hits the ground 4.9 s after launch and at a distance of 69 m from the cliff. Only 2 significant figures have been kept, in agreement with the initial data.

The negative value of t (-0.23 s) given by the binomial theorem is the time at which the stone would have been launched from a level at the base to pass the true launch point with the given velocity (assuming the cliff were not in the way). This path is shown as a dotted line on Fig. 3.4.

As an aside, we can use $v(t) = v_0 + a\,t$ to find the maximum height of the stone. When the stone is at the top of its flight, the y velocity is zero, so

$$0 = 14 \text{ m/s} + (-9.8 \text{ m/s}^2) \times t.$$

That is, the maximum occurs at a time $t = 1.4$ s. At $t = 1.4$ s,

$$y(1.4 \text{ s}) = 50 \text{ m} + 19.6 \text{ m} - 9.6 \text{ m} = 60 \text{ m}$$

and

$$x(1.4 \text{ s}) = 20 \text{ m}.$$

PROBLEM 3.3: A stone is thrown horizontally off a 50 m high building. It hits the ground at a distance of 20 m from the base of the building. What is the initial horizontal velocity of the stone? (Neglect air resistance.) How much longer would it take the stone to hit the ground if it were initially thrown with twice the horizontal speed? (3.5)

Origin for Choice 1

50 m

20 m

Origin for Choice 2

Choice 1

Choice 2

Fig. 3.5:   Sketch of motion and two possible coordinate choices.

Solution:

> **Strategy:** Use the property that the vertical and horizontal motions are independent. The horizontal distance is traveled in the time it takes the stone to fall 50 m.
> To show the independence of the choice of origins, we will solve this problem with two choices of origin for our coordinates.

Let x be the horizontal coordinate and y the vertical coordinate. We will use x_0 for the initial x position, y_0 for the initial y position, v_{xo} for the initial x velocity, $v_{yo} = 0$ for the initial y velocity, and g to represent the acceleration of gravity.

First Coordinate Choice

We will choose a coordinate system with an origin at the top of the building and with x measured positively to the right and y positively downward. This is shown in the box labeled Choice 1 of Fig. 3.5. For this choice,

$$x_0 = 0 \qquad y_0 = 0 \qquad v_{xo} = ? \qquad v_{yo} = 0 \qquad g > 0.$$

Then the general equations of motion are

$$y(t) = y_0 + v_{yo} t + \tfrac{1}{2} a t^2 = \tfrac{1}{2} g t^2 \quad \text{and} \quad x(t) = x_0 + v_{xo} t + \tfrac{1}{2} a t^2 = v_{xo} t$$

with g a positive number (particles are accelerated in the $+y$ direction).

We are given that when $y = 50$ m, $x = 20$ m. This occurs at a time t_f such that

$$y(t_f) = 50 \text{ m} = \tfrac{1}{2} g t_f, \quad \text{and} \quad x(t_f) = 20 \text{ m} = v_{xo} t_f.$$

This gives

$$t_f = \sqrt{\frac{100 \text{ m}}{g}} = 3.2 \text{ s} \quad \text{and} \quad v_{xo} = \frac{20 \text{ m}}{t_f} = 6.3 \text{ m/s}.$$

This coordinate choice is unusual in as much as it is a left-handed coordinate system. That is, if you rotate $+x$ into $+y$ by the smallest angle, the rotation is clockwise rather than counterclockwise. This does not matter for 2-dimensional problems; however, when we deal with 3-dimensional problems, this can get you in trouble. The usual definitions of certain quantities (angular momentum and magnetic field) depend on the right-hand nature of the coordinate system.

Second Coordinate Choice

Choose the coordinate origin at the foot of the building with x increasing positively to the right and y increasing in an upward direction, as shown in Choice 2 of Fig. 3.5. For this choice,

$$x_0 = 0 \qquad y_0 = 50 \text{ m} \qquad v_{xo} = ? \qquad v_{yo} = 0 \qquad g < 0.$$

Then,

$$x(t) = v_{xo} t, \quad \text{and} \quad y(t) = y_0 + \tfrac{1}{2} g t^2.$$

The stone hits the ground at a time t_f given by

$$y(t_f) = 0 = y_0 + \frac{1}{2} g \, t_f^2.$$

This gives

$$t_f = \sqrt{\frac{100 \text{ m}}{-g}} = 3.2 \text{ s}, \quad \text{and} \quad v_{xo} = \frac{20 \text{ m}}{t_f} = 6.3 \text{ m/s}.$$

This is the same result as before (as expected).

Since the x and y motions are independent, the time it takes the stone to hit the ground is independent of the initial horizontal velocity.

PROBLEM 3.4: Galileo's great-great-great-...grandchild drops a Chianti bottle from the top of a vertical 50 m tall tower.
(a)   How long does it take the bottle to fall to the ground?
(b)   What is the velocity when the bottle hits the ground?
(c)   If instead of the bottle being dropped, it was thrown straight out from the same place with a speed of 12 m/s, how far would it land from the base of the tower?
(d)   What is the bottle's horizontal component of velocity as it strikes the ground?
(e)   What is the magnitude of the total velocity of the bottle as its strikes the ground? (3.5)

Solution:

> **Strategy:** Use the property that the vertical and horizontal motions are independent. We will choose the origin at the base of the tower and leave out some of the mathematical steps. You should fill in all missing steps and repeat the calculations taking the top of the tower as the origin.

Let x be the horizontal coordinate and y the vertical coordinate. We will use x_0 for the initial x position, y_0 for the initial y position, v_{xo} for the initial x velocity, $v_{yo} = 0$ for the initial y velocity, and g to represent the acceleration of gravity. A sketch is shown in Fig. 3.6a.

Part a
For the coordinate choice in Fig 3.6a we have

$$x_0 = 0, \ v_{xo} = 0, \ y_0 = 54 \text{ m}, \ v_{yo} = 0, \ g = -9.8 \text{ m/s}^2.$$

Clearly, for part a, there is no x motion.

To solve this part, we need the time, t, when $y = 0$ in the equation of motion

$$y = y_0 + v_0 t + \frac{1}{2} g \, t^2.$$

Using the numbers gives

$$t = \sqrt{\frac{-2 \times 54 \text{ m}}{g}} = 3.4 \text{ s}.$$

Use the coordinate choice in Fig. 3.6b to argue that $y_0 = 0$ and the bottle hits the ground at a time t when $y = -34$ m to obtain the same value of t.

Fig. 3.6:   Coordinates for tower problem: (a) Origin choice at base of tower.
(b) Origin at top of tower.

Part b

The velocity for both choices of origin is given by $v_y = v_{yo} + g\,t$; so when the bottle hits the ground,

$$v_y = 33 \text{ m/s}.$$

Part c

For this part the only change is that $v_{xo} = 12$ m/s for either origin choice. Since the x and y motions are independent, and since no forces are acting in the x direction,

$$x = v_{xo}\,t.$$

Since the y motion is unaffected, the bottle hits the ground 3.3 s after being thrown; it hits 39.6 m from the base of the tower.

Part d

The magnitude of the velocity when the bottle hits the ground will be the magnitude of a vector that is 33 m/s in the negative y direction and 12 m/s in the positive x direction, or 35 m/s.

PROBLEM 3.5: A space explorer fires a gravity testing projectile at an angle of 35° with respect to the horizontal and with a muzzle velocity of 50 m/s and finds that the range is 500 m. What is the acceleration of gravity of the planet? (3.6)

Solution:

> **Strategy:** This is a straightforward range question. Since the launch and impact point are at the same height, we can use the range formula of text Eq. 3.10.

First draw a motion diagram as in Fig. 3.7. Also draw the initial velocity vector at 35°.

Fig. 3.7: Projectile path.

We model the projectile as a point object and measure range along the horizontal. We know that the range of a projectile, R, is related to the launch angle, θ, the initial or muzzle velocity, v_0, and the local gravity, g, by text Eq. 3.10 as

$$R = \frac{v_0^2 \sin(2\theta)}{g}.$$

For this problem,

$$v_0 = 50 \text{ m/s}, \; \theta = 35°,$$

and g is to be found.

So

$$g = \frac{v_0^2 \sin(2\theta)}{R} = \frac{(50 \text{ m/s})^2 \times \sin(70°)}{500 \text{ m}} = 4.698 \text{ m/s}^2.$$

So the acceleration of gravity on this planet is 4.7 m/s^2, or about half that of the earth.

PROBLEM 3.6: What angles give rise to a range of 200 m for a cannonball with an initial speed of 50.0 m/s? (3.6)

Solution:

Strategy: Examine the range formula Eq. 3.10.

The range, R, is given in terms of the initial velocity, v_0, and the launch angle, θ, by

$$R = \frac{v_0^2 \sin(2\theta)}{g},$$

where g is the acceleration due to gravity.

For this problem, the only unknown is θ, so

$$\sin (2\theta) = \frac{9.8 \text{ m/s}^2 \times 200 \text{ m}}{(50 \text{ m/s})^2} = 0.784 .$$

So

$$\theta = 26°.$$

A launch angle of 26° at 50 m/s leads to a range of 200 m.

GENERAL COMMENTS

This chapter introduces both factual knowledge and procedural knowledge. The factual knowledge is the idea of vectors, the parabolic motion of projectiles, and the relation for the range of a projectile. The procedural knowledge is how to combine vectors (both graphically and by components), the independent treatment of the components of motion, and the posing and solving of projectile problems.

This chapter builds on the previous chapter. Since we can treat individual components separately, the equations from Chapter 2 apply to each component.

TRIGONOMETRY REVIEW

Refer to Fig 3.8 to answer these questions.

(a)   **(b)**

FIG. 3.8:   Sample right triangles: (a) Sides given. (b) An angle given.

1. In Fig 3.8a, tan θ = _____ , θ = _____ ,   sin θ = _____ , cos θ = _____ . How would you find the hypotenuse?

2. In Fig. 3.8b, if a = 20 cm, b = _____ .

Answers

1. tan θ = $\dfrac{10 \text{ cm}}{15 \text{ cm}}$ = 0.67, θ = 34°, sin θ = 0.55, cos θ = 0.83. The hypotenuse is given by

$$\sin \theta = \frac{\text{opposite side}}{\text{hypotenuse}} = \frac{10 \text{ cm}}{\text{hypotenuse}} = 0.55,$$

so

$$\text{hypotenuse} = 18 \text{ cm.}$$

You can also use hypotenuse = $\sqrt{(10 \text{ cm})^2 + (15 \text{ cm})^2}$ = 18 cm.

2. Use

$$\tan 20° = \frac{\text{opposite side}}{\text{adjacent side}} = \frac{b}{a} = \frac{b}{20 \text{ cm}} = 0.36,$$

which gives

$$b = 7.2 \text{ cm}.$$

SUGGESTED READINGS

Brancazio, P. J., "Trajectory of a Fly Ball." *The Physics Teacher* , January 1986, p. 20.

Burke, J., *The Day the Universe Changed.* Boston: Little, Brown and Company, 1985. Early attempts to understand projectile motion are described in Chapter 3.

Drake, S., "Galileo's Discovery of the Law of Free Fall." *Scientific American,* May 1973, p. 84.

Drake, S., "Galileo's Discovery of the Parabolic Trajectory," *Scientific American,* March 1975, p.102.

Drake, S., "The Role of Music in Galileo's Experiments." *Scientific American,* June 1975, p. 98.

Egler, R. A., "Student Misunderstanding of the 'Monkey and the Hunter'." *The Physics Teacher,* May 1989, p. 356.

Gardner, M., "Mathematical Games, the Abstract Parabola Fits the Concrete World." *Scientific American,* August 1981, p. 146.

Gingerich, O., "The Galileo Affair." *Scientific American,* August 1982, p. 132.

Chapter 4
Force and Motion

IMPORTANT TERMS

Fill in the blanks with the appropriate word or words.

1. The gravitational force functionally depends upon the _____ of the distance.

2. Newton's laws of motion are valid in a(n) _____ reference frame.

3. A force of one _____ acting on a 1 km mass will give that mass a 1 m/s^2 acceleration.

4. When an object is sitting on a table at rest, the force produced by the table is perpendicular and just the amount needed to balance gravity. This force is called the _____ force.

5. The _____ is the constant of proportionality between the magnitude of the kinetic frictional force and the magnitude of the normal force.

6. Newton's _____ law relates the change in momentum of a body and the applied force.

7. The property of matter that resists changes in motion is called _____.

8. The _____ is a measure of the amount of matter in a body.

9. "In the absence of outside forces a body has constant velocity" is a statement of Newton's _____ law.

10. "If a body exerts a force **F** on another body then the second body exerts a force **−F** on the first" is a statement of Newton's _____ law.

11. _____ equilibrium occurs when a body is at rest and the sum of all forces acting on it is zero.

12. The force produced by two surfaces sliding across each other is named _____.

13. _____ equilibrium occurs when a body has constant velocity and the sum of all forces acting on it is zero.

14. _____ occur(s) when a body falls freely in a gravitational field.

15. The vector diagram for a body showing all forces acting on the body is called a(n) _____.

16. The _____ of a body is the gravitational force exerted upon that body.

17. The product of the mass and the velocity of a body is called the _____ of the body.

18. The vector sum of all the forces acting on a body is called the ___ force on the body.

19. The _____ of gravity on the earth's surface is about 9.0 m/s^2.

20. If a mass is hung on a wire, the force in the wire that opposes the force applied by the mass is termed the _____ in the wire.

Write the definitions of the following:

21. difference between mass and weight (4.5):

22. net force (4.2):

23. Newton's third law (4.6):

24. momentum (4.4):

25. newton (4.4):

26. static equilibrium (4.9):

27. dynamic equilibrium (4.9):

28. normal force (4.5):

29. weight (4.5):

30. mass (4.4):

31. free body diagram (4.7):

32. friction (4.8):

33. acceleration of gravity (4.5):

34. inertial reference frame (4.10):

35. inverse square force (4.1):

36. inertia (4.3):

37. coefficient of friction (4.8):

38. Newton's first law (4.3):

39. Newton's second law (4.4):

40. weightlessness (4.6):

41. tension (4.7):

ANSWERS TO 1–20, IMPORTANT TERMS

1. inverse square; 2. inertial; 3. newton; 4. normal; 5. coefficient of friction; 6. second; 7. inertia; 8. mass; 9. first; 10. third; 11. static; 12. friction; 13. dynamic; 14. weightlessness; 15. free body diagram; 16. weight; 17. momentum; 18. net; 19. acceleration; 20. tension.

SAMPLE QUESTIONS

QUESTION 4.1: A book is at rest on a table top. Does that mean there are no forces acting on the book?

Answer:

No, it only means that the sum of all forces acting on the book is zero. In this case there are two forces acting on the book. Gravity is trying to make the book fall down, while the normal force of the table reacting to the book is equal and opposite.

QUESTION 4.2: A spring scale is placed on an elevator, and a student stands on it. The elevator is moving upward with constant velocity. Will the student weigh more or less than when the elevator is at rest?

Answer:

Since the elevator is moving with constant velocity, the student's weight will be the same as if the elevator were at rest. Since there is no acceleration of the student relative to the earth, the net force is only the gravitational force. Remember, constant velocity is the same as being at rest. As the elevator accelerates to its final velocity, the student will weigh more; but, as the acceleration of the elevator goes to zero, her weight returns to its normal value.

SAMPLE SOLUTIONS

PROBLEM 4.1: A classroom demonstration is done with an Atwood machine. The mass on the right is 1.25 kg. This mass descends a distance of 2.0 m in 3.2 s from rest. What is the other mass? Ignore effects of pulley mass and friction. (4.7)

Solution:

> **Strategy:** To choose our method of solution for this problem we need to analyze the given information. We know from the text that there is a simple relation between the masses and the acceleration of the mass pair. We know the system starts at rest and moves a known distance in a known time. This is sufficient information to determine the acceleration. We then can find the mass.

An Atwood machine is shown in Fig. 4.1a. We have chosen the direction of motion to be positive downward for both masses. Since we are told that the mass on the right descends, the mass on the right must be greater than the mass on the left.

We use the following symbols:

GIVEN:

Quantity	Symbol	Value
mass on right	m_1	1.25 kg
mass on left	m_2	?
acceleration of gravity	g	9.8 m/s^2
acceleration of mass m_i	a_i	? (but equal and opposite)
tension in string	T	?

The forces acting on m_2 are shown in Fig. 4.1b, and the forces on m_1 are shown in Fig. 4.1c. Newton's second law (Σ forces = m a) gives:

$$m_2\, a_2 \ = \ m_2\, g - T,$$

$$m_1\, a_1 \ = \ m_1\, g - T.$$

Combining gives

$$(m_1 + m_2)\, a_1 \ = \ (m_1 - m_2)\, g$$

where use has been made of the constraint

$$a_1 = -a_2.$$

(a)　　　　　**(b)**　**(c)**　　　**(d)**

Fig. 4.1:　　(a) The Atwood machine. (b) and (c) Free force diagrams for the two masses. (d) Net forces on the system.

A constraint is any relation that restricts the possible values that a variable may have.

As an alternative approach, the problem can be considered as shown in Fig. 4.1d. For this method the motion in the *y* direction must be taken as positive for both masses. The force equation is

$$(m_1 + m_2)\, a_1 \ = m_1\, g - m_2\, g,$$

which is the same equation as before. Solving the equation gives

$$m_2 = m_1 \frac{g - a_1}{g + a_1}.$$

To find m_2, we need to know a_1. To get this, use the kinematic law for constant acceleration:

$$y = y_0 + \Delta y = y_0 + v_0\,\Delta t + \frac{1}{2}a\,(\Delta t)^2,$$

where, for this problem, $v_0 = 0$.

This relation gives the acceleration as

$$a = \frac{2\,\Delta y}{(\Delta t)^2} = \frac{2 \times 2.0\ m}{(3.2\ s)^2} = 0.39\ m/s^2;$$

hence

$$m_2 = 1.25\ kg \times \frac{(9.80 - 0.39)m/s^2}{(9.80 + 0.39)m/s^2} = 1.2\ kg.$$

The mass on the left is 1.2 kg.

PROBLEM 4.2: A body of mass 20.0 kg moves with constant velocity of 10 m/s on a horizontal surface under the action of a horizontal force of 15 N. What is the coefficient of friction for the surface and how long will it take the body to stop if the applied force is removed? (4.7)

SOLUTION:

Strategy: Draw a free body diagram. There is no vertical motion so the vertical forces sum to zero. The sum of the horizontal forces will give the acceleration of the body along the surface.

Fig. 4.2: Free body force diagram.

The forces are shown in Fig. 4.2. The mass of the body is m, the acceleration of gravity is g, the horizontal acceleration of the body is a, the normal force is N, the applied force is f, and the frictional force is $F_f = \mu N$.

$$N - mg = 0 \qquad (Vertical\ forces)$$

$$F - \mu N = ma \qquad (Horizontal\ forces)$$

Solving these for the first part when $a = 0$ gives

$$\mu = \frac{F}{mg}.$$

If v_0 is the initial velocity and v is the velocity at a later time, the relation for constant acceleration is

$$v = v_0 + at.$$

Since the applied force is now zero, the acceleration is given by

$$a = -\frac{\mu N}{m}.$$

To find the time it takes the body to stop, set $v = 0$ and solve for t, getting

$$t = \frac{v_0}{-a} = \frac{m v_0}{\mu N}.$$

Entering the numbers gives us $\mu = 0.077$, and the stopping time is about 13 s.

PROBLEM 4.3: A 20 kg block is placed on a frictionless table and connected to a 3 kg block by a string that extends horizontally across the table over a pulley and down to the 3.0 kg block. What is the acceleration of the blocks? Ignore friction in the pulley. (4.7)

SOLUTION:

Strategy: Draw a figure showing the problem. Show the forces, either for the system or as a free body diagram. Since the system is frictionless, the only external force that produces motion is gravity acting on the hanging mass. The gravitational force on the mass on the table is balanced by the normal force. If free body force diagrams are used, include the tension of the string as a force on both bodies.

Fig. 4.3: Problem 4.3: (a) The physical arrangement and coordinates chosen. (b) The motion diagram. (c) Forces on body A. (d) Forces on body B. (e) Forces on the system of both bodies.

We first draw a general figure showing the problem. This is shown in Fig. 4.3a. We will choose coordinates x and y, where x represents the path along which the 20 kg block moves and y the direction that the 3.0 kg block moves. We guess the direction of x and y to be the way the blocks move. If we are wrong, we will get negative velocities.

Known quantities	Symbol
20 kg block	A
mass of the 20 kg block	m_A
3.0 kg block	B
mass of 3.0 kg block	m_B

We also draw a motion diagram in Fig. 4.3b. This diagram is good until one of the blocks hits the pulley. The model assumes the pulley and string are ideal.

If we take the free body force diagram approach, we need force diagrams for both bodies. These are shown in Fig. 4.3c and 4.3d. Since the only forces on block A are either normal or parallel to the surface, we know that

$$N = -m\,g,$$

so no acceleration occurs to A perpendicular to the table. The remaining equations for A and B are

$$T_A = m_A a_A$$

and

$$m_B g - T_B = m_B a_B.$$

Since the string is of fixed length, we have

$$a_A = a_B,$$

and since it is ideal, we have

$$T_A = T_B.$$

Note: There is no minus sign since the directions of the forces have already been taken into account by our drawing.

Applying these two conditions to the preceding equations, we get

$$m_B g = (m_B + m_A)\, a_B$$

or

$$a_B = \frac{m_B}{m_B + m_A}\, g = a_A.$$

If we had chosen to consider the system as a unit, we would have the forces shown in Fig. 4.3e. We could again write

$$N = -mg,$$

and the acceleration is given by

$$(m_A + m_B)\, a = m_B g,$$

which gives the same result.

PROBLEM 4.4: What is the minimum force required to drag a heavy carton across the floor if the force is applied upward at an angle of 40° to the horizontal? Let the mass of the carton be 85 kg and the coefficient of friction be 0.75. (4.8)

SOLUTION:

Strategy: Draw a figure showing the forces acting on the box. These forces are gravity, friction, and the applied dragging force. Choose coordinates along and perpendicular to the floor. This choice will simplify the problem. Write these forces in component notation. Since the box stays on the floor, the net vertical component of the forces must be zero. Use this to find the normal force and substitute this value into the horizontal component equation.

This is posed as a problem in dynamic equilibrium; that is, the assumption is that the force is the minimum needed to maintain a constant velocity of the carton. First draw a diagram showing the physical problem as shown in Fig. 4.4a and a free body force diagram as in Fig. 4.4b. We expect the forces acting to be gravity, the normal force of the floor, friction, and the applied force.

(a)　　　　　　　　　　　　　　　　　　　　　　　**(b)**

Fig. 4.4:　　　Problem 4.4: (a) The physical arrangement.　(b) Force diagram.

We denote the applied force by F_a, the mass of the carton by m, the normal force of the floor on the carton by N, and the frictional force by F_f. We choose coordinates x positive to the right and y positive upward. The conditions for equilibrium are:

$$\text{sum of } x \text{ components of forces} = 0,$$

$$\text{sum of } y \text{ components of forces} = 0.$$

Using the free body diagram Fig. 4.4b, we have

x force components　　　　　　　　　　$F_a \cos\theta - F_f = 0,$

y force components are　　　　　　　　$F_a \sin\theta + N - m g = 0,$

where we have made use of our knowledge that **N** must point in the y direction and F_f must point in the negative x direction.

> **GIVEN:**
>
> $m = 85\,\text{kg}$　　　　$\theta = 40°$　　　　$F_f = \mu N$　　　where $\mu = 0.75$

Note: We cannot assume $N = mg$ since there is a component of F_a in the y direction.

We can put the relation for F_f into the x component equation to get,

$$F_a \cos\theta = \mu N.$$

Then the y component becomes

$$F_a \sin \theta + \frac{F_a \cos \theta}{\mu} - m g = 0$$

or

$$F_a = \frac{m\,g}{\sin\theta + \frac{\cos\theta}{\mu}} = \frac{833\ \text{N}}{0.643 + \frac{0.766}{0.75}} = 501\ \text{N} = 5.0 \times 10^2\ \text{N}.$$

The applied force must be 5.0×10^2 N.

Note: A value of 501 N ignores significant figures, and a value of 500 N implies more significant figures than are known.

PROBLEM 4.5: What is the minimum force required to push a heavy carton across the floor if the force is applied downward at an angle of 40° to the horizontal? Let the mass of the carton be 85 kg and the coefficient of friction be 0.75. (4.8)

SOLUTION:

Strategy: Use the same strategy used for the preceding problem.

This problem is identical with Problem 4.4 except for the direction of the applied force. We will carry the previous analysis through step-by-step with the same detail to show how the analysis is identical except for some sign changes. The numerical results will be very different.

This is a problem in dynamic equilibrium; that is, the assumption is that the force is the minimum needed to maintain a constant velocity of the carton. First draw a diagram showing the physical problem as shown in Fig. 4.5a and a free body force diagram as in Fig. 4.5b. We expect the forces acting to be gravity, the normal force of the floor, friction, and the applied force.

We denote the applied force by F_a, the mass of the carton by m, the normal force of the floor on the carton by N, and the frictional force by F_f. We choose coordinates x positive to the right and y positive upward. The conditions for equilibrium are

Sum of the x components of forces $= 0$ and

sum of the y components of forces $= 0$.

(a)   **(b)**

Fig. 4.5:   Problem 4.5: (a) The physical arrangement. (b) Force diagram.

Using the free body diagram Fig. 4.5b, we have

x force components $\qquad F_a \cos \theta - F_f = 0,$

y force components $\qquad -F_a \sin \theta + N - m g = 0,$

where we have made use of our knowledge that N must point in the y direction and F_f must point in the $-x$ direction.

> **GIVEN:**
>
> $m = 85 \text{ kg} \qquad \theta = 40° \qquad F_f = \mu N \qquad$ where $\mu = 0.75$ is the coefficient of friction.

Also note that we cannot assume $N = mg$ since there is a component of F_a in the y direction.

We can put the relation for F_f into the x component equation to get

$$F_a \cos \theta = \mu N.$$

Then the y component becomes

$$-F_a \sin \theta + \frac{F_a \cos \theta}{\mu} - m g = 0$$

or

$$F_a = \frac{m g}{\dfrac{\cos \theta}{\mu} - \sin \theta} = \frac{833 \text{ n}}{\dfrac{0.766}{0.75} - 0.643} = 2.2 \times 10^3 \text{ N}.$$

This is a much larger force than we found in Problem 4.4. Clearly it is better to pull up on the box to move it than to push down.

PROBLEM 4.6: What is the minimum force required to push an 80 kg carton up an incline of 25° if the coefficient of friction between the incline and the box is 0.70 and if the force is applied upward at an angle of 20° with respect to the horizontal? (4.8)

SOLUTION:

> **Strategy:** Draw a force diagram. Analyze the forces acting on the box. Since there is no motion perpendicular to the plane, a choice of coordinates parallel and perpendicular to the plane will simplify the solution. Then a component expression has no motion and hence no acceleration perpendicular to the plane. The perpendicular component can be used to eliminate the normal force. The parallel component is then analyzed to determine the minimum size force needed for the acceleration to be zero or directed up the plane.

Fig. 4.6: Problem 4.6.

Note: The problem posed is sketched in Fig. 4.6. A choice of coordinates has been made with the axes parallel and perpendicular to the plane. This choice is not essential, but it has the good feature that any motion occurs along only one of the coordinate axes.

A choice of vertical and horizontal axes would have involved motion in both directions, and we would have had to determine that the motion was confined to the surface of the inclined plane. To determine that the motion was on the plane, we would have had to require that for any motion Δ(horizontal) and Δ(vertical) that

$$\frac{\Delta(\text{vertical})}{\Delta(\text{horizontal})} = \tan(25°).$$

This would make a relatively easy problem appear very hard.

Advice: **Pick your coordinate axes so as to minimize the number of coordinates that change value.**

Since the force is given as at an angle of 20° upward with respect to the horizontal, it will be downward at an angle of 20° − 25° = − 5° (the minus sign tells us that it is downward) with respect to the surface of the inclined plane. We will call this angle ϕ. We will also let f represent the magnitude of the applied force and N represent the magnitude of the normal force. The inclination of the plane will be represented by θ.

Newton's second law then gives us

　　sum of forces in y direction $= 0 = N - mg\cos\theta - f\sin\phi$

and

　　sum of forces in x direction $= ma = f\cos\phi - mg\sin\theta - \mu N$

where a is the acceleration in the x direction and μ is the coefficient of friction.

First, solve the y equation for N and put the results into the x equation, giving:

$$ma = f[\cos\phi - \mu\sin\phi] - mg[\sin\theta + \mu\cos\theta].$$

For the box to move up the inclined plane we must have $a \geq 0$ with $= 0$ giving constant velocity. Thus,

$$f \geq mg\frac{\sin(\theta) + \mu \cos(\theta)}{\cos(\phi) - \mu \sin(\phi)} = 8.1 \times 10^2 \text{ N.}$$

So a force of at least 8.1×10^2 N must be applied to the box.

PROBLEM 4.7: Four coplanar forces act on a 10 kg mass. Three of them are as follows (directions are measured clockwise relative to the reference direction):

7.00 N at 0°; 20.0 N at 100°; 25.0 N at 250°.

What are the magnitude and direction of the fourth force if the 10 kg mass is to be in translational equilibrium? (4.9)

SOLUTION:

Strategy: We will consider two methods:
1. Graphical: Make an accurate drawing of the three given forces, placing tail to head for the force vectors. The fourth force must bring you back to to the tail of the first force since the system is in equilibrium.
2. By components: Calculate the components of each of the three forces and add like components. The fourth force must have components equal to the resultant of the first three and with the opposite sign.

The term "coplanar" means all the forces act in a plane, so the problem is two-dimensional. We are looking for equilibrium, so there is no information in a motion diagram. Since there is only one body, the only force diagram we can draw is a free body force diagram. This is shown in Fig. 4.7a. The fourth force has been drawn at an angle ϕ with respect to the reference direction.

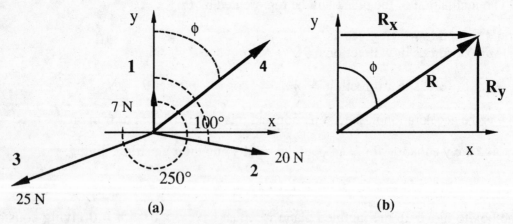

FIG. 4.7: (a) Four forces acting on a 10 kg mass. (b) Components of the vector **R**.

In the figure, we have chosen the reference direction for angle measurement as the y-axis and have also chosen an x-axis. This is done so we can take components to calculate the resultant force of the three given forces. That is, for equilibrium,

$$F_1 + F_2 + F_3 + F_4 = 0,$$

where

$$F_1 = 7.00 \text{ N at } 0°, \qquad\qquad F_2 = 20.0 \text{ N at } 100°,$$
$$F_3 = 25.0 \text{ N at } 250°, \text{ and} \qquad F_4 = R \text{ N at } \phi°$$

represents the unknown fourth force with a magnitude of R newtons.

Writing the component equations gives:

$$(F_1 + F_2 + F_3 + F_4)_x = 0$$

and

$$(F_1 + F_2 + F_3 + F_4)_y = 0.$$

The components are

$$(F_1)_x = 0, \ (F_1)_y = 7.00 \text{ N};$$

$$(F_2)_x = 20.0 \text{ N} \times \cos(10°) = 20.0 \times 0.985 \text{ N} = 19.7 \text{ N};$$
$$(F_2)_y = 20.0 \text{ N} \times (-\sin(10°)) = -20.0 \times 0.174 \text{ N} = -3.47 \text{ N};$$

$$(F_3)_x = 25.0 \text{ N} \times (-\cos(20°)) = -25.0 \times 0.940 \text{ N} = -23.5 \text{ N};$$

$$(F_3)_y = 25.0 \text{ N} \times (-\sin(20°)) = -25.0 \times 0.342 \text{ N} = -8.55 \text{ N};$$

where we have used Fig. 4.7a to get the proper angles less than 90° and the proper signs. This is not necessary if you are careful with the sign of the trig functions in each quadrant.

From Fig. 4.7b we see that for a vector R that $R_x = R \sin \phi$ and $R_y = R \cos \phi$. Then, for example,

$$(F_2)_x = 20.0 \text{ N} \times \sin(100°) = 20.0 \times (0.985) \text{ N} = 19.7 \text{ N}$$

$$(F_2)_y = 20.0 \text{ N} \times \cos(100°) = 20.0 \times (-0.174) \text{ N} = -3.47 \text{ N}.$$

The signs of the trig functions should be correctly generated by your calculator.

We can now calculate the needed force

$$F_4 = -(F_1 + F_2 + F_3)$$

or

$$(F_4)_x = -(0 \text{ N} + 19.7 \text{ N} - 23.5 \text{ N}) = 3.80 \text{ N},$$

and

$$(F_4)_y = -(7.00 \text{ N} - 3.47 \text{ N} - 8.55 \text{ N}) = 5.02 \text{ N}.$$

Fig. 4.7b also shows us that

$$R = \sqrt{R_x^2 + R_y^2} \quad \text{and} \quad \tan \phi = \frac{R_x}{R_y},$$

so

$$R = \sqrt{(3.80)^2 + (5.02)^2} \text{ N} = 6.30 \text{ N}$$

and

$$\tan \phi = \frac{3.80 \text{ N}}{5.02 \text{ N}} = 0.75697.$$

Hence,

$$\phi = 37°.1,$$

where we need to know that both components of R are positive to take the inverse tangent.

The entire problem could have been done in radians by restating the initial conditions. This is done by dividing the angle in degrees by 2π. Most calculators will compute trig functions with angles in either degrees or radians and can convert degrees to radians and conversely. Just be sure to be consistent in your choice and be sure that your calculator is using the units that you input.

Note: Information that is not needed was given. The mass of the body was not used in this solution. Let this be a warning that the given information for a problem may be more than you need, so don't insist on using all information. This is a better situation than in the real world, where you may leave the laboratory without all the needed information.

PROBLEM 4.8: A 5.0 kg weight is suspended with a rope (string) and pulley as shown in Fig. 4.8a. What is the magnitude of the force needed for equilibrium? The pulley has a mass of 0.50 kg.(4.9)

Solution:

Strategy: Draw a force diagram and analyze the forces acting on the system. Use components, preferably vertical and horizontal, and require that the sum of the forces is zero. If you use free body diagrams, include tension where appropriate.

Fig. 4.8:   Problem 4.8: (a) The proposed problem. (b) The idealized problem. (c) Free body diagram (pulley). (d) Free body diagram (5 kg mass). (e) External force diagram for the problem.

We will let F be the magnitude of the force. We idealize the rope as a massless inextensible string. Thus, the force desired must be in the same direction the string points, 30°, and directed outward from the end of the string. For this problem, we must deal with the mass of the pulley since it is given; but, other than that, the pulley is ideal in the sense that it changes only the direction of the string. The connection between the pulley and the 5.0 kg mass is assumed to be an ideal string.

Fig. 4.8b shows the idealized problem and the external forces acting on the system. We will measure vertical directions + downward from the supporting surface.

GIVEN or to be FOUND:

Quantity	Symbol
5.0 kg mass	M
0.50 kg mass	m
tension in the string supporting the pulley	T
tension in the string between the pulley and the 5 kg mass	T_d
applied force	F
angle between left support and horizontal	θ

Since the string holding the pulley is ideal, we know the tension in the string is equal in magnitude to the applied force F. Since the string is ideal, the tension in the string on the right-hand side of the pulley has to be the same tension as on the left-hand side. Since we are in equilibrium, the horizontal forces must add to zero. Hence,

$$T \cos \theta = T \cos 30° \quad \text{so} \quad \theta = 30°.$$

The free body force diagrams for the pulley and the mass are shown in Figs. 4.8c and 4.8d. We know the horizontal forces cancel since the system is in equilibrium. So only the vertical components need be considered. The equations of equilibrium for the **pulley** are

$$-T \sin \theta - T \sin \theta + m g + T_d = 0;$$

while for the **mass**, they are

$$-T_d + M g = 0.$$

The unknown force is equal in magnitude to the tension, T, and directed at an angle of 30° along the string. Solving the equations, we have

$$2 T \sin 30° = m g + T_d$$

and

$$M g = T_d.$$

This gives

$$T = \frac{(m + M) g}{2 \sin \theta} = \frac{5.5 \text{ kg} \times 9.8 \text{ m/s}^2}{2 \times 0.5} = 54 \text{ N} = F.$$

The force diagram for the system of pulley plus mass is shown in Fig. 4.8e. For equilibrium, the vertical forces add as

$$-T \sin \theta - T \sin \theta + (m + M) g = 0,$$

which gives, as it must, the same result.

COMMENTS ON FRICTION

In the model for friction used in the text, we take the frictional force as a given expression without attempting to understand how it arises in the real world. As stated in the text, the frictional force is understood in terms of molecular forces and Newton's laws. In the context of the point object model used at this stage in the course, we will have to take the friction simply as a given property. The frictional force is a **force** that **always opposes** motion along the surface upon which the body is moving or resting. For Fig. 4.9 we take the direction x to be along the surface.

If you use the free body diagram approach, you must consider the tension in the string as a force applied to each body. The force is horizontal in Fig. 4.9a and at an angle θ in Fig. 4.9b. Even though we are modeling the body as a point, figures usually show it as extended for visibility.

In Fig 4.9a the surface is perpendicular to the weight of the body, so the normal force, N, is opposite and equal to the weight, **mg** (note the weight of an object is a vector). This equality follows from Newton's second law and the assumption that the body does not move off the surface. As a vector expression,

$$N = -mg.$$

If no other forces act ($F_{applied} = 0$ in Fig 4.9a) then the frictional force is zero. If there is an applied force, there will be a frictional force opposing the applied force. If the component of the applied force parallel to the surface is large enough, the body will begin to accelerate and the frictional force opposing the motion is given by μN. Note that any component of the applied force perpendicular to the surface will change the value of the normal force and hence affect the size of the frictional force. See Problem 4.4 for an example.

Fig. 4.9: Frictional force: (a) Horizontal surface with an external force $F_{applied}$. (b) tilted surface. (c) Force components resolved along x and y.

In Fig. 4.9b, we show the same body on an inclined plane and the forces that act. In Fig. 4.9c, the gravitational force is decomposed into its x and y components. The normal force, N, is in the $+y$ direction and has magnitude $mg \cos \theta$. The body is in equilibrium provided the frictional force, $F_{friction}$, can equal the x component of gravity. This is possible if

$$F_{friction} = - (-mg \sin \theta) = mg \sin \theta \leq \mu N = \mu\, mg \cos \theta,$$

which means that

$$\mu \geq \tan \theta.$$

If the preceding inequality is not satisfied, then the block is accelerated down the incline by a force $-mg \sin \theta + \mu N < 0$ (the $-$ sign indicates acceleration down the incline).

In no case does the frictional force encourage acceleration. It can only oppose applied forces and reduce the resulting acceleration. For the case shown in Fig. 4.9b–c, the applied force along the plane is $-mg \sin \theta$, provided by the force of gravity. Only when the applied force is larger than the maximum frictional force (μN), does the body move.

GENERAL COMMENTS

This chapter introduces a number of new **variables** into our model of the real world. The mass is an intrinsic property of each body. When we have problems where the mass changes, we can break the body into smaller pieces. Each of these pieces has its own intrinsic mass. Alternatively, we can relax the intrinsic nature of mass and let it be a variable, just as the velocity and position are. Variables that change are position, velocity, and acceleration, which we saw in Chapter 3. Finally, there are the forces that are variables representing interactions among the bodies of the problem.

We have finally introduced some **equations of the model**, Newton's laws. These laws provide relationships among the variables both for single bodies (Newton's first law) and among the bodies themselves (Newton's second and third laws). Although the three laws of Newton as given in this text and in most others use the word **body**, we must remember that so far we have a model where all bodies are point objects. We will later see how to extend or modify our model to include extended bodies.

We have also introduced several new objects into our model. One such object is the surface. The surface is an idealized plane or curved object that has the property of being impenetrable and perfectly rigid. The surface of our model is clearly an object that does not exist in the real world; however, we will interpret the surface as our modeling of a real but simplified surface.

Our model surface can also have a property that we call the coefficient of friction. This is our modeling of the true frictional properties of surfaces. The coefficient of friction depends both on the surface and on the object resting or moving upon the surface. For any problem we have a given body resting on a given surface, so there is a number (usually denoted by the Greek letter μ), which gives the coefficient of friction. We will discuss friction more fully under its own separate heading.

Other objects introduced are the light inextensible string and the massless pulley. The string, whose mass can be neglected and whose length does not change, is a model of the real world string. Such a string is assumed for now to transmit tensional forces along its length without loss and not to support any compressional forces. That is, if you pull on the string with a force, F, this force appears without loss at the opposite end.

Likewise, the ideal pulley is the model for the real world pulley that is used to change the direction of forces without introducing the complication of friction. Both of these idealizations can be considered a good approximation when the masses of the attached bodies are large compared with the string and pulley masses and for the pulley when the friction of the bearings can be neglected. In later chapters, both the string and pulley will be given additional properties.

COMMENTS ON FORCE DIAGRAMS

The drawing of one or more force diagrams is essential to the understanding and the solution of problems. Force diagrams should supplement motion diagrams; they should not replace them. Of course, if the problem is a statics problem (sum of forces equals zero), a motion diagram will be uninformative. Along with the force diagrams, it is often useful to draw a separate figure showing the physical arrangement.

There are two kinds of force diagrams introduced in physics texts — the **external force diagram** and the **free body diagram**. The use of each of these is shown at the start of text Section 4.7 with the Atwood machine example. We cannot replace the real Atwood machine with a single point body. The best we can do is to model each mass as a point and replace the string with an idealized massless inextensible string. If we ask only about the motion of the system of two masses and string, then use the external force diagram, it is possible to solve for the motion of the system and not find the tension in the string connecting the masses.

If you use the **free body diagram** approach, you must consider the tension in the string as a force applied to each body and also have the tension at each end of the string be equal and opposite. The free body diagram approach has the advantage of forcing you to consider all forces that are acting, even forces that hold separate parts (nonpoint) of the system together. However, this method has the disadvantage of forcing you to find forces that need not be known to solve the posed problem.

Although the free body diagram approach may take a few extra steps, studies have shown that a consistent free body diagram approach leads to better student understanding of the posed problems. We will use the free body approach for all sample solutions as well as for the external force approach when appropriate.

SUGGESTED READINGS

Escobar, C., "Amusement Park Physics." *The Physics Teacher,* October 1990, p. 446.

McFarland, E., "Newton's Third Law at a Traffic Intersection." *The Physics Teacher,* May 1990, p. 315.

Moore, D. F., *The Friction of Pneumatic Tyres.* Amsterdam: Elsevier, 1975.

Chapter 5
Uniform Circular Motion and Gravitation

IMPORTANT TERMS

Fill in the blank with the appropriate word or words.

1. The _____ is the constant that appears in Newton's law of gravitation.

2. The quantity 2π divided by the period of rotation of an object is called the object's _____.

3. The force needed to keep a body in a circular orbit is called the _____.

4. The SI unit of frequency is the _____.

5. The time it takes to complete one revolution about any closed path is called the _____ of the motion.

6. Newton proposed the _____, which tells us the gravitational force between two bodies.

7. The _____ of a substance is the mass per unit volume.

8. The reciprocal of the period of motion is the _____.

9. Motion in a circular path with constant speed is called _____.

10. The technical name for the acceleration needed to move in a circle at constant speed is the _____ acceleration.

11. The _____ for circular motion is the rate of change of the angular position as seen from the center of the orbit.

12. The _____ is the ratio of the gravitational force on a body to the mass of the body.

13. Drawing _____ of force can help us visualize the gravitational field.

Write the definitions of the following:

14. angular frequency (both concept and equation with symbols defined) (5.1):

15. hertz (5.1):

16. gravitational field strength (both concept and equation with symbols defined) (5.5):

17. uniform circular motion (5.1):

18. law of universal gravitation (both concept and equation with symbols defined) (5.3):

19. universal gravitational constant (both concept and value with units) (5.3):

20. frequency (5.1):
21. angular velocity (5.1):
22. centripetal acceleration (5.1):
23. centripetal force (5.2):
24. density (5.3):
25. period (5.1):
26. lines of force (5.5):

ANSWERS TO 1–13, IMPORTANT TERMS

1. gravitational constant; 2. angular frequency; 3. centripetal force; 4. hertz; 5. period;
6. law of universal gravitation; 7. density; 8. frequency; 9. uniform circular motion;
10. centripetal; 11. angular velocity; 12. gravitational field strength; 13. lines.

BINOMIAL EXPANSION

We will often see an expression of the form

$$(1 + x)^n$$

where $x \ll 1$ and n is any number positive or negative.
If n is a positive integer we know that

$$(1 + x)^1 = 1 + x,$$

$$(1 + x)^2 = 1 + 2x + x^2,$$

$$(1 + x)^3 = 1 + 3x + 3x^2 + x^3,$$

or in general

$$(1 + x)^n = 1 + \frac{nx}{1} + \frac{n(n-1)x^2}{1(2)} + \frac{n(n-1)(n-2)x^3}{1(2)(3)} + \dots + x^n.$$

This is an increasing power series in x. If $x \ll 1$, then each successive term is smaller than the preceding one, so we can often approximate the series as

$$(1 + x)^n \approx 1 + nx.$$

This approximation works even if n is not a positive integer provided $x \ll 1$.

SAMPLE QUESTIONS

QUESTION 5.1: How can a body in uniform circular motion be moving at constant speed and at the same time be constantly accelerated?

Answer:

This is a good example of the difference between speed and velocity. The speed (which is the magnitude of the velocity) stays constant, but the velocity is changing. This is possible if the acceleration is perpendicular to the velocity. If the acceleration is not perpendicular to the velocity, the speed will also change.

QUESTION 5.2: What are the restrictions on the orbit of a communications satellite if it is to remain motionless as viewed in the sky from a location on the earth's surface?

Answer:

The satellite must be in a circular orbit with the same period as the rotation of the earth. This is commonly referred to as a "24 hour" orbit; but, since the earth is rotating about the Sun, both periods are completed in slightly less than a day (about 4 minutes less). Look up the words sidereal and synodic.

SAMPLE SOLUTIONS

PROBLEM 5.1: A merry-go-round (carrousel) has a diameter of 50 feet and a period of 25 s. What is its frequency in hertz? A ticket taker hangs on an outer pole. What is her angular velocity in rad/s, her speed in m/s and her radial acceleration in m/s²? A rider is 15 feet from the center. What are his angular velocity, speed and radial acceleration in the same units? Using text Eq. 5.3, verify that both have the same angular velocity . (5.1)

Solution:

> **Strategy:** For this problem we are given a period of rotation and several different radii and are asked to find linear velocities and accelerations. This suggests using the relation between angular and linear velocity and the formula for centripetal acceleration. We choose to use meters and radians as our units.

We first draw a sketch showing the idealized problem. The ticket taker is idealized as a point mass 25 feet from the center and the rider as a point 15 feet from the center. The merry-go-round is a disk rotating with a frequency ω that corresponds to a 30 s period. Numbers have apparently been given to 2 significant figures.

The frequency, f, is given by text Eq. 5.5 as

$$f = \frac{1}{T}$$

where T is the period, so

$$f = \frac{1}{25\ \text{s}} = \frac{0.040}{\text{s}} = 0.040\ \text{hertz}.$$

The ticket taker is at a distance of

$$25\ \text{feet} = 25\ \text{feet} \times \frac{12\ \text{in.}}{\text{foot}} \times \frac{0.0254\ \text{m}}{\text{in.}} = 7.62\ \text{m}$$

from the center of rotation. Her angular frequency is then

$$\omega = 2\pi \times f = 0.2513272\ \text{rad/s} = 0.25\ \text{rad/s}.$$

The factor of 2π in the conversion from frequency to angular frequency is a result of choosing radians as the unit of angular frequency.

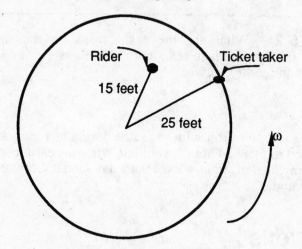

Fig. 5.1: Merry-go-round with rider and ticket taker.

The circumference of a circle of radius $r = 7.62$ m is

$$2\pi r = 2\pi \times 7.62\ \text{m} = 47.8778316\ \text{m} = 48\ \text{m}$$

to 2 significant figures. Thus the speed of the ticket taker, v, is given by

$$v = \frac{\text{distance}}{\text{time}} = \frac{48\ \text{m}}{25\ \text{s}} = 1.92\ \text{m/s} = 1.9\ \text{m/s}.$$

This means the ticket taker has a centripetal acceleration, $(a_c)_{tt}$, given by

$$(a_c)_{tt} = \frac{v^2}{r} = \frac{(1.9\ \text{m/s})^2}{7.6\ \text{m}} = 0.475\ \text{m/s}^2 = 0.48\ \text{m/s}^2.$$

The rider is 4.572 m $= 4.6$ m from the center, so he is moving on a circle of radius 29 m. His speed is then 1.2 m/s and he has a centripetal acceleration, $(a_c)_r$, given by

$$(a_c)_r = \frac{v^2}{r} = \frac{(1.2\ \text{m/s})^2}{4.6\ \text{s}} = 0.31\ \text{m/s}^2.$$

The angular frequency, ω_{tt}, of the ticket taker is

$$\omega_{tt} = \frac{v}{r} = \frac{1.9 \text{ m/s}}{7.6 \text{ m}} = 0.25 \text{ Hz}$$

while the angular frequency, ω_r, of the rider is

$$\omega_r = \frac{1.2 \text{ m/s}}{4.6 \text{ m}} = 0.26 \text{ Hz}.$$

If numbers differ only by one in the last significant figure, they must be considered equal to the given significance. With 2 significant figures these numbers (0.25 Hz and 0.26 Hz) are effectively equal.

PROBLEM 5.2: A belt passes over two pulleys. The diameter of the smaller pulley is 6.0 cm while the diameter of the larger is 24 cm. The centers of the pulleys are separated by 2.0 m. If the smaller pulley is turning at a frequency of 100 hertz, what is the frequency of the larger pulley? (5.1)

Solution:

> **Strategy:** Since the pulleys are connected by a belt, the linear distance a point on one pulley travels must be the same as the distance a point on the other pulley travels.

We let A indicate the smaller pulley and B the larger one. A sketch of the problem is shown in Fig. 5.2.

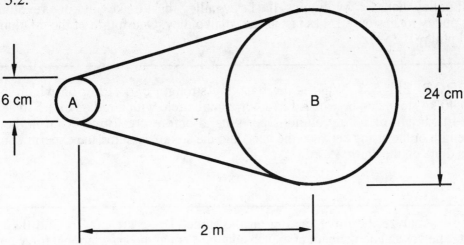

Fig. 5.2:   Two pulleys connected by a belt.

The smaller pulley has a circumference given by:

small pulley circumference $= \pi \times (6.0 \text{ cm}) = 18.84954 \text{ cm} = 19 \text{ cm}$

while the larger pulley has a circumference given by:

large pulley circumference = 75 cm.

When the smaller pulley makes one revolution, the belt moves 19 cm. When the belt moves 19 cm, the larger pulley turns through an angle θ given by

$$\theta = \frac{\text{distance}}{\text{radius}} = \frac{19 \text{ cm}}{12 \text{ cm}} = 1.5833333 \text{ rad} = 1.6 \text{ rad.}$$

This angle is $1.6/(2\pi) = 0.245$ of a complete revolution. The small pulley makes a complete revolution in

$$\frac{1}{f} \text{ s} = 0.01 \text{ s,}$$

so the larger pulley makes 0.25 of a turn in 0.01 s. Thus, it will take

$$\frac{0.01 \text{ s}}{0.25} = 0.040 \text{ s}$$

for a complete revolution. The frequency of the larger pulley is then

$$\frac{1}{0.040 \text{ s}} = 25 \text{ Hz.}$$

Another approach is to note that the ratio of the diameters of the pulleys is

$$\frac{24 \text{ cm}}{6 \text{ cm}} = 4.0.$$

This is also the ratio of the circumferences. The belt constrains the system such that when the small pulley makes 4 revolutions, the large pulley must make only one revolution. So the large pulley rotates only 1/4 as fast as the small pulley. One-fourth of the rotation rate of the small pulley is 25 Hz.

PROBLEM 5.3: What is the breaking strength of a string on which a 0.40 kg stone can be whirled in a horizontal circle with a radius of 1.5 m at a maximum of 2.0 revolutions per second before breaking? What is the length of the string between the hand and the stone? Assume the experiment is done on the earth. (5.2)

Solution:

Strategy: Analyze the forces acting on the stone. These are gravity and the string tension. The vertical component of the tension must equal the gravitational force and the horizontal component must be the same size as the centrifugal force needed for circular motion. The tension in the string for our conditions is assumed to be the maximum tension allowed without breaking the string.

We let the stone be a point of mass m, and the string be an idealized massless string whose radius of motion is 1.5 m. The frequency $f = 2.0 / \text{s}$. Our problem is to find the breaking strength of the string. The statement of the problem implies that if we had used a more massive rock or if we had spun the rock faster, the string would break. The force which could break the string is the tension in the string. First we make a sketch of the problem which also serves as a motion diagram as shown in Fig. 5.3a.

Fig. 5.3: (a) Preliminary drawing of stone on the end of a whirled string. (b) Forces on the mass. (c) Perspective view of problem. (d) Forces on string. (e) Force on hand from the string

Fig. 5.3 shows the forces on the mass. There are only two forces, gravity acting downward and the tension. Clearly for dynamic equilibrium the string cannot be exactly horizontal since there must be a vertical component of the tension to balance the force of gravity. The path of the stone can be horizontal as given in the problem, but the string must lie above this path as shown in Fig. 5.3c. Other possible free body force diagrams are shown for the string in Fig. 5.3d and for the hand holding the string in Fig. 5.3e.

Let the angle between the horizontal and the string be θ. The vertical component of the tension must balance the force of gravity on the stone and the horizontal component of the tension must provide the centripetal force on the stone. The centripetal force is given by Eq. 5.6.

GIVEN:

radius of motion	r	1.5 m
frequency	f	2.0 hertz
mass on end of string	m	0.40 kg
tension in string	T	?

The force equations are:

$$m g - T \sin \theta = 0 \qquad \text{vertical component of force}$$

$$T \cos \theta = m \frac{v^2}{r} \qquad \text{horizontal component of force,}$$

where v is the circular velocity corresponding to the radius r and frequency f.

The vector sum of the two components of the tension $[T \sin(\theta), T \cos(\theta)]$ must be the tension, T. So we can find the tension without knowing θ by squaring and summing; that is,

$$T^2 = (T_{\text{horizontal}})^2 + (T_{\text{vertical}})^2$$

or

$$T^2 = T^2 \cos^2 \theta + T^2 \sin^2 \theta = (m g)^2 + m^2 \frac{v^4}{r^2} \; .$$

We will need the velocity,

$$v = \frac{\text{circumfrence}}{\text{period}} \; ,$$

$$v = \frac{2 \times \pi \times 1.5 \text{ m}}{\dfrac{1}{2.0/s}} = 18.85 \text{ m/s} \; .$$

Then

$$T = m \sqrt{g^2 + \frac{v^4}{r^2}} = 0.40 \text{ kg} \sqrt{(96 + 56,113) \text{ m}^2/\text{s}^4} = 94.8 \text{ N}.$$

The tension in the string is 95 N.

To find the length of the string we will need θ. From the vertical component of force we get

$$\sin \theta = \frac{m g}{T} = \frac{3.9 \text{ N}}{95 \text{ N}} = 0.041 \quad \text{or} \quad \theta = 2.4°.$$

The string and the radius form the hypotenuse and one side of a right triangle, so

$$\text{length of string} = \frac{\text{radius of circle}}{\cos \theta} = \frac{1.5 \text{ m}}{0.9992} = 1.5 \text{ m}.$$

The string must be longer than the radius of the circle, but for this problem the difference is lost in the significant figures.

PROBLEM 5.4: Consider the preceding problem (Problem 5.3). The speed is increased slightly and the string breaks. Which of the following sketches (Fig. 5.4) of motion in the horizontal plane properly shows the prior motion in that plane? What will be the motion in the vertical plane? Assume there are 15 revolutions before the string breaks. (5.2)

Solution:

Look at Fig. 5.4a-c and choose the answer you feel is correct.

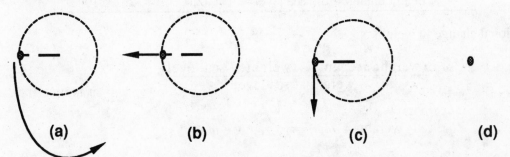

(a)　　　　　　**(b)**　　　　　　**(c)**　　　　　　**(d)**

Fig. 5.4:　　Three possible motions of the stone in the horizontal plane when the string breaks: (a) Spiral. (b) Radially outward. (c) Tangential. (d) Space for force diagram.

The proper way to answer this question is to draw a force diagram for the stone when the string has broken. Space has been left for you to draw this in Fig. 5.4d. Please do so before proceeding.

The only force is that of gravity acting vertically downward. There are no forces in the horizontal plane. Newton's second law then tells us that the velocity must continue with the value it had when the string breaks. Figure 5.4c must be the correct answer since the velocity was tangential before the string broke. The number of revolutions before the string breaks is not relevant information.

The vertical motion would correspond to uniform acceleration vertically downward under the force of gravity. The total motion would then be the same as if the stone had been fired horizontally from a cannon with velocity v.

PROBLEM 5.5: What would be the fractional change in the weight of a 120 lb woman when she moves from sea level to a mountain 2.0 km high? Assume that she is at the equator. (5.3)

Solution:

> **Strategy:** Consider what forces might act on the woman. The major force on the woman is gravity; however, the rotation of the earth will make her appear to weigh less. So we must subtract the centripetal force from the gravitational force to get her apparent weight.

The only externally applied force we will consider as acting on her is gravity as shown in Fig. 5.5. Let m be her mass, g_1 the acceleration of gravity at sea level, and g_2 the acceleration of gravity at the top of the mountain.

(a)　　　　　　　　**(b)**

Fig. 5.5:　　(a) Weight at sea level. (b) Weight at 2 km above sea level.

GIVEN:

sea level weight	120 lbs
height change from sea level	$2.0 \text{ km} = 2.0 \times 10^3 \text{ m}$

The fractional change in weight is

$$\frac{(\text{weight at 2 km}) - (\text{weight at sea level})}{\text{weight at sea level}}.$$

Now

$$\text{weight} = mg - \frac{mv^2}{r} = mg - m\,\omega^2 r,$$

where m is her mass, g the acceleration due to gravity, r the distance from the center of the earth and ω is the earth's rotation frequency. If she had not been at the equator, r would be the distance to the earth's rotation axis rather than the distance to the earth's center. We let r_1 and r_2 be the respective distances from the earth's center.

Then

$$\text{fractional change} = \frac{(mg_2 - m\,\omega^2 r_2) - (mg_1 - m\,\omega^2 r_1)}{mg_1 - m\,\omega^2 r_1}$$

$$= \frac{g_2 - \omega^2 r_2}{g_1 - \omega^2 r_1} - 1.$$

The first thing we note is that the actual weight does not matter when we calculate the fractional change in weight. The acceleration due to gravity we can get from Newton's second law and the law of universal gravitation:

$$F = mg = \frac{GMm}{r^2},$$

where m is the woman's mass, M the mass of the earth and r the distance from the earth's center. The first equality is from the second law and the second equality is from the gravitational law. Thus we have that

$$g = \frac{GM}{r^2},$$

and the fractional weight change, f, is

$$f = \frac{g_2 - \omega^2 r_2}{g_1 - \omega^2 r_1} - 1 = \frac{GM/r_2^2 - \omega^2 r_2}{GM/r_1^2 - \omega^2 r_1} - 1$$

$$= \frac{\dfrac{r_1^2}{r_2^2} - \dfrac{\omega^2}{GM} r_2 r_1^2}{1 - \dfrac{\omega^2}{GM} r_1^3} - 1 = \frac{\dfrac{r_1^2}{r_2^2} - 1 + \dfrac{\omega^2}{GM} r_1^2 (r_2 - r_1)}{1 - \dfrac{\omega^2}{GM} r_1^3}.$$

The first term

$$\left(\frac{r_1^2}{r_2^2} - 1 \right)$$

is the correction due to the change in gravity resulting from the change in distance. The second term,

$$\left(\frac{\omega^2}{GM} r_1{}^2 (r_2 - r_1) \right),$$

is the correction for the change in the centripetal force. The denominator of the fraction is not one since we used the measured weight at sea level instead of the true weight.

Now, using the values in text examples 5.12 and 5.13:

radius of earth	$= 6.38 \times 10^6$ m
mass of earth	$= 5.98 \times 10^{24}$ kg
gravitational constant	$= 6.67 \times 10^{-11}$ N m/kg^2.

Since the earth rotates with a 24 hr period,

$$\omega = \frac{2\pi}{T} = \frac{2\pi}{24 \text{ hr} \times 3600 \text{ s/hr}} = 7.27 \times 10^{-5} \text{ /s}.$$

We have

$$\frac{\omega^2 r_1{}^2}{GM} = 5.41 \times 10^{-10}/\text{m},$$

so

$$1 - \frac{\omega^2}{GM} r_1{}^3 = 1 - 3.45 \times 10^{-3} = 0.997,$$

while

$$\frac{\omega^2 r_1{}^2}{GM} (r_2 - r_1) = 1.08 \times 10^{-6},$$

and if we let $\Delta r = r_2 - r_1$,

$$\frac{r_1{}^2}{r_2{}^2} - 1 = \frac{r_1{}^2}{r_1{}^2 (1 + \Delta r/r_1)^2} - 1 = (1 + \Delta r/r_1)^{-2} - 1.$$

Now use

$$(1 + x)^n = 1 + nx + \ldots \approx 1 + nx \text{ for } x \ll 1$$

and get

$$\frac{r_1{}^2}{r_2{}^2} - 1 = -2\frac{\Delta r}{r_1} = -6.3 \times 10^{-4}.$$

We have only two significant figures in the last term and hence the other corrections can be neglected and the fractional change is -6.3×10^{-4}.

We conclude that the centripetal force could have been neglected from the beginning; however, it helps to do the complete calculation at least once to know this.

PROBLEM 5.6: What is the orbital velocity of an earth satellite that is in orbit at a radius of 1.02 times the earth's radius? (5.3)

Solution:

> **Strategy:** Consider the forces. There is only gravity. Since the orbit is circular, the gravitational acceleration must be equal to the centripetal acceleration.

From the preceding problem the radius of the earth is 6.38×10^6 m. Make a motion sketch in the space labeled Fig. 5.6a. We take R to be the radius of the earth, m the mass of the satellite, M the mass of the Earth.

(a)   (b)

Fig. 5.6:   (a) Space to sketch circular orbit of space satellite and forces acting for Problem 5.6. (b) Space to sketch circular orbit of Titan around Saturn for Problem 5.7.

The only force present is the gravitational force given by

$$\text{gravitational force} = \frac{GmM}{(1.02\,R)^2},$$

where M is the mass of the Earth, m is the mass of the satellite and R is the radius of the Earth.

Since the orbit is circular, this force must be the centripetal force:

$$\text{centripetal force} = \frac{mv^2}{1.02\,R},$$

where v is the orbital velocity of the satellite.

Equating these gives
$$v^2 = \frac{G\,M}{1.02\,R}$$
$$= \frac{(6.67 \times 10^{-11}\text{N m}^2/\text{kg})(5.98 \times 10^{24}\text{ kg})}{1.02 \times 6.38 \times 10^6\text{ m}} = 6.13 \times 10^7\text{ m}^2/\text{s}^2.$$

So
$$v = 7.8 \text{ km/s}.$$

This result is independent of the mass of the satellite (as expected).

PROBLEM 5.7: Calculate the mass of Saturn from the orbital data of Titan. Titan orbits Saturn at 1.221×10^6 km with a period of 15.594 days. (5.3)

Solution:

> **Strategy:** This is the same problem as 5.6 except for the numbers used. Follow the same procedure.

First sketch the orbit of Titan about Saturn in the space left in Fig. 5.6b. This will be a circle with Saturn in the center. Both Saturn and Titan are nearly at the same distance from the sun, so we can treat them is being in the same gravitational field due to the sun. This allows us to ignore the effect of the sun on the motion and only consider the gravitational interactions of the satellite and the planet. It is also reasonable to neglect the motion of Saturn produced by Titan if we can assume that the mass of Saturn is much larger than the mass of Titan. These assumptions allow us to claim that the gravitational attraction between Saturn and Titan is equal to the centripetal force needed to keep Titan in orbit about Saturn.

GIVEN:

Quantity	Symbol
mass of Saturn	M
mass of Titan	m
center distance between Saturn and Titan	r
velocity of Titan about Saturn	v
period of Titan about Saturn	T
gravitational force between objects	F_g
centripetal force	F_c

KNOWN:

$$r = 1.221 \times 10^6 \text{ km}, \quad T = 15.594 \text{ days}$$

The force diagram which you have drawn in Fig. 5.6b gives

$$m\frac{v^2}{r} = F_c = F_g = G\frac{m\,M}{r^2}.$$

The velocity can be found from the period by the relation

$$\text{velocity} = \frac{\text{circumfrence}}{\text{period}} = 2\pi\frac{r}{T} = v.$$

Solving these relations will give you the mass of Saturn, M.

PROBLEM 5.8: In Problem 5.3, how long is the the string between the rock and the pivot point? Assume the radius of rotation is 1.500 m. (5.2)

Solution:

> **Strategy:** We have done most of the work in this problem. It is only necessary to determine the angle θ between the plane of rotation and the string. Then simple trig will give us the solution.

Examine Fig. 5.3. The length of the string is the hypotenuse of a right triangle with long side equal to the radius of the circle of rotation and included angle θ. Thus

$$\cos \theta = \frac{r}{L},$$

where $r = 1.5$ m as given in the problem, and L is the desired length.

We found the vertical and horizontal forces to be

$$mg - T \sin \theta = 0$$

and

$$T \cos \theta - m\frac{v^2}{r} = 0,$$

so

$$\tan \theta = \frac{\sin \theta}{\cos \theta} = \frac{T \sin \theta}{T \cos \theta} = \frac{mg}{m\frac{v^2}{r}} = \frac{r\,g}{v^2} = 0.0407.$$

This gives

$$\theta = 0.041 \text{ rad} = 2.3°.$$

Since this is such a small angle, it is easy to see why many people assume that the pivot point is in the plane of rotation.

The length of the string is then

$$L = \frac{r}{\cos \theta} = \frac{1.500 \text{ m}}{0.999} = 1.502 \text{ m}.$$

The radius is only slightly different from the string length. The purpose of this problem is to make sure you understand that these two quantities (radius, length) are different.

PROBLEM 5.9: Compute the gravitational force between the sun ($M = 1.99 \times 10^{30}$ kg) and the planet Mars ($m = 0.108\ M_e$, $R = 1.52$ AU). Express the answer in Newtons. (5.4)

Solution:

> **Strategy:** Use Newton's gravitational law and convert masses and distances to SI units.

The mass of the earth has been previously given as 5.98×10^{24} kg and the AU as 1.50×10^{11} m. Newton's gravitational law is

$$F = G \frac{m_1 m_2}{R^2} \ ,$$

where F is the force, R is the center to center separation, G is Newton's gravitational constant, and m_1, m_2 are the masses.

Converting units:

$$m_1 = \text{mass of sun} = 1.99 \times 10^{30} \text{ kg},$$
$$m_2 = \text{mass of Mars} = 0.108 \times 5.98 \times 10^{24} \text{ kg},$$
$$R = 1.52 \times 1.50 \times 10^{11} \text{ m}, \ G = 6.673 \times 10^{-11} \text{ N·m}^2/\text{kg}^2$$

This gives a force of 1.65×10^{21} N.

PROBLEM 5.10: Masses of 5.0 kg are located at two of the corners of an equilateral triangle with sides of 10.0 m. Determine the magnitude and direction of the gravitational field at the third corner due to these masses. Neglect the effects of all other masses, including the earth. (5.5)

Solution:

Strategy: The gravitational force between two masses acts along the line joining the two masses. Find the magnitude and direction of the gravitational field due to each of the masses and vectorially add these forces.

The physical arrangement is shown in Fig. 5.7. The two masses are denoted by "1" and "2", while the triangle is shown by the light lines. The direction of the gravitational field due to each mass is shown by the arrow. From the symmetry of the problem, we know that the resultant field bisects the triangle and has a magnitude of

$$2 \times \cos (30°) \times \text{magnitude of gravitational field due to one mass.}$$

The factor of two comes from combining the equal field strengths of the two masses. For either mass the gravitational field strength is given by:

$$\Gamma = G \ \frac{m_1}{R^2} = 6.673 \times 10^{-11} \text{ N·m}^2/\text{kg}^2 \times$$

$$\frac{5.00 \text{ kg}}{10^2 \text{ m}^2} = 3.3 \times 10^{-12} \text{ N/kg}.$$

So the total gravitational field is

5.71×10^{-12} N/m

to the right in the figure.

FIG. 5.7: Field caused by two masses.

COMMENTS ON CENTRIPETAL FORCE

When labeling a force diagram for an object moving on a circular path, it is best never to label a force only as the "centripetal force." All forces drawn must have a physical existence independent of the particular path the moving body takes. Forces such as that of a rope, the normal force, friction, and gravity act on the body. If we are told that the body is moving on a circular path, then we can identify one or more of the physical forces as those causing the circular motion. We can then set the sum of these forces equal to the value of the centripetal force needed for circular motion. For example, consider Figure 5.8 of the text. The automobile is moving on a circle whose center lies to the right. The car's velocity changes direction, so we know there is a centripetal force. The only physical force that can provide this force is the frictional force $\mathbf{F}_{fr}$ between the tires and the road. Thus the total frictional force (that is, the force provided by the 4 tires) in the radial direction must be identified as the centripetal force. So:

$$\sum_{4 \text{ tires}} (\mathbf{F}_{fr})_{radial} = \frac{m\,v^2}{r}$$

where m is the mass of the auto, v the speed of the auto, and r is the radius of the turning circle. This allows us to calculate the frictional forces that must exist for the motion to be circular.

What may be unclear to you is how we knew that frictional forces existed. Consider what happens when you try to turn a car on glare (smooth) ice. When you try to change direction by turning the steering wheel, you may start to spin but you continue in a straight line. On glare ice there is little or no friction between the tires and the ice. On a normal roadway, changing the steering causes the direction of motion of the car to change. This tells us that there must be friction between the tires and the road for a car to turn.

When analyzing the physical forces, always choose a frame of reference that is not rotating or at least rotates only a small amount during the time under consideration. For this course, unless you are explicitly told otherwise for a particular problem, neglect the rotation of the earth. When you cannot neglect the rotation of your reference frame, there will be other coordinate dependent forces, such as the Coriolis force.

SUGGESTED READINGS

Bartlett, A. A., and J. P. Lamb, "The Train Left the Track." *The Physics Teacher*, December 1990, p. 586.

Beams, J. W., "Finding a Better Value for G." *Nature*, Vol. 310, 1984, p. 723.

Clodfelter, B. E., "The Cavendish Experiment as Cavendish Knew It." *American Journal of Physics*, March 1987, p. 210.

Cohen, I.B., "Newton's Discovery of Gravity." *Scientific American*, March 1981, p. 166.

Hawking, S. W., and W. Israel, eds., *Three Hundred Years of Gravitation*. New York: Cambridge, 1987. A contemporary view of gravitational theory.

Chapter 6
Work and Energy

IMPORTANT TERMS

Fill in the blank with the appropriate word or words.

1. The _____ is the unit of work in the SI system.

2. The time rate of doing work is called _____.

3. _____ is the energy of translational motion.

4. The ability to do work is called _____.

5. The energy that a body has because it is in a gravitational field is called the body's _____ energy.

6. A body's _____ energy is the sum of its kinetic energy and its potential energy.

7. _____ is defined as the product of a force and the distance through which the force acts.

8. The _____ states that the work done on a body is equal to its change in kinetic energy.

9. A quantity that remains constant during a change in a system is said to be _____.

10. A(n) _____ diagram is a plot of the potential energy against the displacement.

11. The unit of power is called the _____.

12. The sum of kinetic and potential energy is called _____.

13. Our first conservation law is the _____.

14. Another name for the energy of translational motion is _____ energy.

15. _____ laws tell us which quantities are the same before and after an interaction.

16. For a(n) _____ force, the work done depends upon the starting and ending points and is independent of the path.

Write the definitions of the following:

17. energy (6.3):

18. conservation laws (6.8):

19. potential energy diagram (6.8):

20. gravitational potential energy (6.5):

21. work-energy theorem (6.1):

22. total mechanical energy (6.8):

23. watt (and its definition in terms of joule/s) (6.9):

24. instantaneous power (6.9):

25. translational kinetic energy (6.4):

26. joule (6.1):

27. conservation of mechanical energy (6.8):

28. power (6.9):

29. conservative force (6.8):

30. mechanical energy (6.8):

31. kinetic energy (6.4):

32. work (6.1):

33. horsepower (and its definition in terms of the watt) (6.9):

34. distinguish between work and power

ANSWERS TO 1–16, IMPORTANT TERMS

1. joule; 2. power; 3. translational kinetic energy; 4. energy; 5. gravitational potential; 6. mechanical; 7. work; 8. work-energy theorem; 9. conserved; 10. potential energy; 11. watt; 12. total mechanical energy; 3. conservation of mechanical energy; 14. kinetic; 15. conservation; 16. conservative.

COMMENTS ON WORK AND POWER

The following questions should always be asked:

> **Who or what does the work?**
>
> **Who or what has work done on it?**
>
> **Is there motion with respect to the force?**

If you properly ask and answer these questions you will have few difficulties in expressing the physics of a problem. Failing to ask these questions will make the problems difficult. In the sample solutions we will show examples of how to apply properly these questions.

The work done by a variable force can be found by adding up the area under a force versus displacement curve. This follows from the fundamental definition of Work = Force × displacement. For example, Fig. 6.1a shows a force versus displacement plot for a force similar to the gravitational force $F \propto 1/r^2$. The problem is how to calculate the total work done while moving from one value of r to another. In Figure 6.1b we show an enlarged portion of the curve of Fig. 6.1a between the two values r_1 and r_2. We let $F(r)$ be the force at r and let the separation between r_1 and r_2 be small enough that the straight line between

$F(r_1)$ and $F(r_2)$ is a good approximation to the curve $F(r)$. Then the area under the curve is approximately the sum of the rectangle and the triangle:

$$\text{area} \quad = (r_2 - r_1) \times F(r_2) + \frac{1}{2} \times (r_2 - r_1) \times [F(r_1) - F(r_2)]$$

$$= (r_2 - r_1) \times \frac{F(r_1) + F(r_2)}{2}$$

$$= \text{displacement} \times \text{average force} = \text{work}.$$

Fig. 6.1:   (a) Force versus displacement curve for an inverse square law. (b) A small section of the curve in part a. (c) A power versus time curve.

This procedure can be done for a larger displacement by taking the total displacement as a sum of small displacements. Then the area under a work versus displacement curve is the total work done. The same arguments can be made for a power versus time curve such as shown in Fig. 6.1c to allow us to conclude that the total work done is the area under the power versus time curve.

When analyzing curves such as in Fig. 6.1, if a positive displacement indicates work done on the body then a negative displacement would indicate work done by the body.

SAMPLE QUESTIONS

QUESTION 6.1: A golf ball and a Ping-Pong ball have the same speed. Which one has the larger kinetic energy and why?

Answer:

The kinetic energy of a body is proportional to the mass of the object and its speed squared. The speeds are the same, but the golf ball will be more massive than the Ping-Pong ball, so the kinetic energy of the golf ball will be greater.

QUESTION 6.2: Why do small cars generally get better mileage than larger cars?

Answer:

A small car has less mass than a large car, So less work is done to move it at the same velocity as a larger and hence more massive car. Small cars also present a smaller cross-section and so have less air resistance to overcome at moderate speeds.

SAMPLE SOLUTIONS

PROBLEM 6.1: How much work is done by a man who weighs 550.0 N and who carries a box whose mass is 3.0 kg up a 20.0 m hill? Hint: Assume the motion is done with constant velocity. (6.1)

Solution:

> **Strategy:**
> 1. Force method: Examine the forces. Determine which are exerted **by** the man rather than **on** the man. Choose the component of force in the direction of motion and calculate the work.
> 2. Work–energy method. The work done must be the change in potential energy.

1. Force Method

A sketch of the problem is given in Fig. 6.2a. We let M be the mass of the man, m the mass of the box, g be the acceleration of gravity and h the height of the hill.

If we only had to worry about the work done on the box, we would argue that he applies an upward force of magnitude mg on the box. This force acts directly upwards for a distance h, so the work done by the man on the box is $m\,g\,h$.

Fig. 6.2: (a) Sketch of man and box moving uphill. (b) Forces involved. (c) How height moved is related to distance traveled.

The analysis of the work done by the man to move up the hill is much more complicated. The problem is that the object being moved is, in some sense, the source of the work. Let us analyze the problem by examining the forces acting on the man as shown in Fig. 6.2b. Gravity acts downward on the man–box system with a force of

$$\text{total gravitational force} = (M + m)\, g,$$

and the plane also provides a force which balances out the force of gravity. This force by the plane can be broken into two components, a tangential force, F_f, and a normal force, N. The normal force is directed upwards from the plane and balances the downward gravitational force. That is,

$$N = mg \cos \theta,$$

where θ is the slope of the hill at that location. There is also a frictional force directed up the hill which opposes the downward directed tangential gravitational force. Since the velocity is constant we must have

$$F_f = mg \sin \theta,$$

This frictional force is produced by the interaction of the man and the hill. In order to walk up the hill, the man must contract and release his muscles. The net effect of this is an effective force of

$$\text{force} = (M + m)g \cos \theta,$$

where θ is the slope angle. The distance moved as seen in Fig. 6.2c is just

$$\text{distance} = \frac{\Delta h}{\cos \theta},$$

where Δh is the vertical height traveled at the angle θ. Adding these together and summing over the path gives

$$\text{work} = \text{force} \times \text{distance} = (M + m)g \cos \theta \times \frac{\Delta h}{\cos \theta} = (M + m)g\, \Delta h$$

$$= (M g + m g)\, h = (550\,\text{N} + 3\,\text{kg} \times 9.8\,\text{m/s}^2) \times 20\,\text{m}.$$

$$= 11588\ \text{N·m} = 1.2 \times 10^4\ \text{N·m}$$

This analysis is incomplete, since there are chemical and biological reactions involved in holding a box at a fixed height and in the process of walking. That analysis can not be completed until a study of thermodynamics (see Chapter 11) is done in depth. Then we find that problems involving living organisms are much more complicated than a simple work picture and hence produce a different answer than the simple one given. However, if we ignore internal changes in the organism, the simple work picture is correct and easily applied provided you use the work-energy approach.

2. Work-Energy Method

Another way to approach this problem is by the work-energy theorem. The difference in energy between the top and the bottom of the hill is just the added potential energy

$(M + m) g h$. This energy change can only be provided by the effort of the man, so that must be the work done:

$$\text{work} = (M + m) g h.$$

Another simplification that results from the use of the work-energy theorem is that we need not deal with the details of human motion and work.

Our definition of work (at this point) does not include the internal processes of living organisms.

PROBLEM 6.2: A 50 N force is applied to a box at a downward angle of 30° to the horizontal. The box moves with constant velocity against a frictional force of 43.3 N. How much work is done for a travel distance of 20 m? What is the coefficient of friction? If this takes 10 s, what is the power? (6.1, 6.9)

Solution:

> **Strategy:** This problem involves sliding friction. Such a force cannot be treated with the work-energy theorem, so use the force equations.

The first step is to draw a motion diagram and a free body force diagram. This is done in Fig. 6.3a for the motion and 6.3b for the forces.
We let

d	distance traveled,
F_{applied}	applied force,
F_{friction}	frictional force,
m	mass of the box,
θ	30° angle,
g	acceleration due to gravity,
N	normal force
μ	coefficient of friction.

Fig. 6.3: (a) Motion diagram for the problem. (b) Free body force diagram. Compare with Fig. 4.4b. They are the same except for the angle (30° versus 40°).

Since the velocity is constant, the system is in dynamical equilibrium and the sum of the forces is zero:

x component $F_{\text{applied}} \cos \theta - F_{\text{friction}} = 0,$

y component $F_{\text{applied}} \sin \theta + mg - N = 0.$

The motion is in the x direction, so only the x component forces need be considered to calculate the work done.

The work done by the applied force is given by

$$\text{work} = F_{\text{applied}} \times \cos\theta \times d$$

$$= 50 \text{ N} \times 0.8660254 \times 20 \text{ m} = 8.7 \times 10^2 \text{ J}.$$

Friction affects the amount of work done. If the frictional forces were smaller, a smaller applied force would be needed to keep the box moving at a constant velocity, so less work would be done. In the absence of friction no work would be needed to move the box once it reached the velocity of travel. That is

$$\text{work by friction} = - \text{ work by applied force}.$$

The coefficient of friction is given by

$$F_{\text{friction}} = \mu N.$$

We are given the magnitude of the frictional force, which is not sufficient to find the coefficient of friction. That is, we have two equations, which we can solve for μ as

$$\mu = \frac{F_{\text{applied}} \cos\theta}{F_{\text{applied}} \sin\theta + mg},$$

We will need the mass of the box to find the coefficient of friction. Since this was not given, the coefficient of friction cannot be found.

We can get the power. The total work done is

$$\text{work} = F_{\text{applied}} \times \cos(\theta) \times d.$$

This work is done in 10 s so the power is:

$$\text{power} = \text{rate of work} = \frac{\text{work}}{\text{time}} = \frac{870 \text{ J}}{10 \text{ s}} = 87 \text{ W}.$$

Another way to get the power is by

$$\text{power} = \text{force} \times \text{velocity}.$$

The box moves 20 m in 10 s, so its speed, v, is 2 m/s. Thus, the power is:

$$\text{power} = F \times v = F_{\text{applied}} \cos\theta \times v$$

$$= 50 \text{ N} \times 0.8660254 \times 2 \text{ m/s} = 87 \text{ W}.$$

PROBLEM 6.3: 30 J is required to extend a spring 5.0 cm from its equilibrium position. How much force is required to hold the spring there? (6.2)

Solution:

> **Strategy:** The work done is given. This work should be determined by the force applied and the distance, or since this is a spring we can use the work-energy theorem to relate the work done to the spring constant and displacement. This determines the distance and then, by Hooke's law, the force.

First draw a motion diagram as in Fig. 6.4a. Once the spring is extended there are two forces acting: F_s, the force of the spring to the left and F_a, the applied force to the right. Draw these forces in the space left for Fig. 6.4b.

(a)	**(b)**

Fig. 6.4:   (a) Motion diagram. (b) Space for force diagram at final equilibrium.

The work done, W, to extend a spring a distance x is given by text Eq. 6.3 as

$$W = \frac{1}{2} k x^2$$

where k is the spring constant.

The force, F, needed to keep the spring extended a distance x is

$$F = kx.$$

From Fig. 6.4b we see that

$$F_a = F_s = k x.$$

GIVEN:

Quantity	Symbol	Value
work done	W	30 J
displacement	x	5.0 cm.
spring constant	k	?
applied force	F_a	?

We can use the work equation to find k and then use the force equation to find F_a; that is,

$$W = \frac{1}{2} k x^2 \text{ is solved for } k = \frac{2\,W}{x^2}.$$

This value for k can be substituted into $F_a = k x$ to get

$$F_a = \frac{2\,W}{x} = \frac{2 \times 30 \text{ J}}{0.050 \text{ m}} = 1.2 \times 10^3 \text{ N}.$$

So it requires 1.2×10^3 N to hold the spring.

PROBLEM 6.4: What is the kinetic energy of a 15 g bullet which is fired at 1200 m/s from a 5 kg rifle? Assume the velocity is given relative to the ground. (6.4)

Solution:

Strategy: The kinetic energy is given by the mass and the velocity of an object, so a straight forward application of the kinetic energy law should suffice.

Kinetic energy is given as $1/2\ m\ v^2$ where m is the mass and v is the velocity. To answer this question we do not need to know the mass of the rifle or even how the bullet reached the speed v.

$$v = 1200 \text{ m/s}$$
$$m = 15 \text{ g}$$

Fig. 6.5:   Bullet leaving rifle.

GIVEN:

mass	m	$15 \text{ g} = 1.5 \times 10^{-2} \text{ kg}$
velocity	v	$1.200 \times 10^{3} \text{ m/s}$

Then

$$\text{kinetic energy} = \frac{1}{2} m\, v^2$$

$$= \frac{1}{2} \times 1.5 \times 1.44 \times 10^{-2+6} \text{ kg m}^2/\text{s}^2$$

$$= 1.1 \times 10^4 \text{ J}$$

Usually the velocity of a bullet is given relative to the gun (the muzzle velocity). In that case we must also consider the recoil of the gun in order to calculate the velocity of the bullet relative to the ground. This problem can be considered after you complete the next chapter.

PROBLEM 6.5: A ball is thrown upwards with sufficient speed to reach a height of 12.0 m. What is its speed when it reaches a height of 10 m? (6.5)

Solution:

Strategy: This problem involves one-dimensional motion in the gravitational field. Since we know the maximum height, we have sufficient information to get the motion at any other height.
1.   Energy conservation method: The sum of kinetic energy and potential energy is constant.
2.   Kinematic approach. Write the equation for constant acceleration. Use the value of the maximum height to determine constants and then solve for time at 10 m.

There are two ways of approaching this problem: the energy conservation methods of this chapter and the kinematic equations of Chapter 2. This problem provides a good comparison of the methods. We will choose h as the vertical coordinate and x as the horizontal coordinate and make a motion diagram sketch as shown in Fig. 6.6a. The only force acting is gravity, vertically downward as shown in Fig. 6.6b.

GIVEN:

velocity at 12.0 m	0.0 m/s
maximum height, h_O	12 m

Fig. 6.6: (a) Motion of ball in a gravitational field. (b) Free body force diagram for ball.

1. Energy Approach

Use the energy conservation theorem:
$$\text{kinetic energy + potential energy = constant.}$$
Let us call this constant k.

Since the x and y components can be treated separately and since there are no x forces, we will treat the problem as if only y motion occurs and represent the y velocity by v. Picking the potential energy origin at $h = 0$, we have

$$\text{kinetic energy} \left(\frac{1}{2} m \, v^2 \right) + \text{potential energy} \, (m \, g \, h) = k .$$

We can evaluate this constant if we know the kinetic energy and the potential energy at any one height.

We know the form of the potential energy at all heights and that the kinetic energy at $h_O =$ 12 m is zero, so

$$\frac{1}{2} m \, (0 \text{ m/s})^2 + m \, g \, h_O = k.$$

Thus $k = m \, g \, h_O$, and at any height

$$\frac{1}{2} m \, v^2 + m \, g \, h = k = m \, g \, h_O,$$

giving

$$v^2 = 2 \times g \, (h_O - h).$$

Then at $h = 10$ m,

$$v^2 = 2 \times 9.8 \text{ m/s}^2 \times (12 \text{ m} - 10 \text{ m}),$$

so

$$v = \pm 6.26 \text{ m/s} = \pm 6.3 \text{ m/s},$$

considering significant figures. The ± is there since the ball is at 10 m on both its upward and its downward paths. Note that the original height and the original velocity are not needed or even mentioned .

2. Kinematic Approach

The kinematic equations of Chapter 2 of the text can be used to solve this problem. These equations were summarized in Table 2.1 of the text. There are two events discussed, the ball at 10 m and the ball at 12 m. Since we are not concerned with when the two events occurred, the last of the kinematical equations is most likely the one we want:

$$v^2 = v_0^2 + 2\,a\,(x - x_0)$$

where v_0 is the velocity at x_0.

For our problem we replace x by y, x_0 by h_0, and a by $-g$. The minus sign is needed since y is measured positively upwards and the gravitational force is downwards. Then for $h_0 = 12$ m, $v_0 = 0$,

$$v^2 = -2\,g\,(y - h_0) = 2\,g\,(h_0 - y),$$

which is the same result as obtained with the energy conservation method. In this simple case there isn't much difference in effort, but often the energy method will be simpler. This is especially true in cases where the acceleration is not uniform or where rotational motion is involved (see Problems for Chapter 9).

PROBLEM 6.6: A box of mass m slides on a frictionless loop-the-loop starting from rest at a distance h_0 above the bottom. Let the loop radius be R. What is the minimum value that h_0 can have so that the box will not fall off the the loop? (6.5)

Solution:

Strategy: Since the normal force does no work, energy is conserved. Calculate the kinetic energy at the top of the loop and thus find the centrifugal force at that point. The centrifugal force must be larger than the gravitational force at that point.

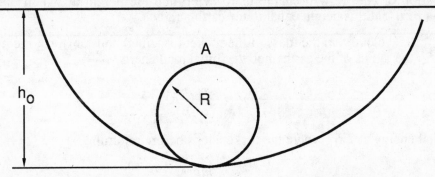

Fig. 6.7: Loop-the-loop with a loop radius of R. A box enters the path from a height h_0.

We can use conservation of energy since the normal forces do no work. The total energy of the box is equal to its initial energy, which is

$$mgh_0.$$

At the top of the loop, the box is moving horizontally to the left with a velocity v. The total energy at that point is

$$\frac{1}{2}mv^2 + mg(2R), \qquad \text{so} \qquad v^2 = 2g(h_0 - 2R).$$

The centrifugal force needed for this motion is

$$\frac{mv^2}{R} = 2mg\left(\frac{h_0}{R} - 2\right).$$

The centrifugal force is provided by the normal force minus the gravitational force,

$$\text{centrifugal force} = \text{normal force} - mg,$$

so

$$\text{normal force} = \frac{2mgh_0}{R} - 5mg.$$

The condition that the box not leave the loop is that the normal force be greater than zero, that is

$$\frac{2mgh_0}{R} > 5mg \qquad \text{or} \qquad h_0 > \frac{5}{2}R.$$

PROBLEM 6.7: What velocity must a rocket have at the surface of the earth if it is to rise at a height equal to the earth's radius before it starts to descend? Use energy conservation. (6.7)

Solution:

Strategy: Since this problem involves heights about the earth's surface which are comparable to the earth's size, we cannot assume that gravity is constant. The conservation of energy will enable us to determine the height at which the kinetic energy is zero. Set this height equal to the earth's radius.

If v is the initial velocity, m the mass of the rocket (assumed constant), m_E the mass of the earth and R_E the radius of the earth, then the initial total energy is

$$\frac{1}{2}mv^2 - G\frac{m\,m_E}{R_E},$$

and the total energy at $2R_E$, where the rocket stops before returning, is

$$- G\frac{m\,m_E}{2R_E}.$$

Conservation of energy makes these two equal. This gives

$$v = \sqrt{G\frac{m_E}{R_E}} = \sqrt{gR_E},$$

where g is the acceleration of gravity at the Earth's surface.

PROBLEM 6.8: A 1200 kg automobile, going 100 m/s, collides head-on with a 1400 kg automobile which was traveling in the opposite direction at 50 m/s. During the collision the two cars get stuck together. What is the total kinetic energy immediately after the collision, assuming the cars are not spinning? (6.8)

Solution:

Strategy: Since the two colliding cars stick together, this is clearly not an elastic collision. That leaves us with conservation of momentum as our only conservation law. Since we know the masses and velocities before the collision and since there is only one object after the collision, that is sufficient to give us the solution.

We let

Object	Symbol	Value
mass of car moving to right	m_1	1200 kg
velocity of car moving to right	v_1	100 m/s
mass of car moving to left	m_2	1400 kg
velocity of car moving to left	v_2	– 50 m/s
final velocity	v_f	?

A choice of positive direction has been made for car #1 so $v_2 < 0$; and, if $v_f > 0$ the final combination will be moving to the right in Fig. 6.8b.

Fig. 6.8: (a) The two cars before collision. (b) The two cars moving as a unit after collision.

Conservation of momentum is

initial momentum = final momentum,

which is

$$m_1v_1 + m_2v_2 = (m_1 + m_2)v_f.$$

This gives

$$v_f = \frac{m_1v_1 + m_2v_2}{m_1 + m_2} = 19 \text{ m/s.}$$

Since there is no rotation after the collision, the entire kinetic energy is translational and

$$\text{final kinetic energy} = \tfrac{1}{2}(m_1 + m_2)v_f^2 = 4.8 \times 10^5 \text{ J.}$$

If there had been some final rotation, the value of v_f we calculated would have been correct and the final kinetic energy would have been correct for the translational energy, but there would also have been rotational kinetic energy.

The initial kinetic energy was about 6.8×10^6 J. The difference between this value and our calculated final value represents the work done to deform the cars and heat the materials.

PROBLEM 6.9: A 20 kg suitcase falls from an airplane at a height of 500 m. If it loses 85% of its initial potential energy through friction with the air, what kinetic energy does it have just before it strikes the ground?

Solution:

> **Strategy:** Calculate the kinetic energy the suitcase would have without friction and reduce that value by 85%.

If a mass, m, falls a distance, d, its kinetic energy will be:

$$KE = m g d,$$

where g is the acceleration due to gravity.

For a fall of 500 m, the KE without air friction will be 9.8×10^4 J. With friction there will be only 15% of this energy; so the final kinetic energy will be 1.5×10^4 J.

PROBLEM 6.10: A 40 kg box is pushed at a constant velocity of 0.80 m/s with a horizontal force on a level floor. The coefficient of friction between the box and the floor is 0.500. (6.9)
(a) What is the applied force?
(b) At what rate is power expended?
(c) How much work is done in 1.0 minute?

Solution:

> **Strategy:** Since there is friction and we are not told how much energy is lost, we will use forces. Since there is no vertical motion the vertical forces sum to zero, and since the horizontal velocity is constant those forces also sum to zero.

The forces acting are gravity, the normal force and friction as shown in Fig. 6.9. We use the following notation:

Quantity	Symbol	Value
applied force	F	?
frictional force	F_f	?
mass of box	m	40 kg
acceleration of gravity	g	9.8 m/s^2
normal force	N	mg
velocity	v	0.80 m/s
coefficient of friction	μ	0.500

Since the box moves with constant horizontal velocity and has no vertical velocity:

$$N - mg = 0 \qquad \text{and} \qquad F - F_f = 0.$$

We assume that the friction force is related to the normal force via the coefficient of friction; hence

$$F - F_f = F - \mu N = F - \mu\, mg = 0,$$

so

$$F = \mu\, mg = 196\,\text{N} = 2.0 \times 10^2\,\text{N},$$

while

$$\text{power} = \frac{\text{work}}{\text{time}} = \frac{\text{force} \times \text{distance}}{\text{time}}$$

$$= \text{force}\,\frac{\text{distance}}{\text{time}} = \text{force} \times \text{velocity}$$

$$= (2.0 \times 10^2\,\text{N}) \times 0.80\,\text{m/s}$$

$$= 1.6 \times 10^2\,\text{W}.$$

Fig. 6.9: Forces acting on box.

The work in one minute is

$$\text{work} = \text{power} \times \text{time} = 1.6 \times 10^2\,\text{N} \times 60\,\text{s}$$

$$= 9.6 \times 10^3\,\text{J}.$$

Or, if you prefer, calculate the work in 1.0 minute by

$$\text{distance} = \text{velocity} \times \text{time} = 48\,\text{m}$$

then

$$\text{work} = 2.0 \times 10^2\,\text{N} \times 48\,\text{m} = 9.6 \times 10^3\,\text{J}$$

and

$$\text{power} = \frac{\text{work}}{\text{time}} = \frac{9.6 \times 10^3\,\text{J}}{60\text{s}} = 1.6 \times 10^2\,\text{W}.$$

SUGGESTED READINGS

Alexander, R. McN., "Walking and Running." *American Scientist,* July 1984, p. 348.

Brooks, A. N., A. V. Abbott, and D. Gordon, "Human Powered Watercraft." *Scientific American*, December 1986, p.120.

Gray, C. L. Jr., and F. von Hipple, "The Fuel Economy of Light Vehicles." *Scientific American*, May 1981, p. 48.

Keeports, D., "Numerical Calculation of Model Rocket Trajectories." *The Physics Teacher,* May 1990, p. 274.

Nadel, E. R., and S. R. Bussolari, "The Daedalus Project: Physiological Problems and Solutions." *American Scientist,* July 1988, p. 351.

Scientific American, Issue on Energy and Power, *Scientific American,* September 1981.

Chapter 7
Linear Momentum

IMPORTANT TERMS

Fill in the blanks with the appropriate word or words.

1. Momentum associated with translational motion is called _____ momentum.

2. A collision occurring between two bodies which stick together is called a(n) _____ collision.

3. The total linear momentum before a collision is equal to the total linear momentum after the collision. This is a statement of the _____.

4. The product of a force and the time that force acts is called a(n) _____.

5. In a(n) _____ collision kinetic energy is conserved.

Write the definitions of the following:

6. impulse (7.2):

7. elastic collision (7.4):

8. conservation of linear momentum (7.3):

9. perfectly inelastic collision (7.4):

10. linear momentum (7.1):

ANSWERS TO 1–5, IMPORTANT TERMS

1. linear; 2. perfectly inelastic; 3. conservation of linear momentum; 4. impulse; 5, elastic.

SAMPLE QUESTIONS

QUESTION 7.1: Under what conditions can a compact car have the same momentum as a large car?

Answer:

The momentum of an object depends upon both the mass and the velocity of the object. A compact car at high velocity can have the same momentum as a large car at a lower velocity.

QUESTION 7.2: Which has a larger momentum, a bug flying at 10 mi/hr or a bird flying at the same speed?

Answer:

The bird has a larger mass, so it will have the larger momentum for the same speed.

SAMPLE SOLUTIONS

PROBLEM 7.1: A 0.14 kg baseball was thrown with an average speed of 30 m/s. The ball is hit with a bat and leaves the batting area with a speed of 35 m/s and at an angle of 45° upwards with respect to the ground and in the direction opposit to its approach. What is the impulse given to the ball? If the duration of the contact between bat and ball is 20 ms, what is the average force? Assume the ball was traveling horizontally when hit. (7.2)

Solution:

Strategy: The baseball changes direction. For that to happen an impulse must act on the ball. Use the law that the change in momentum is equal to the applied impulse.

The simplest model will ignore all details of the ball except its mass, which we will represent by m. The initial direction of the ball will be taken as $+x$ and its final direction is 45° to the initial and with negative x and positive y components as shown in the motion diagram Fig. 7.1a. We let v_0 be the initial velocity and v_1 be the velocity after impact with associated momenta p_0 and p_1. The relation we need is the definition of impulse:

$$\textbf{impulse} = \textbf{F} \, \Delta t = \Delta \textbf{p}.$$

Fig. 7.1:   (a) Motion diagram for ball of Problem 7.1. (b) Velocity components of final velocity.

GIVEN:

initial velocity	v_0	30 m/s
final velocity	v_1	35 m/s
final angle	θ	45°
duration of impulse	Δt	20 ms
mass of ball	m	0.14 kg

FIND:
    impulse   **impulse**
    average force   **F**

The component equations are

$$\text{impulse}_x = p_{x\ initial} - p_{x\ final}$$

$$= m\, v_0 - m(-v_1 \times \cos(\theta)) = m\,(v_0 + v_1 \times \cos(\theta))$$

$$= 0.14\ \text{kg} \times (30\ \text{m/s} + 0.707 \times 35\ \text{m/s}) = 7.6643\ \text{N·s} = 7.7\ \text{N·s}$$

$$\text{impulse}_y = 0 - m(-v_1 \times \sin(\theta)) = m\, v_1 \sin(\theta)$$

$$= 0.14\ \text{kg} \times 35\ \text{m/s} \times 0.707 = 3.4643\ \text{N·s} = 3.5\ \text{N·s}.$$

So the total impulse is 7.7 N·s in the $-x$ direction and 3.5 N·s in the $+y$ direction. These components are shown in Fig. 7.1c.

The total impulse is

$$\text{impulse} = \sqrt{(7.7\ \text{N·s})^2 + (3.5\ \text{N·s})^2} = 8.5\ \text{N·s}$$

at an angle of

$$\phi = \tan^{-1}(3.5/7.7) = 24°$$

to the $-x$-axis. The average force is

$$\frac{8.5\ \text{N·s}}{20 \times 10^{-3}\text{s}} = 4.3 \times 10^2\ \text{N at } 24° \text{ up from } -x.$$

PROBLEM 7.2: A sniper is prone on smooth ice. Her mass and that of her equipment total 91 kg. She uses a rifle to fire a 47-g bullet with a velocity of 900 m/s towards her chemistry professor. What is her recoil velocity? (7.4)

Solution:

Strategy: Use conservation of momentum. The initial momentum is zero , so her final momentum is equal and opposite to the momentum of the bullet.

We assume that the sniper was initially at rest on the ice. The momentum before she fires her rifle is zero, so her final momentum must be equal and opposite to the momentum of the bullet. Remember to convert all units to SI units.

v_s v_b

Fig 7.2:   Velocities of sniper and bullet.

We let m_b be the mass of the bullet, v_b be the velocity of the bullet, m_s be the mass of the sniper, and v_s be the velocity of the sniper. Conservation of momentum gives

$$\text{total momentum} = 0 = \text{momentum of bullet} + \text{momentum of sniper},$$

so

$$0 = m_b \times v_b + m_s \times v_s.$$

Solving, we get

$$v_s = -\frac{0.047 \text{ kg} \times 900 \text{ m/s}}{91 \text{ kg}} = 0.46 \text{ m/s}.$$

Her recoil velocity is 0.46 m/s.

PROBLEM 7.3: A bird of mass 0.10 kg flying horizontally at 1.5 m/s is hit by a blob of tar of mass 0.050 kg falling vertically downward at 3 m/s. The blob sticks to the bird. What is the velocity of the bird immediately after the collision? (7.5)

Solution:

> **Strategy:** This is an inelastic collision. For such a collision energy is not conserved but momentum is conserved. Only one body (bird + tar) exists after the collision, so there is sufficient information in the one conservation law.

We will choose +x to be the initial direction of the bird and +y to be vertically downward. Since all information is given just before and just after the collision, we can neglect the effects of gravity, air resistance and the bird's wing movement. The three velocities are shown in Fig. 7.3a while the components and direction of the final velocity are shown in Fig. 7.3b.

(a)   **(b)**

Fig. 7.3:   (a) Initial velocities and final velocity. (b) Components of final velocity.

GIVEN:

Quantity	Symbol	Value
initial velocity of bird	v_0	1.5 m/s in +x direction
initial velocity of blob	v_1	3.0 m/s in +y direction
mass of bird	M	0.10 kg
mass of tar blob	m	0.050 kg

FIND:

final velocity v_f

final direction θ

Conservation of momentum gives us

initial momentum = final momentum

$$M v_0 + m v_1 = (M + m) v_f.$$

In terms of components:

$$M v_0 = (m + M) v_f \times \cos(\theta),$$

$$m v_1 = (m + M) v_f \times \sin(\theta).$$

Using

$$\sin^2(\theta) + \cos^2(\theta) = 1 \qquad \text{and} \qquad \tan(\theta) = \frac{\sin(\theta)}{\cos(\theta)},$$

we get:

$$v_f = \frac{\sqrt{M^2 v_0^2 + m^2 v_1^2}}{m + M} = \frac{0.212 \text{ kg m/s}}{0.15 \text{ kg}} = 1.4 \text{ m/s}$$

and

$$\theta = \tan^{-1}\left(\frac{m \, v_1}{M \, v_0}\right) = \tan^{-1}\left(\frac{0.15}{0.15}\right) = 45°.$$

The bird is now moving forward and downward at an angle of 45°.

PROBLEM 7.4: A plastic "flying bird" consists of a plastic body on a string. The string passes through a hollow tube (like a soda straw) and is held at the other end. The tube is held vertically and moved rapidly in a small horizontal circle to make the "bird" fly as shown in Fig. 7.4. The "bird" has a mass of 80 g. The "bird" is set "flying" in a 50 cm radius and the straw is held fixed. If the string is pulled down in the straw so that the radius is now 15 cm, what is the fractional change in the period of the "bird" about the straw? Assume the "bird" is moving slowly enough that we can neglect the lift provided by the air. This assumption makes for poor flying, but it gives us a simpler problem to solve. (7.5)

Solution:

Strategy: Since the string provides only a radial force the angular momentum is conserved when the string is shorted, so we will use conservation of angular momentum. Since we are asked to find only the fractional change, it is not necessary to solve for the frequency.

We will consider the "bird" to be a small mass of mass m and the string to be ideal. The motion of the straw is just to get the bird flying so we will consider the straw as fixed while the string length is changed. The physical arrangement is shown in Fig. 7.4a.

We can answer the question without finding the frequency of motion. For the "bird" the initial angular momentum, L_i, was

$$L_i = m R_i^2 \omega_i,$$

while the final angular momentum, L_f, is

$$L_f = m R_f^2 \omega_f,$$

where R and ω are the radii and frequency, respectively.

Fig. 7.4: (a) The "flying bird" on a straw. (b) Forces in the horizontal plane. (c) Forces in the vertical plane.

Since the string exerts only a radial force, the initial and final angular momenta are equal, giving

$$m\,R_i^2\,\omega_i = m\,R_f^2\,\omega_f$$

or

$$\frac{\omega_f}{\omega_i} = \frac{R_i^2}{R_f^2} = \frac{(50\text{ cm})^2}{(15\text{ cm})^2} = 11.$$

The final angular frequency is 11 times the initial angular frequency.

PROBLEM 7.5: Suppose in the preceding problem the string makes an angle of 10° below the horizontal. What is the force that is exerted to the string to keep the "bird" at 15 cm radius and what is the frequency at which the "bird" revolves? Neglect lift provided by the air. (7.5)

Solution:

Strategy: Now we must use the equality of the horizontal tension of the string to the centripetal force and that of the vertical tension to the gravitational force.

We can answer this question with material from Chapter 5. Text Eq. 5.9 relates the centripetal force, F_C, and the resulting circular motion:

$$F_C = m\frac{v^2}{r} = m\,\omega^2 r \qquad \text{(centripetal force)}$$

where r is the radius of the circle, v is the tangential velocity, and ω is the angular frequency.

As we see from Fig. 7.4c, letting θ be the angle from the horizontal,

$$m\,g = T\sin(\theta) \qquad\qquad \text{(vertical forces)}$$

and from Fig. 7.4b,

$$F_C = T\cos(\theta) = m\,\omega^2 r. \qquad\qquad \text{(horizontal forces)}$$

Hence,

$$T = \frac{m\,g}{\sin(\theta)} = \frac{0.080 \times 9.8}{0.178}\,\text{N} = 44\,\text{N},$$

while

$$\omega = \sqrt{\frac{T\cos(\theta)}{m\,r}} = \sqrt{\frac{g\cos(\theta)}{r\sin(\theta)}} = \sqrt{\frac{9.8 \times 0.9848}{0.15 \times 0.1736}}\,\text{rad/s} = 19\,\text{rad/s}.$$

$$= \sqrt{\frac{9.8 \times 0.9848}{0.15 \times 0.1736}}\,\text{rad/s} = 19\,\text{rad/s}.$$

A force of 44 N is exerted and the "bird" rotates with a frequency of 19 rad/s = 3.0 rev/s.

PROBLEM 7.6: Two cars approach each other along streets that meet at a right angle. They collide at the intersection, sticking together. If one car has a mass of 1500 kg and an initial speed of 3.00 m/s and the other has a mass of 2000 kg and an initial speed of 5.00 m/s, what will be their speed and direction immediately after impact? How much energy is converted to heat, work in bending material, etc.? (7.5)

Solution:

Strategy: Since the cars stick together we do not expect the collision to be elastic; therefore, energy is not conserved. The collision conserves momentum. Conserving momentum will enable us to calculate the final velocity of the resulting single body. Calculate the final energy from the total masses and final velocity.

We will take the initial velocity of the 1500 kg car (denoted by m_1) as being in the $+x$ direction and that of the 2000 kg car (m_2) as $+y$ as shown in Fig. 7.5.

FIG. 7.5:   Cars of masses m_1 and m_2 colliding at right angles and sticking together.

Conservation of momentum applies to each coordinate direction separately. The x component is

$$m_1\,v_1 = (m_1 + m_2)\,v_f\cos(\theta),$$

while the y component is

$$m_2 \, v_2 \; = \; (m_1 + m_2) \, v_f \sin(\theta).$$

Solving gives $v_f \cos(\theta) = 1.29$ m/s and $v_f \sin(\theta) = 2.86$ m/s.

Square and sum to use $\sin^2(\theta) + \cos^2(\theta) = 1$, and take the ratio to use $\tan(\theta) = \dfrac{\sin(\theta)}{\cos(\theta)}$. This gives

$$v_f \; = \; 3.63 \;\; \text{m/s} \qquad \text{and} \qquad \theta \; = \; 65.7°.$$

The change in kinetic energy is given by

$$\Delta \text{KE} \; = \; \text{KE}_{final} - \text{KE}_{inital}$$

$$= \; \tfrac{1}{2} \, (m_1 + m_2) \, v_f^2 \; - \; \tfrac{1}{2} \, m_1 \, v_1^2 \; - \; \tfrac{1}{2} \, m_2 \, v_2^2 \; .$$

$$= \; 2.31 \times 10^4 \, \text{J} \; - \; 6.75 \times 10^3 \, \text{J} \; - \; 2.50 \times 10^4 \, \text{J} \; = \; -1.77 \times 10^4 \, \text{J}.$$

That is, 1.77×10^4 J of kinetic energy is lost, being converted to heat and work.

PROBLEM 7.7: A rocket with an initial mass of 1.0×10^4 kg is fired in a vertical direction. Its exhaust gases are ejected at the rate of 50 kg/s with a relative velocity of 3.0×10^3 m/s. What is the initial acceleration of the rocket? What is the acceleration after 30 seconds have passed? What is the acceleration after 2 minutes have passed? Why do you think the question does not ask for the velocity or the position at 30 s and 120 s? Assume that the rocket was initially at rest on the earth's surface and that the rocket does not run out of fuel during the time considered. (7.6)

Solution:

Strategy: The first parts of this problem are a simple application of the thrust relation of the text. Identify the forces acting on the rocket and use Newton's second law.

Fig. 7.6: Rocket with thrust upwards and gravity pulling downwards.

We model the rocket as a point and consider the forces acting on the rocket. These are gravity acting downward and the rocket thrust acting upwards as shown in Fig. 7.6. We let F_R represent the rocket thrust and F_G be the gravitational force. The thrust is given in the text as

$$F_R = v_r \frac{\Delta m}{\Delta t}$$

where

$$\frac{\Delta m}{\Delta t}$$

is the rate that the rocket ejects gases and v_r is the ejection velocity relative to the rocket. The gravitational force will be $F_G = -mg$ where m is the mass of the rocket plus remaining fuel and g is the acceleration due to gravity. The minus sign shows that F_R and F_G are in opposite directions. We let m_0 be the initial mass of rocket plus fuel.

GIVEN:

Quantity	Symbol	Value
ejection rate	$\frac{\Delta m}{\Delta t}$	50 kg/s
ejection speed	v_r	3.0×10^3 m/s
initial mass	m_0	1.0×10^4 kg

If we let $t = 0$ be the time of launch, then the mass of rocket plus remaining fuel at any time t, $m(t)$, will be given by:

$$m(t) = m_0 - \frac{\Delta m}{\Delta t} t.$$

The physical law we need is Newton's third law that the net force is equal to the mass times the acceleration. We let a represent the acceleration in the x direction, taken positively upwards, and can now write

$$F_{total} = F_R + F_G = v_r \frac{\Delta m}{\Delta t} - m(t)g = m(t) a.$$

This gives

$$a = \frac{v_r}{m(t)} \frac{\Delta m}{\Delta t} - g$$

$$= \frac{v_r}{m_0 - \frac{\Delta m}{\Delta t} t} \frac{\Delta m}{\Delta t} - g.$$

For our problem this is

$$a = \left(\frac{15}{1 - 5.0 \times 10^{-3} \times t} - 9.8 \right) \text{m/s}^2,$$

where t is in seconds. Hence

$$a(0 \text{ s}) = 5.2 \text{ m/s}^2,$$

$$a(30 \text{ s}) = 7.8 \text{ m/s}^2,$$

and

$$a(120 \text{ s}) = 28 \text{ m/s}^2.$$

Since the acceleration is not constant we cannot use the relations in Chapter 2 to calculate the velocity and the displacement. Calculus is needed to exactly solve problems with varying acceleration. You have the background to get numerical answers to the position and velocity for problems such as this. The following steps would be needed:

1. Plot the acceleration as a function of time.
2. For any time t, the total area under the acceleration curve, up to time t, is the velocity at that time t.
3. Plot the calculated velocities against time. The displacement is the total area under the velocity curve up to the time t.

GENERAL COMMENTS

All the text equations that involve acceleration are valid only when the acceleration is constant. Often we can model motion with changing acceleration provided the motion has relatively long periods of constant acceleration and only relatively short periods when the acceleration changes. The model is to neglect the motion that occurs when the acceleration is changing.

In the preceding chapter we dealt with perfectly elastic collisions where the total kinetic and potential energy is conserved. In perfectly elastic collisions, momentum is conserved. This chapter introduces inelastic collisions that conserve only momentum. Momentum is always conserved while the total of kinetic and potential energy may not be conserved. We will later see that this missing energy shows up as heat or other forms of energy. With this broader definition of energy, we may also say that energy is always conserved; however, at this point in the course, you will be able to use energy conservation only when you know that the collisions are perfectly elastic.

SUGGESTED READINGS

Priest, J., and D. F. Griffing, "Model for an Undergraduate Physics Problem." *The Physics Teacher,* April 1990, p. 210.

Spradley, J. L., "Meter-Stick Mechanics." *The Physics Teacher,* May 1990, p. 312.

Chapter 8
Combining Conservation of Energy and Momentum

IMPORTANT TERMS

Fill in the blanks with the appropriate word or words. Since this chapter consists of applications of the material in the last two chapters, some important terms from those chapters are included here again.

1. The _____ is the unit of work in the SI system.

2. The time rate of doing work is called _____.

3. _____ is the energy of translational motion.

4. The ability to do work is called _____.

5. The energy that a body has because it is in a gravitational field is called the body's _____ energy.

6. _____ is defined as the product of a force times the distance through which the force acts.

7. The _____ states that the work done on a body is equal to its change in kinetic energy.

8. The idealized spring produces a force proportional to its stretching. This is called _____.

9. The SI unit of power is called the _____.

10. The sum of kinetic and potential energy is called _____ .

11. Our first conservation law is the _____ .

12. The product of a force and the velocity with which it is applied is the _____.

13. A collision which conserves energy is said to be a(n) _____ collision.

14. The _____ of a rocket is the force produced by ejecting gases.

15. Momentum associated with translational motion is called _____ momentum.

16. A collision occurring between two bodies which stick together is called a(n) _____ collision.

17. A collision which does not conserve kinetic energy is called a(n) _____ collision.

18. The total linear momentum before a collision is equal to the total linear momentum after the collision. This is a statement of the _____ .

19. The total velocity corresponding to the initial upward motion necessary for a body not to fall back to earth is called the _____ velocity.

Write the definitions of the following:

20. work (6.1):

21. energy (6.2):

22. joule (6.1):

23. Hooke's law (6.1):

24. gravitational potential energy (6.4):

25. inelastic collision (8.1):

26. work-energy theorem (6.1):

27. total mechanical energy (6.7):

28. perfectly elastic collision (8.1):

29. instantaneous power (6.8):

30. translational kinetic energy (6.8):

31. perfectly inelastic collision (8.1, 7.4):

32. power (6.8):

33. distinguish between work and power (6.8):

34. inelastic collision (8.1):

35. conservation of mechanical energy (6.7):

36. conservation of linear momentum (7.4):

37. escape velocity (8.7):

38. linear momentum (7.4):

39. watt (and its definition in terms of joule/s) (6.8):

40. thrust (7.6):

ANSWERS TO 1–19, IMPORTANT TERMS

1. joule; 2. power; 3. translational kinetic energy; 4. energy; 5. gravitational potential; 6. work; 7. work-energy theorem; 8. Hooke's law; 9. watt; 10. total mechanical energy; 11. conservation of mechanical energy; 12. instantaneous power; 13. elastic; 14. thrust; 15. linear; 16. perfectly inelastic; 17. inelastic; 18. conservation of linear momentum; 19. escape.

SAMPLE QUESTIONS

QUESTION 8.1: A cosmonaut wishes to dock her spacecraft with another craft several hundred meters ahead of him in the same orbit. The two spacecraft are moving with the same speed at the same radius in the same circular orbit. The cosmonaut can use thruster rockets directed fore, aft, up, or down. Which should she use and in what order? Describe the subsequent motion of his craft.

Answer:

The cosmonaut cannot simply speed up (use forward thruster). If she does that, the orbit becomes an ellipse and the cosmonaut will have to use the down thruster rockets to correct the radial distance. Several corrections will be needed. If the cosmonaut quickly fires the down thruster rockets properly, she can move to a smaller circular orbit. This orbit will have a shorter period, so she will catch up with the leading spacecraft. A quick upward thrust will bring them together. In simplistic terms, she must move inwards to catch up.

QUESTION 8.2: Can an object have, at the same time, more kinetic energy but less momentum than another object?

Answer:

Yes. The kinetic energy depends upon the square of the velocity while the momentum depends upon the velocity to the first power. Let the subscript "1" denote the first body and the subscript "2" the second body. Suppose, $m_1v_1 > m_2v_2$; that is, the momentum of body "1" is bigger than the momentum of body "2". Let us call the ratio of kinetic energies R. Then

$$R = \frac{\frac{1}{2}m_2v_2^2}{\frac{1}{2}m_1v_1^2} = \frac{m_2v_2}{m_1v_1}\frac{v_2}{v_1}.$$

By assumption, the first ratio is less than one. We can choose the velocities and masses so that the second fraction is larger than the inverse of the first. This is clearly possible since we have two fractions v_2/v_1 and m_2/m_1.

SAMPLE SOLUTIONS

PROBLEM 8.1: When a glass marble bounces off a steel plate 80.0% of the kinetic energy is conserved. If the marble is dropped from an initial height of 3.000 m, how high will it bounce after the second impact? (8.2)

Solution:

Strategy: The momentum of the marble is not conserved upon collision, but upon each collision 80% of the kinetic energy is retained. The total energy (kinetic + potential) is conserved between collisions. This means that at the top of each bounce the potential energy is 80% of what it was at the top of the last bounce.

Fig. 8.1: Marble bouncing off a steel plate.

A motion plot is shown in Fig. 8.1. We let h_0 be the initial height, h_1 be the height of the first bounce, and h_2 be the maximum height after the second bounce. Potential energy is measured from the plate. Since the system is interacting with the earth, we need not consider the conservation of momentum. Energy will be conserved except at the collisions when 20.0% of the energy is lost. We label events as follows:

Event label	Event	Energy and type
A	start at h_0	mgh_0 -- potential
B	just before first impact	mgh_0 -- kinetic
C	just after first impact	$0.8\ mgh_0$ -- kinetic
D	at h_1	$mgh_1 = 0.8\ mgh$ -- potential
E	just before second impact	mgh_1 -- kinetic
F	just after second impact	$0.8\ mgh_1$ -- kinetic
G	at h_2	$mgh_2 = 0.8\ mgh_1$
		$= 0.64\ mgh_0$

From this analysis, we see that the final height after two bounces is just 0.64 of the initial height or 1.920 m. Notice that three significant figures were given.

A shorter analysis is to state that 80% of the initial energy remains after each collision, so that after two collisions, only $64\% = 80\% \times 80\%$ of the initial energy remains.

PROBLEM 8.2: An air track glider of 100 g mass with an initial speed of 6.0 m/s collides head on with another glider at rest. The other glider is 5.0 times as massive as the first one. What are the final speeds and directions of the gliders if the collision is 90% elastic? (8.2)

Solution:

Strategy: Assume conservation of momentum. After the collision, only 90% of the initial kinetic energy remains. These two laws give us sufficient information.

(a)   **(b)**

Fig. 8.2:   (a) Two gliders before collision. (b) Two gliders after collision.

Since we are told that these bodies are air track gliders, we will assume that the collision conserves momentum. The arrangement before and after collision is shown in Fig. 8.2.

We choose the initial direction to be positive and let

Symbol	Quantity represented
m	mass of first glider
v	initial velocity of first glider = 6.0 m/s
v_1	velocity of first glider after collision
v_2	velocity of second glider after collision

Momentum conservation gives:
$$m\,v = m\,v_1 + 5\,m\,v_2, \tag{8.1}$$

while energy conservation (keep 90% after collision) gives:

$$0.90 \times [\text{initial kinetic energy}] = \text{final kinetic energy}$$

$$0.90 \times [\tfrac{1}{2}\,m\,v^2] = \tfrac{1}{2}\,m\,v_1^2 + \tfrac{1}{2}\,(5.0\,m)\,v_2^2. \tag{8.2}$$

Since m is a common factor in all terms, we can rewrite the equations as

$$v = v_1 + 5.0\,v_2 \tag{8.1'}$$
and
$$0.90\,v^2 = v_1^2 + 5.0\,v_2^2. \tag{8.2'}$$

The problem is reduced to how best to solve these two equations.

First Method
Using Eq. 8.1' we can eliminate v from Eq. 8.2', to obtain

$$17.5\,v_2^2 + 9\,v_1 v_2 - 0.1\,v_1^2 = 0 \quad \text{giving} \quad 0 = v_2^2 + 0.51 v_1 v_2 - 0.0057 v_1^2.$$

This quadratic has two solutions: $v_2 = 0.01\,v_1$ and $v_2 = -0.51\,v_1$.

NOTE: In the first solution we lose a significant figure since we must subtract two nearly equal answers.

Since we know v, we can use Eq. 8.1' to find the two possible sets of v_1 and v_2:

$$v_1 = 5.7 \text{ m/s} \qquad \text{and} \qquad v_2 = 0.06 \text{ m/s} \qquad (\text{using } v_2 = 0.1\,v_1)$$
and
$$v_1 = -3.9 \text{ m/s} \qquad \text{and} \qquad v_2 = 2.0 \text{ m/s} \qquad (\text{using } v_2 = -0.51\,v_1).$$

The first solution would correspond to the two bodies passing through each other (v_1 is bigger than v_2), so we must discard this solution.

Second Method
Use the value of v to rewrite Eq. 8.1' as
$$v_1 = 6.0 \text{ m/s} - 5.0\,v_2.$$
Substitute this into Eq. 8.2', giving

$$0.9 \times 36 \text{ m}^2/\text{s}^2 \quad = (6.0 \text{ m/s} - 5.0 \ v_2)^2 + 5.0 \ v_2{}^2$$

$$= 36 \text{ m}^2/\text{s}^2 - (60 \text{ m/s})v_2 + 30 \ v_2{}^2 \ ,$$

or

$$v_2{}^2 - (2.0 \text{ m/s})v_2 + 0.12 \text{ m}^2/\text{s}^2 = 0.$$

The two roots are 1.9 m/s and 0.06 m/s, which are close to the numbers we found in the first method. Then Eq. 8.1' can be used to find the corresponding values for v_1 as before. As before, we lose a significant figure in the subtraction for one of the roots of the quadratic.

A corresponding argument could be made by solving Eq. 8.1' for v_2 and using that in Eq. 8.2' to find the pairs of v_1. From any of these approaches we conclude that the solutions are that the first glider reverses its direction and has a speed of 3.9 m/s while the second glider moves at a speed of 2.0 m/s in the direction that the first glider originally had.

PROBLEM 8.3: After colliding elastically with a stationary puck of the same mass, an air hockey puck is observed to have a momentum whose magnitude is only 1/3 the magnitude of the initial momentum. What angle does it make with its initial direction? (8.3)

Solution:

Strategy: Since these are air hockey pucks, we will consider this as a two dimensional scattering problem without friction. Since the collisions are elastic, we can conserve both energy and momentum. Write the equations showing conservation of each component of momentum and an equation showing conservation of energy. Solve for the unknown angle.

Fig. 8.3: (a) Before the collision. (b) After the collision.

We sketch the problem in Fig. 8.3. The symbols used are:

Symbol	Meaning
m	mass of either puck
v	initial velocity of puck A
v_1	final velocity of puck A
v_2	final velocity of puck B

We orient our coordinates so the collision occurs in the x-y plane and choose x and y as shown in Fig. 8.3a. In terms of the symbols adopted, this means that

$$|m\,v_1| = |\tfrac{1}{3}\,m\,v| \qquad \text{or} \qquad v_1 = \tfrac{1}{3}\,v.$$

Using the angles shown in Fig. 8.3b, momentum conservation gives:

(x component) $\qquad m\,v = m\,v_1 \cos\theta + m\,v_2 \cos\phi \qquad$ (8.4)

and

(y component) $\qquad m\,v_1 \sin\theta = m\,v_2 \sin\phi,\qquad$ (8.5)

while energy conservation gives:

$$\tfrac{1}{2}\,m\,v^2 = \tfrac{1}{2}\,m\,v_1^2 + \tfrac{1}{2}\,m\,v_2^2.\qquad (8.6)$$

Divide out the common factor of m and use $v_1 = \tfrac{1}{3}\,v$ to get

$$v = \tfrac{1}{3}\,v\cos(\theta) + v_2 \cos\phi,\qquad (8.4')$$

$$\tfrac{1}{3}\,v\sin(\theta) = v_2 \sin\phi,\qquad (8.5')$$

and

$$\tfrac{8}{9}\,v^2 = v_2^2.\qquad (8.6')$$

Rewrite Eq. 8.4' as

$$v_2 \cos(\phi) = v\,(1 - \tfrac{1}{3}\cos\theta).$$

Square this and add the result to the square of Eq. 8.5' so we can eliminate ϕ by using the trig identity

$$\sin^2(\phi) + \cos^2(\phi) = 1$$

to get

$$v_2^2 = (\tfrac{10}{9} - \tfrac{2}{3}\cos(\theta))\,v^2.$$

Substituting this into Eq. 8.6' gives

$$\tfrac{8}{9}\,v^2 = (\tfrac{10}{9} - \tfrac{2}{3}\cos(\theta))\,v^2,$$

so

$$\cos(\theta) = \tfrac{1}{3} \qquad \text{or} \qquad \theta = 71°.$$

The puck is deflected by 71° from its original direction.

PROBLEM 8.4: An air hockey puck of mass m, which was moving at 50 cm/s, strikes a stationary puck of mass 2.0 m. The collision is perfectly elastic. The first puck moves off at an angle of 60° with respect to its initial direction. What is the velocity of the initial puck after the collision? (8.3)

Solution:

> **Strategy:** The collision is elastic so energy and momentum are conserved. Write the momentum conservation laws component by component and write the conservation law of energy. Solve for the final velocity.

We can use Fig. 8.3 again provided we choose

$$m_B = 2.0 \, m_A = 2.0 \, m$$
$$v = 50 \text{ cm/s}$$
$$\theta = \text{final angle of puck A} = 60°$$
$$\phi = \text{final angle of puck B.}$$

The x component of momentum conservation is

$$mv = mv_1 \cos \theta + 2.0 \, mv_2 \cos \phi \; ;$$

for the y component $(p_{Ay} + p_{By}) = 0$,

$$0 = mv_1 \sin \theta - 2.0 \, mv_2 \sin \phi;$$

and energy conservation gives

$$\tfrac{1}{2}mv^2 = \tfrac{1}{2}mv_1^2 + \tfrac{1}{2}2.0 \, mv_2^2.$$

The most direct way to find v_1, is to solve the first two equations for v_2 and use it to find v_1 with the third equation; that is, solve the x momentum and the y momentum equations as

$$v_2 \cos \phi = \tfrac{1}{2}(v - v_1 \cos \theta)$$

and

$$v_2 \sin \phi = \tfrac{1}{2}v_1 \sin \theta \, .$$

Squaring and adding these gives us

$$v_2^2 = \tfrac{1}{4}\left(v^2 - 2vv_1 \cos \theta + v_1^2\right).$$

Then energy conservation gives

$$v^2 = v_1^2 + 2 \, v_2^2 = v_1^2 + \tfrac{1}{2}[v^2 - 2vv_1 \cos \theta + v_1^2],$$

$$= \tfrac{3}{2}v_1^2 + \tfrac{1}{2}v^2 - vv_1 \cos \theta,$$

which, rewritten, is

$$v_1^2 - \tfrac{2}{3}vv_1 \cos \theta - \tfrac{1}{3}v^2 = 0.$$

The solution of this is

$$v_1 = \frac{v}{3}\left(\cos \theta \pm \sqrt{3 + \cos^2 \theta}\right)$$

and, since $\cos 60° = \frac{1}{2}$,

$$v_1 = \frac{v}{3}\left(\frac{1}{2} \pm \sqrt{\frac{13}{4}}\right) = \frac{v}{6}(1 \pm 3.6),$$

so

$$v_1 = 0.77\,v \qquad \text{or} \qquad v_1 = -0.43\,v.$$

If we assume that the scattering was forward as shown in Fig. 8.3, we conclude that

$$v_1 = 0.77\,v = 38 \text{ cm/s}.$$

The other root corresponds to $\theta = 120°$ and comes from

$$\cos 120° = -\cos 60°.$$

PROBLEM 8.5: A driver in an automobile traveling at 35 miles per hour sees a child in the road. Assuming the frictional coefficient is 0.72 for this road and that the car is 50 ft away from the child when the brakes are applied, will the child be hit? How much time elapses before the car stops? (8.4)

Solution:

> **Strategy:** Use the stopping distance formula or the kinematic laws to calculate stopping distance and time.

We can use the results of Problem 8.23 of the text to quickly get the first answer. If s is the stopping distance in feet, then

$$s(\text{ft}) = \frac{[v(\text{mph})]^2}{30\mu},$$

where v is the speed in mph and μ is the coefficient of friction. Using $\mu = 0.72$ and $v = 35$ mph, we get s = 57 ft, so the child will be hit.

This formula for stopping distance follows from the basic kinematic laws. The forces acting are shown in Fig. 8.4, which is a free body diagram for the car.

$$\text{The frictional force} = -\mu\,N = -\mu\,m\,g = ma$$

where a is the acceleration; so

$$a = -\mu\,g.$$

The law of motion relating velocity and acceleration is

$$v^2 = v_0^2 + 2\,a\,x.$$

Setting $v = 0$ gives x as the stopping distance, so

$$\text{stopping distance} = \frac{v_0^2}{\mu g}.$$

Choosing English units gives the stopping distance formula used above.

Fig. 8.4: Free-body force diagram for automobile sliding with locked wheels.

To answer the second part of the question, use another kinematic law:

$$v = v_0 + a\,t.$$

Setting $v = 0$ with the a above gives us the stopping time,

$$t = \frac{-v_0}{a} = \frac{v_0}{\mu\,g},$$

where $v_0 = 35\ \text{mph} = 35 \times \frac{1\ \text{hr}}{3600\ \text{s}} \times \frac{5280\ \text{ft}}{\text{mile}} = 51\ \text{ft/s}$ and $g = 32\ \text{ft/s}^2$.

The stopping time is

$$t = \frac{51\ \text{ft/s}}{0.72 \times 32\ \text{ft/s}^2} = 2.2\ \text{s}.$$

PROBLEM 8.6: A meteor is detected heading toward the earth on a path that is nearly a straight head-on collision. The meteor was detected at a distance of $R = 7 \times R_e$ and a relative speed of 1.0×10^4 m/s. What will be the velocity when the meteor reaches the top of the atmosphere $R \approx R_e$? Why can't we easily answer the question, "How much time will elapse between detection and collision?" (8.5)

Solution:

Strategy: Use conservation of energy.

$$\text{total energy} = \text{kinetic} + \text{potential} = \frac{1}{2}m\,v^2 - \frac{G\,m}{R}.$$

GIVEN:

initial distance	$7 \times R_e$
initial velocity	1.0×10^4 m/s
final distance	R_e

Fig. 8.5: Motion diagram for meteor falling on the earth.

This gives

$$\frac{1}{2} m \, v_i^2 - \frac{G \, m}{7 \, R_e} = \frac{1}{2} m \, v_f^2 - \frac{G \, m}{R_e} \; ,$$

so

$$v_f^2 = v_i^2 + \frac{6 \, G}{7 \, R_e} = v_i^2 + \frac{6}{7} g \, R_e$$

$$= (10^4 \text{ m/s})^2 + \frac{6}{7} \times 9.8 \text{ m/s}^2 \times 6.38 \times 10^6 \text{ m} = 1.54 \times 10^8 \text{ m}^2/\text{s}^2$$

$$v_f = 1.2 \times 10^4 \text{ m/s}.$$

We cannot easily answer the question about the elapsed time because the acceleration is not constant and because frictional forces will be important as the meteor passes through the atmosphere. Either advanced mathematics or graphical analysis will be needed.

PROBLEM 8.7: What is the period of a geostationary earth satellite? By what factor would this period need to be increased in order for the satellite to escape? (8.5)

Solution:

> **Strategy:** A satellite can escape only if its total energy is nonnegative; thus the critical escape condition is total energy (kinetic + potential) equal to zero.

A geostationary earth satellite has a 24 hr period in which to remain over the same location. Hence its velocity must satisfy

$$\text{period} = P = \frac{2\pi R}{v} = 24 \text{ hr} \frac{3600 \text{ s}}{\text{hr}} = 8.784 \times 10^5 \text{ s, where } R \text{ is the orbital radius.}$$

For a circular orbit the centripetal force equals the gravitational force or

$$\frac{m \, v^2}{R} = \frac{G \, M \, m}{R^2} \; ,$$

where M is the mass of the earth. We can use this last equation to find v, giving

$$v = \sqrt{\frac{GM}{R}} \; .$$

If the satellite had escape velocity, the total energy (kinetic plus potential) would be zero, so the minimum velocity for escape at the same R would satisfy

$$\frac{1}{2} m \, v_{esc}{}^2 = \frac{GmM}{R}, \quad \text{(kinetic} = -\text{potential); which is} \quad v_{esc} = \sqrt{\frac{2GM}{R}}.$$

So the ratio is

$$\frac{v_{esc}}{v} = \sqrt{2},$$

independent of R and v.

PROBLEM 8.8: Compute the escape velocity from the surface of Venus which has a radius of 0.949 that of the earth and a mass of 0.814 that of the earth.

Solution:

> **Strategy:** Use the text Eq. 8.11 applied to Venus, expressing the numbers in terms of the earth values.

The text gives the relation for escape velocity as

$$v_{esc} = \sqrt{\frac{2\,G\,M}{R}}, \text{ where } M \text{ is the mass and } R \text{ is the radius of the planet.}$$

Since we know the escape velocity for the earth (1.12×10^4 m/s), we can write

$$v_{esc} = \sqrt{\frac{2\,G\,(0.814M)}{0.949R}} = \sqrt{\frac{0.814}{0.949}}\,\sqrt{\frac{2\,G\,M_{earth}}{R_{earth}}},$$

$$v_{esc} = 0.926 \times (v_{esc})_{earth} = 1.04 \times 10^4 \text{ m/s}.$$

The escape velocity for Venus is 0.926 that of earth, or 1.04×10^4 m/s.

SUGGESTED READINGS

Brody, H., "The Tennis-Ball Bounce Test." *The Physics Teacher,* September 1990, p. 407.

Damask, A. C., "Forensic Physics of Vehicle Accidents." *Physics Today,* March 1987, p. 36.

Frolich, C., ed., *Physics of Sports: Selected Reprints.* College Park, Md.: American Association of Physics Teachers,1986.

Rees, M. J., "Black Holes in Galactic Centers." *Scientific American,* November 1990, p. 56.

Townes, C. H., and R. Genzel, "What Is Happening at the Center of Our Galaxy." *Scientific American,* April 1990, p. 46.

Chapter 9
Rigid Bodies and Rotational Motion

IMPORTANT TERMS

Fill in the blanks with the appropriate word or words.

1. A body behaves as if all its mass were concentrated at a point called the _____.

2. The _____ takes into account both the force acting and the placement of this force.

3. The _____ of a body is the time rate of change of the body's angular velocity.

4. A body whose angular motion is constant is said to be in _____.

5. The _____ is the product of the linear momentum and the distance through which it acts.

6. The quantity analogous to mass in angular motion is called _____.

7. The total angular momentum before a collision is equal to the total angular momentum after the collision. This is a statement of the _____.

8. A(n) _____ rotates with little or no deformation.

9. A(n) _____ consists of a pair of forces (not lying along the same line) of equal magnitude but opposite directions.

10. The ratio of the applied force to the area upon which that force acts is called the tensile _____.

11. The _____ is a measure of the energy of rotation.

12. The tensile _____ is defined as the ratio of the change in length to the initial length of an object when a force is applied which causes the object to stretch.

13. The ratio of stress to strain is called _____.

14. The _____ strain is defined as the ratio of the change in length to the initial length of an object when a force is applied which causes the object to shorten

Write the definitions of the following:

15. center of mass (9.2):

16. torque (9.2):

17. angular momentum (9.6):

18. conservation of angular momentum (9.6):

19. rotational equilibrium (9.2):

20. angular acceleration (9.1):

21. moment of inertia (9.5):

22. compressive strain (9.4)

23. rigid body (9.1)

24. couple (9.3)

25. tensile stress (9.4)

26. rotational kinetic energy (9.8)

27. tensile strain (9.4)

28. Young's modulus (9.4)

ANSWERS TO 1–14, IMPORTANT TERMS

1. center of mass; 2. torque; 3. angular acceleration; 4. rotational equilibrium; 5. angular momentum; 6. moment of inertia; 7. conservation of angular momentum; 8. rigid body; 9. couple; 10 stress; 11. rotational kinetic energy; 12. strain; 13. Young's modulus; 14. compressive.

EQUATION ANALOGIES

The number of equations that you must remember can be reduced by considering the linear and the rotational equations as analogies. To help you remember these you should write the analogous angular equation for each of the given linear equations; however, you should first write the interpretation of the symbols used (fill in missing entries).

SYMBOLS USED FOR CONSTANT LINEAR OR ANGULAR ACCELERATION

Linear		Angular	
Symbol	Interpretation	Symbol	Interpretation
x	linear position	θ	angular position
v	linear velocity	ω	angular velocity
a	linear acceleration	α	angular acceleration
F	force	τ	torque
m	mass	I	moment of inertia
p	linear momentum	L	angular momentum
KE_{trans}	linear kinetic energy	KE_{rot}	angular kinetic energy
Δ	indicates a change	Δ	indicates a change
t	time	t	time

The subscript "o" indicates the value of the quantity at $t = t_0$.

ANALOGOUS EQUATIONS

Linear	Rotational
$\bar{v} = \dfrac{\Delta x}{\Delta t}$	
$\bar{a} = \dfrac{\Delta v}{\Delta t}$	
$v^2 = v_o^2 + a(x - x_o)$	
$v = v_o + a (t - t_o)$	
$x = x_o + v_o(t - t_o)$	
$x = x_o + v_o(t - t_o) + \dfrac{1}{2} a (t - t_o)^2$	
$\Sigma \mathbf{F_i} = 0$	
$\mathbf{p} = m \mathbf{v}$	
$KE_{trans} = \dfrac{1}{2} m v^2$	

Use Table 9.1 of the text to check and correct your entries.

PROCEDURES FOR EQUILIBRIUM PROBLEMS INVOLVING TORQUE

The following general procedures will usually make equilibrium problems with torques reasonably easy to solve. These usually are problems in which extended bodies are supported by wires, pivots, brackets, and/or studs. A wire can provide a force only along its length while a pivot can provide a force in any direction but allows rotation. A bracket provides support and prevents rotation. A stud is assumed to be able to provide a force only perpendicularly to its surface.

1. Identify all forces acting. There will be gravity on the center of the object and forces at each support point.
2. RESOLVE all forces into horizontal and vertical components. Some of these components are known and others are not. Depending upon the questions asked you may need to find only some of the unknown force components. You will need at least as many equations among the force components as there are needed unknown components.
3. Translational equilibrium gives two equations that you should always use relating the forces:

    sum of horizontal forces = 0,
    sum of vertical forces = 0.
4. Write as many additional torque equations as are needed. Choose the points about which you calculate torques carefully. Choose them to be as simple as possible; and, if possible, not to involve the force components that you neither know nor need to know. Usually the best points to use are those for which one or more force components have a zero length action arm. Such points have fewer terms in the torque equation.

5. If a careful choice of points for torques does not eliminate unnecessary unknowns, additional torque equations may be needed.
6. Solve the equations for the unknowns.

SAMPLE QUESTIONS

QUESTION 9.1: What would happen to the planets if the gravitational force had a tangential component as well as a radial component?

Answer:

The angular momenta of the planets would no longer be constants of the motion. This would mean that Kepler's laws would no longer be true.

QUESTION 9.2: Explain how ice skaters can quickly go from a slow to a fast spin and vice versa.

Answer:

A skater can change her angular velocity by changing her moment of inertia. The moment can be decreased by bring her arms close to the trunk of her body. Since angular momentum is conserved, this increases her angular velocity.

SAMPLE SOLUTIONS

PROBLEM 9.1: An electric motor is used to lift a 20 kg bucket of water at constant speed from a well. If the diameter of the pulley is 10.0 cm, what must be the torque of the motor? (9.2)

Solution:

> **Strategy:** At constant speed the force in the support rope must be the weight of the bucket. Calculate the torque for this force.

To lift the 20 kg mass at constant speed, we need a force equal and opposite to the force of gravity. This is a force

$$F = mg = 20 \text{ kg} \times 9.80 \text{ m/s}^2 = 196 \text{ N}.$$

Text Eq. 9.9 applies here with $\theta = 90°$, so

$$\tau = rF \sin 90° = rF = 0.050 \text{ m} \times 196 \text{ N} = 9.8 \text{ N·m}.$$

The motor must be able to provide a torque of 9.8 N·m.

PROBLEM 9.2: A sign weighing 980 N is suspended from the end of a uniform rod 5.0 m long and weighing 1470 N. There is a support cable on the outer end which makes an angle of 45° with respect to the rod (as shown in Fig. 9.1a). What is the tension in the support cable? (9.3)

Solution:

> **Strategy:** There is one unknown so one equation should solve this problem. If we take torques about the wall support for the wall, we can determine the tension.

For the rod to be in equilibrium the translational forces and rotational torques must each add to zero. There are four forces on the rod. Its own weight, the weight of the sign, the tension in the support cable, and the force provided by the wall-rod support. This is shown in Fig. 9.1b. We are asked to find the tension in the wire. This is one unknown, which suggests that possibly only one equation is needed.

Since we are not asked for the rod-wall support force, we can get the desired equation by taking torques about the rod-wall point. This equation, which will not involve the forces at that point, is

$$\text{(weight of rod)} \times \frac{L}{2} + \text{(weight of sign)} \times L - T \sin \theta \times L = 0,$$

where L is the length of the rod, θ is the angle between the rod and the cable, and we have assumed the rod acts as if all its mass is concentrated at the center. Thus,

$$T = \frac{\dfrac{\text{weight of rod}}{2} + \text{weight of sign}}{\sin \theta} = 2.4 \times 10^3 \text{ N}.$$

The cable has a tension of 2.4×10^3 N.

(a)   **(b)**

Fig. 9.1:   (a) A sign hanging on a rod with a 45° tension wire. (b) Forces on the rod.

PROBLEM 9.3: A square sign one meter on a side and of mass 50 kg is suspended from a wall by a bracket at the upper corner, a stud at the lower corner, and a support wire as shown in Fig. 9.2a. The wire makes an angle of 30° with respect to the horizontal at the outer edge of the sign and has a tension of 40 N. What is the force provided by the bracket and how much force does the stud exert on the sign? (9.3)

Solution:

Strategy: Follow the procedures for equilibrium problems involving torque. The first two steps are to identify the forces and resolve them into components.

(a)   (b)

(c)

Fig. 9.2: (a) The physical arrangement of sign, wall, brackets, and tension wire. (b) A free body force diagram for the sign. The dotted lines are the diagonals of the sign and the bold arrows are the forces. Each force is shown acting at its proper location. (c) Torque produced about C by force F may be calculated as either $F \times R \times \sin\theta$ or $F \times R \times \sin\phi$.

Refer to Fig 9.2a. We label the tie point for the wire by the letter A and the upper and lower supports by B and C, respectively. The length of each side of the sign is denoted by the symbol d. The center of mass of the sign will be taken to be in the geometric center of a perfect square one meter on a side. The bracket will be assumed to be able to provide any force needed. Details of the bracket will be ignored.

In Fig. 9.2b we show the free body diagram for the sign. The bracket can provide an upward or downward force and a horizontal force. Rather than putting in an arbitrary force and angle, it is usually easier to put in the vertical and horizontal components and calculate the force from these at the last step or just give these components as the answer. The positive direction of the forces has been arbitrarily chosen as shown. If the magnitude of any force comes out negative, then the force points in the opposite direction to that shown.

The unknown forces we want are V_B, F_B, and F_C; thus we need three equations among the force components.

The sign should be in both translational and rotational equilibrium. If we let F_i denote any of the forces and τ_i be the torques calculated about any point then these equilibrium conditions are:

$$\sum_i F_i = 0 \qquad \text{and} \qquad \sum_i \tau_i = 0.$$

Summing the forces will present no problems; however, to sum the torques, you must decide which and how many centers to use. The following rules provide a good guide:

1. Determine how many unknown forces need to be determined. (For this problem there are three forces that are not known, V_B, F_B, and F_C, since m and T are given.)
2. Since translational equilibrium will give us two equations (considering two-dimensional problems), subtract two from the number of unknown forces (leaving one for this problem).
3. The remaining number is the number of torque equations needed (for this problem there is one).
4. Choose the centers needed so as to produce the simplest equations to solve. This is the hardest part to do without experience. Usually the best choices are points at which forces act, since these choices will eliminate one or more forces from the torque equation(s); however, any choice(s) will work. (For this problem we could choose point A or B or C.)

First we draw the force diagram for the sign as shown in Fig. 9.2b. There are five forces, two of which are known.

The second step is to write the equations for translational equilibrium. These are:

$$\text{for } x \qquad T\cos\theta - F_B - F_C = 0, \qquad (9.1)$$

$$\text{for } y \qquad T\sin\theta + V_B - mg = 0. \qquad (9.2)$$

These two equations involve all of the unknowns, so we will need one torque equation.

For this particular example there are four locations at which forces act. They are locations A, B, C, and the center of mass. These locations are among the centers which should be used for calculating torques. In general, the center of mass is a poor choice since it usually maximizes the number of terms present.

A set of torque equations are as follows.

Torques about B (Eliminating V_B and F_B)

$$T \times d \times \sin(\theta) + F_C \times d - m g \frac{\sqrt{2}\, d}{2} \times \sin(135°) = 0 \qquad (9.3)$$

where torques are taken as positive if clockwise (arbitrary choice).

At this point some additional comments on writing torques are in order. First, since $\sin(\Omega) = \sin(90° + \Omega)$ we could have written Eq. 9.3 as

$$T \times d \times \sin(\theta) + F_C \times d - m g \frac{\sqrt{2}\, d}{2} \times \sin(45°) = 0. \qquad (9.3')$$

This choice is shown in Fig. 9.2c. Secondly, we could have also projected the gravitational force on the dotted line and immediately written Eq. 9.3 as

$$T \times d \times \sin(\theta) + F_C \times d - m g \frac{d}{2} = 0. \qquad (9.3'')$$

All of these are equivalent; the latter obviously is simplest.

Torques about A (Eliminating T and F_B)

Another possible choice is

$$- V_B \times d + F_C \times d = 0. \qquad (9.4)$$

Eq. 9.4 gives

$$F_C = V_B \qquad (9.5)$$

while 9.3 gives

$$F_C = \frac{mg}{2} - T \sin \theta. \qquad (9.6)$$

Eqs. 9.1 and 9.2 give, respectively

$$T \cos \theta = F_B + F_C \qquad (9.7)$$

and

$$T \sin \theta = mg - V_B. \qquad (9.8)$$

The simplest choice is to use Eq. 9.8 to find V_B; then Eq. 9.5 gives F_C and Eq. 9.7 gives F_B. That is,

$$V_B = mg - T \sin \theta$$

$$F_C = V_B, \text{ and } F_B = T \cos \theta - F_C.$$

From the numbers given for this problem, we find

$$V_B = 4.7 \times 10^2 \text{ N}, F_C = 4.7 \times 10^2 \text{ N, and } F_B = 4.4 \times 10^2 \text{ N}.$$

The force F_B is in the opposite direction from that shown in Fig. 9.2b.

PROBLEM 9.4: A wheel consists of a hub of mass 0.10 kg and diameter 10 cm, a set of 8 spokes and an outer rim 1.2 m in diameter. The outer rim is of iron and has a mass of 50 kg. The wheel is initially spinning on an axis about the hub at a rate of 100 revolutions per minute without friction. If a force of 10 N is applied tangentially to the wheel at the rim, how long will it be before the wheel stops and through how many revolutions will it turn? Neglect the mass of the spokes. (9.5)

Solution:

Strategy: This is an application of the equations relating angular motion and angular acceleration. First calculate the moment of inertia of the wheel. There is torque due to the applied force. This torque causes an angular acceleration. The angular acceleration stops the wheel.

Fig. 9.3: Wheel of radius 0.60 m with constant stopping force. The angular position, θ, is measured positively in a clockwise sense as shown.

The problem is illustrated in Fig. 9.3. We will assume the hub is a uniform cylinder of radius $r = 5.0$ cm and mass, $m = 0.10$ kg, while the rest of the wheel is a ring of radius $R = 0.60$ m and mass $M = 50$ kg.

We will represent the initial rotation of the wheel by $\omega_0 = 2\pi \times$ frequency $= 2\pi \times 100$ rev/min $\times$ 1min/60 s $= 10.5$ rad/s. The applied torque is constant and will produce a constant angular acceleration, α, which will reduce the rotation rate of the wheel. Hence we have the following.

GIVEN:

Quantity	Symbol	Value
radius of hub	r	0.05 m
mass of hub	m	0.10 kg
radius of rim	R	0.6 m
mass of rim	M	50 kg
applied force	F	10 N
applied torque	$\tau = -FR$	-6 N•m
initial rotation rate	ω_0	10.5 rad/s

Calculate the Moment of Inertia, I_T, for the Wheel

The net moment of inertia of the system is

$$I_T = I_{hub} + I_{rim},$$

where

$$I_{hub} = \frac{1}{2} m \, r^2$$

and

$$I_{rim} = M \, R^2$$

as can be found in Table 9.3 of the text.

Calculate the Angular Acceleration

We choose the angular position to increase positively in the clockwise sense, so the angular acceleration, α, as given by Eq. 9.4 of the text is

$$\alpha = \frac{\tau}{I_T} = \frac{\tau}{\frac{1}{2} m \, r^2 + M \, R^2}$$

$$= \frac{R \, F}{\frac{1}{2} m \, r^2 + M \, R^2} = -0.33 \text{ rad/s}^2,$$

where τ is the torque.

The effect of the hub is too small to consider with our significant figures.

Calculate the Rotation Angle

The rotational equation from Table 9.1 of the text is

$$\omega = \omega_0 + \alpha \, t.$$

The wheel will stop at a time given by

$$t = -\frac{\omega_0}{\alpha}$$

$$= -\frac{10.5 \text{ rad/s}}{-0.33 \text{ rad/s}^2} = 32 \text{ s}.$$

The wheel will stop in 32 s.

Angle of Turn

This is most easily calculated by using the relation

$$\omega^2 = \omega_0^2 + 2 \, \alpha \, (\theta - \theta_0) \, .$$

We know $\omega = 0$ and $\omega_0 = 10.5$ /s, and where θ is the angle of rotation this gives

$$\Delta\theta = \theta - \theta_0 = \frac{\omega^2 - \omega_0^2}{2\alpha} = -\frac{\omega_0^2}{\alpha} = 167 \text{ rad.}$$

Each revolution is 2π radians so this is about 27 revolutions.

PROBLEM 9.5: A physics teacher stands on a freely rotating platform. She holds a dumbbell in each hand of her outstretched arms while a student gives her a push until her angular velocity reaches 2.0 rad/s. When the freely spinning professor pulls her hands in close to her body, her angular velocity increases to 7.0 rad/s. What is the ratio of her final rotational kinetic energy to her initial rotational kinetic energy? Where does this energy originate? (9.8)

Solution:

Strategy: The term freely spinning suggests that friction can be neglected. Thus the total energy and the total angular momentum should be conserved. There is no translation so we need not consider linear momentum. Write the expression for angular momentum and conserve the angular momentum when the arms are moved towards her body.

The situation is shown in Fig. 9.4, with 9.4a showing the initial arrangement of arms out and angular speed $\omega_0 = 2.0$ rad/s and 9.4b showing the final arrangement with arms in and angular speed $\omega_1 = 7.0$ rad/s.

Since the platform is freely rotating, we assume momentum conservation. Let I_i be the initial moment of inertia (arms out) and I_f be the final moment of inertia (arms in); then

$$I_i \, \omega_0 = I_f \, \omega_1,$$

and so

$$I_f = I_i \, \frac{\omega_0}{\omega_1}.$$

(a) (b)

Fig. 9.4: (a) Starting with arms out and speed ω_0. (b) With arms in and speed ω_1.

The initial kinetic energy is $$KE_i = \frac{1}{2} I_i \, \omega_0^2,$$

while the final kinetic energy is $\qquad KE_f = \frac{1}{2} I_f \omega_1{}^2.$

This gives:

$$\frac{KE_f}{KE_i} = \frac{I_f}{I_i} \frac{\omega_1{}^2}{\omega_0{}^2} = \frac{\omega_1}{\omega_0} = 3.5.$$

So momentum is conserved but kinetic energy increases. The kinetic energy must come from the work done by the teacher in moving her arms in to her sides. That is, food energy has been converted to work which was converted to kinetic energy.

PROBLEM 9.6: A cylinder of mass 18 kg and radius 75 cm rolls down an inclined plane whose slope is 30° from a height of 1.5 m. Assume the cylinder has an initial horizontal velocity of 2.0 m/s and that it rolls without slipping. What is the translational speed at the bottom of the ramp? (9.9)

Solution:

> **Strategy:** Use conservation of energy. Since there is no slipping, the angular velocity of the cylinder is related to its translational velocity.

A sketch of the problem is shown in Fig. 9.5.

Fig. 9.5: Cylinder rolling, without slipping, down an inclined plane.

We will use the following symbols

Quantity	Symbol	Value
initial translational velocity	v_i	2.0 m/s
initial rotational velocity	ω_i	?
final translational velocity	v_f	?
final rotational velocity	ω_f	?
height of inclined plane	h	1.5 m
angle of inclination	θ	30°
radius of cylinder	r	0.75 m
mass of cylinder	m	1.8 kg

The relation between the rotational speed and the translational speed is given by the requirement that there is no slipping. This gives:

$$\omega r = v \qquad \text{or} \qquad \omega = \frac{v}{r}.$$

Choose the zero of potential energy at the bottom of the plane and let I be the moment of inertia of the cylinder about the contact point. Note that the rotational frequency of the cylinder about the contact point is the same as its rotational frequency about its center. Energy conservation requires

(kinetic energy + potential energy)$_{initial}$ = (kinetic energy + potential energy)$_{final}$,

so

$$\tfrac{1}{2} m \, v_i^2 + \tfrac{1}{2} I \omega_i^2 + mgh = \tfrac{1}{2} m \, v_f^2 + \tfrac{1}{2} I \omega_f^2.$$

Using the relation between ω and r and the value of I from text Table 9.3, we have

$$I \omega^2 = \tfrac{3}{2} m r^2 \omega^2 = \tfrac{3}{2} m r^2 \frac{v^2}{r^2} = \tfrac{3}{2} m v^2 \ .$$

The energy equation is then

$$\tfrac{1}{2} m \, v_i^2 + \tfrac{3}{2} m v_i^2 + mgh = \tfrac{1}{2} m \, v_f^2 + \tfrac{3}{2} m v_f^2 \ .$$

The result is independent of m and θ, $\qquad v_f^2 = v_i^2 + \dfrac{gh}{2} \ ,$

so the translational speed of the cylinder at the bottom of the ramp is 3.4 m/s.

PROBLEM 9.7: A drum full of hardened cement is rolled down an incline at a warehouse for removal and disposal. The drum has a diameter of 0.50 m, a length of 1.0 m, and a mass of 750 kg. The incline is at 30° and lowers the drum 1.2 m. What is the translational velocity of the drum just after it leaves the ramp? Assume the drum rolls without slipping and that it is rolling at a translational velocity if 1.0 m/s just as it starts down the ramp. (9.9)

Solution:

(a)	(b)

Fig. 9.6: (a) Drum just before it starts down the ramp. (b) Drum just after leaving the ramp.

Strategy: Since there is no slipping, we can relate the angle through which the drum has rotated to the distance it has traveled. Also, since no work is done by frictional forces, we can use conservation of energy. Total energy = kinetic + potential is a constant of the motion.

This problem is most easily solved by energy conservation. By use of the conservation theorem we can eliminate the need to know all the forces. In Fig. 9.6a we show the initial conditions and in Fig. 9.6b the drum just after leaving the ramp. We use the following information.

GIVEN:

Quantity	Symbol used	Value given
radius of drum	r	0.25 m
length of drum	b	1.0 m
mass of drum	m	750 kg
initial height	h	1.2 m
incline angle	θ	30°
horizontal distance moved	d	$h/\tan(\theta)$
initial translational velocity	v_0	1.0 m/s
initial rotational velocity	ω_0	?
final translational velocity	v	?
final rotational velocity	ω	?

Since the drum rolls without slipping there is a relation between the angle through which the drum has turned and the distance it has moved:

$$\text{angle turned} = \frac{\text{distance}}{2\pi \text{ radius}} \times 2\pi,$$

where the ratio tells us the number of rotations and the last 2π converts this to radians. (2π radians = one circle.) Dividing both sides of this relation by time we can convert this to a relation between translational velocity and angular velocity:

$$\text{angular velocity} = \frac{\text{translational velocity}}{r}.$$

This is just Eq. 5.3 of the text, which gives

$$\omega_0 = \frac{v_0}{r}$$

and

$$\omega = \frac{v}{r}.$$

Choose the Zero for Potential Energy

We choose the zero of potential energy to be the height of the center of the drum after it leaves the ramp. Any other choice is just as good.

The conservation of mechanical energy theorem gives

$$\text{initial translational kinetic energy} \left(\frac{1}{2} m \, v_0^2\right)$$

$$+ \text{ initial rotational kinetic energy} \left(\frac{1}{2} I \, \omega_0^2\right) + \text{initial potential energy } (m \, g \, h)$$

$$= \text{final translational kinetic energy} \left(\frac{1}{2} m \, v^2\right)$$

$$+ \text{final rotational kinetic energy} \left(\frac{1}{2} I \, \omega^2\right) + \text{final potential energy } (0),$$

where we have chosen the reference of potential energy to be the center of the drum's final height.

Writing the energy conservation in terms of the symbols used above gives

$$\frac{1}{2} m \, v_0^2 + \frac{1}{2} I \, \omega_0^2 + m \, g \, h = \frac{1}{2} m \, v^2 + \frac{1}{2} I \, \omega^2 + 0.$$

Using the relations between ω and v (and ω_0 and v_0) allows the elimination of the ω, giving

$$\frac{1}{2}\left(m + \frac{I}{r^2}\right) v_0^2 + m \, g \, h = \frac{1}{2}\left(m + \frac{I}{r^2}\right) v^2,$$

which gives us

$$v^2 = v_0^2 + \frac{m \, g \, h \, r^2}{m \, r^2 + I} = v_0^2 + \frac{gh}{\left(1 + \dfrac{I}{mr^2}\right)}.$$

The moment of inertia to use is just that for a disk rotating about one edge parallel to the axis of symmetry, which is given in Table 9.3 of the text as

$$I = \frac{3}{2} m \, r^2.$$

So

$$v^2 = v_0^2 + \frac{2}{5} g \, h$$

and

$$v = 2.4 \text{ m/s}.$$

Another Choice for the Zero of Potential Energy

Had you chosen the reference for potential energy differently, the results would have been the same. For example, suppose we choose the zero of potential energy to be the center of the drum at the top of the ramp. Then the conservation of mechanical energy theorem gives us:

initial translational kinetic energy $\left(\frac{1}{2} m v_0^2\right)$

+ initial rotational kinetic energy $\left(\frac{1}{2} I \omega_0^2\right)$ + initial potential energy (0)

= final translational kinetic energy $\left(\frac{1}{2} m v^2\right)$

+ final rotational kinetic energy $\left(\frac{1}{2} I \omega^2\right)$ + final potential energy $(-m g h)$.

This gives us the same results:

$$\frac{1}{2} m v_0^2 + \frac{1}{2} I \omega_0^2 + 0 = \frac{1}{2} m v^2 + \frac{1}{2} I \omega^2 - mgh.$$

It is also interesting to note that the result is independent of the length and of the mass of the drum. This means that a drum of styrofoam would travel at the same translational velocity as the cement. This is not too surprising since they would fall at the same rate if dropped. The independence of the result on the mass is then a rough check of our results.

PROBLEM 9.8 An Atwood machine has a mass of 10 kg hanging on the right side and a mass of 5.0 kg on the other. The pulley is a disk with a mass of 4.0 kg and a radius of 8.0 cm. Assume the rope moves over the pulley without slipping and that the objects are not moving initially. What is the motion of the 10 kg mass? (9.9)

Solution:

> **Strategy:** Choose the force or energy method. We choose the energy approach. Require the conservation of total energy.

Fig. 9.7a shows the Atwood machine for this problem. This problem will force us to revise our model of an ideal rope. We are told that the rope does not slip on the pulley, so the pulley itself must turn. If the pulley starts from rest and accelerates, there must be forces on the pulley. Since the rope is the only thing attached to the edge of the pulley, the rope must provide these forces. This is only possible if the tension in the right side of the rope is different than the tension in the left-hand side. Obviously, for a real rope, if we carefully analyzed the frictional forces we could understand this behavior. For our model of a rope we must allow the tension to change across those regions for which the rope is in contact with another surface across which it does not slip.

The best approach to problems is to first decide if a force or an energy approach is most appropriate. We will first solve this problem using energy methods; we will then solve this problem again using forces.

We need to look first at the variables used and their interpretations. The center of the pulley can serve as a reasonable reference point. We will let

x indicate the distance of mass A from this point,

y be the distance of mass B as shown in Fig. 9.7a,

θ be the angle through which the pulley has turned, and

R represent the radius of the pulley.

Fig. 9.7:   (a) The physical arrangement of the Atwood machine. (b) Free body force diagram for body A. (c) Free body force diagram for body B. (d) Free body force diagram for the pulley.

Since the rope has not stretched and does not slip on the pulley:

$$y - y_0 = -(x - x_0) \qquad \text{and} \qquad \theta - \theta_0 = \frac{x - x_0}{R},$$

where y_0, x_0, and θ_0 represent the initial positions of the masses and pulley. We will represent the time by t and take the initial time to be $t = 0$ and let

v_y and a_y represent the velocity and acceleration of mass B,

v_x and a_x be the velocity and acceleration of mass A, and

ω and α be the velocity and acceleration of the pulley, respectively.

The relations between y, θ, and x give us

$$v_y = -v_x, \qquad a_y = a_x,$$

$$\omega = \frac{v_x}{R}, \text{ and } \alpha = \frac{a_x}{R}.$$

These definitions and relations are given in Table 9.1.

Table 9.1: Variables and Relations for Problem 9.8

Object	Position	Velocity	Acceleration	Relation
mass A	x	v_x	a_x	
mass B	y	v_y	a_y	$v_y = -v_x$ $a_y = -a_x$
pulley	θ	ω	α	$\omega = v_x/R$ $\alpha = a_x/R$

GIVEN:

Object	Symbol	Value
mass of body A	m_A	10.0 kg
mass of body B	m_B	5.0 kg
mass of pulley	m_C	4.0 kg
radius of pulley	R	8.0 cm
moment of inertia of pulley	I	$\frac{1}{2} m_C R^2$

NOT KNOWN:

Quantity	Symbol	Quantity	Symbol
position of mass A	x	tension in rope on right	T_A
velocity of mass A	v_x	tension in rope on left	T_B
acceleration of mass A	a_x	force on support (tension in support rope)	F

Energy Method

We will take the zero of potential energy at the pulley. Then the

$$\text{initial potential energy} = -(m_A\, g\, x_0 + m_B\, g\, y_0)$$

and

$$\text{initial kinetic energy} = 0.$$

At any time the energy conservation theorem gives us:

$$-(m_A\, x_0 + m_B\, y_0)\, g = -(m_A\, x + m_B\, y)\, g + \frac{1}{2} m_A\, v_x^2 + \frac{1}{2} m_B\, v_y^2 + \frac{1}{2} I\, \omega^2 .$$

We also have

$$y - y_0 = -(x - x_0)$$

since as x increases y decreases by the same amount. Using this and the relations among v_x, v_y and ω, we get:

$$v_x^2 = 2 \times \frac{(m_A - m_B)\, g}{m_A + m_B + \dfrac{m_C}{2}} \times (x - x_0) = 2\, a_x\, (x - x_0),$$

where the last equality follows from Eq. 2.7 of the text; that is, we can identify the coefficient of $(x - x_0)$ as the acceleration in the x direction. If you enter the numbers for this problem, you will get the velocity of the 10 kg mass.

This result was obtained without reference to the forces on the masses or the pulleys.

Force Method

This problem is most simply solved by energy considerations; however, it is instructive to consider a force approach. Consider the variables used and their interpretation. The center of the pulley can serve as a reasonable reference point. We will let x indicate the distance of mass A from this point and y be the distance of mass B as shown in Fig. 9.7a. We will also let θ be the angle through which the pulley has turned. R will represent the radius of the pulley.

Since the rope has not stretched and does not slip on the pulley:

$$y - y_0 = -(x - x_0)$$

and

$$\theta - \theta_0 = \frac{x - x_0}{R},$$

where y_0, x_0, and θ_0 represent the initial positions of the masses and pulley. We will represent the time by t and take the initial time to be $t = 0$. We will also let v_y and a_y represent the velocity and acceleration of mass B, v_x and a_x be the velocity and acceleration of mass A, and ω and α the velocity and acceleration of the pulley. The relations between y, θ and x give us:

$$v_y = -v_x, \qquad a_y = a_x,$$

and

$$\omega = v_x/R, \qquad \alpha = a_x/R.$$

GIVEN:

Quantity	Symbol	Value
mass of body A	m_A	10 kg
mass of body B	m_B	5.0 kg
mass of pulley	m_C	4.0 kg
radius of pulley	R	8.0 cm

UNKNOWN:

Quantity	Symbol
position of mass A	x
velocity of mass A	v_x
acceleration of mass A	a_x
tension in rope on right	T_A
tension in rope on left	T_B
force on support (tension in support rope)	F

Figs.9.7b–d show the forces and torques acting on the three bodies of the problem. We let F be the force on the string holding the pulley up. Then, from Newton's second law for translational and rotational motion, we have

Translational:

$$m_A g - T_A = m_A a_x,$$

$$m_B g - T_B = m_B a_y = -m_B a_x,$$

$$F - T_B - T_A = 0.$$

Rotational:

$$(T_A - T_B) \times R = I\alpha = I\frac{a_x}{R} = \frac{1}{2}m_C R^2\frac{a_x}{R} = \frac{m_C \times R \times a_x}{2},$$

where we have used the moment of inertia, I, for a disk about its axis as given in Table 9.3 of the text. The first, second and last equations together give us:

$$a_x = \frac{m_A - m_B}{m_A + m_B + \dfrac{m_C}{2}} g = 2.9 \text{ m/s}^2.$$

Note that these results are independent of R.

Since the initial velocity was zero and the initial time was taken as zero,

$$x(t) = x_0 + \frac{1}{2} a_x t^2 = x_0 + 1.5 \text{ m/s}^2 \times t^2.$$

This allows us to find the position at any time.

PROBLEM 9.9: An Atwood machine has a mass of 12 kg hanging on one side and a mass of 7.0 kg on the other side. The pulley is a disk with a mass of 9.0 kg and a radius of 8.00 cm. Assume the string connecting the two masses does **not** slip on the pulley. (a) If the masses start from rest, how fast will the 12 kg mass be moving when it has fallen 80 cm? (b) How long will this take? (9.9)

Solution:

Strategy: Use energy considerations to find the velocity after the masses have moved 80 cm. From this, and the basic kinematic laws, the acceleration and time can be determined.

Fig. 9.8: (a) Atwood machine when the masses are released. (b) The system after a time t.

The initial state is shown in Fig. 9.8a and the system after it has fallen 80 cm is shown in Fig. 9.8b. We denote the two masses by A and B. Since they are connected by a string, the amount A moves down is equal to the amount B moves up. At any point in the motion the energy is conserved; that is, if PE is gravitational potential energy and KE is kinetic energy:

$$PE_A + KE_A + PE_B + KE_B + KE_{\text{pulley}} = \text{constant.}$$

Since the system starts from rest the initial kinetic energy is zero and if we measure the positions of the two masses by x, the displacement from the starting position, then we may choose this constant to be zero:

$$PE_A + KE_A + PE_B + PE_B + KE_{\text{pulley}} = 0.$$

In terms of x and the instantaneous angular velocity of the pulley this becomes

$$-m_Agx + \frac{1}{2}m_Av_A^2 + m_Bgx + \frac{1}{2}m_Bv_B^2 + \frac{1}{2}I\,\omega^2 = 0,$$

where v_A and v_B are the velocities of mass A and mass B, respectively, and I is the moment of inertia of the pulley. The first and third terms are of opposite sign since A and B move in opposite directions.

Since the string is assumed not to slip on the pulley,

$$\omega r = v_A = v_B$$

where r is the radius of the pulley, ω is the rotational velocity of the pulley, and the last equality follows from the fixed length of the string. Thus the energy equation becomes

$$-(m_A - m_B)\,gx + \frac{1}{2}(m_A + m_B)v_A^2 + \frac{1}{2}I\left(\frac{v_A}{r}\right)^2 = 0$$

or

$$v_A^2 = \frac{2(m_A - m_B)gx}{m_A + m_B + \dfrac{I}{r^2}} = \frac{2(m_A - m_B)gx}{m_A + m_B + \frac{1}{2}m_{\text{pulley}}}.$$

For $x = 0.80$ m, this gives

$$v_A = 1.9 \text{ m/s},$$

and since the mass started from rest, we can use $v^2 = 2ax$ to find the acceleration,

$$a = \frac{v^2}{2} = 2.3 \text{ m/s}^2.$$

Hence, since $x = \frac{1}{2}a\,t^2$,

$$t = \sqrt{\frac{2x}{a}} = 0.83 \text{ s}.$$

GENERAL COMMENTS

With the introduction of rotation we must modify our model of bodies. Instead of points, they must now be extended bodies. There are two models we can take:

1. Deal with the extended body by dividing it into small pieces and sum over the pieces with the appropriate forces. ALSO add internal forces to keep these pieces in a fixed relationship to each other (rigid body). This approach is used in advanced courses.

2. Treat the body as a point mass, but also give a set of numbers that describes the body's rotational properties about different axes. External forces, except for gravity, act on the surface of the body.

For this course we will adopt the latter approach: The body has finite size but acts as if all the mass were concentrated at one point of the body (center of mass). The gravitational force will act on the body as if all the mass were at this point. Other forces will act on points of the surface of the body. The body will respond with translation and rotation and change gravitational potential energy as given by text Eq. 9.20. The moment of inertia is then just another number, like the mass, of the set of numbers that give the properties of a body.

Some moments of inertia are given in text Table 9.3. Notice that the moment depends not only upon the mass and the shape of the body but also upon the location of the axis about which the rotation occurs. It would seem that the moment of inertia is not just characteristic of the body, but also depends upon the axis of rotation. It is possible to define a set of numbers that depend only upon the body (the characteristic moments of inertia for the body) and then give rules allowing one to calculate the moment of inertia for any given axis. Assigned problems will either use the quantities in Table 9.3 or will give you the necessary moment of inertia.

SUGGESTED READINGS

Brody, H., "The Moment of Inertia of a Tennis Racket." *The Physics Teacher,* April 1985, p. 312.

Burger, W., "The Yo-Yo: A Toy Flywheel." *American Scientist,* March 1984, p. 137.

Edge, R. D., "Murphy's Law or Jelly-Side Down." *The Physics Teacher,* September 1988, p. 392.

Frederickson, J. E., "The Tail-less Cat in Free-Fall." *The Physics Teacher,* November 1989, p. 620.

Frohlich, C., "The Physics of Somersaulting and Twisting," *Scientific American,* March 1980, p. 154.

Laws, K., "Physics and Dance." *American Scientist,* September 1985, p. 426.

Salvadori, M., *Why Buildings Stand Up: The Strength of Architecture.* New York: W. W. Norton & Co., 1990. A very interesting and readable survey of construction including the pyramids, buildings, and bridges.

Scaglia, G., "Building the Cathedral in Florence." *Scientific American,* January 1991, p. 66.

Walker, J., "The Amateur Scientist, The Physics of the Follow, the Draw, and the Massé (in Billiards and Pool)." *Scientific American,* July 1983, p. 124.

Chapter 10
Fluids

IMPORTANT TERMS

Fill in the blank with the appropriate word or words.

1. A substance which cannot maintain its own shape is called a(n) _____.

2. The SI unit of pressure is called the _____. It is equivalent to $1 N/m^2$.

3. Pressure measured relative to the surrounding atmosphere is called _____ pressure.

4. "A body, either submerged or floating, is buoyed up by a force that is equal to the weight of the displaced fluid" is a statement of _____.

5. The _____ is a parameter which tells us if fluid flow is laminar or turbulent.

6. _____ flow is a smooth streamline flow.

7. _____ action, which results from surface tension and adhesive forces between a fluid and the walls of a small tube, shows when the fluid rises in the small tube above the level of the surrounding fluid.

8. The pressure difference between two points in a pipe containing a flowing viscous fluid is given by _____ law.

9. A(n) _____ is a device for measuring the velocity of a fluid flowing through a pipe.

10. The upward force exerted on a body in a fluid is called the _____ force.

11. A falling body entering a fluid slows its velocity until it reaches _____ velocity, when the upward fluid force equals the gravitational force.

12. _____ is the force per unit area. Its SI unit is the pascal (Pa).

13. _____ law gives us the relation between the drag force exerted by a fluid on a moving body and the velocity of the body.

14. _____ principle tells us that the pressure exerted at any point of an enclosed fluid is transmitted to every part of the fluid.

15. The _____ of a fluid is a measure of the fluid's internal friction.

16. The equation of _____ must hold if the total amount of fluid crossing any surface is constant.

17. The _____ of a fluid is a measure of the force per unit length needed to remove a rigid body from the fluid.

17. The _____ of a fluid is a measure of the force per unit length needed to remove a rigid body from the fluid.

18. _____ equation is the form that the conservation of energy takes for fluids.

19. If neighboring layers of a fluid flow smoothly past each other, the resulting paths are called _____.

20. _____ flow is characterized by irregular, complex motion of the fluid.

21. The _____ of a substance is defined as the ratio of its density to the density of water at 4°C

Write the definitions of the following:

22. Reynolds number (10.9):

23. fluid (10.0):

24. laminar flow (10.5):

25. equation of continuity (10.5):

26. pressure (10.1):

27. terminal velocity (10.8):

28. surface tension (10.4):

29. turbulent flow (10.5):

30. pascal (10.1):

31. Stokes's law (10.8):

32. Bernoulli's equation (10.6):

33. Poiseuille's law (10.6):

34. Pascal's principle (10.2):

35. viscosity (10.7):

36. Archimedes' principle (10.3):

37. specific gravity (10.1):

38. buoyant force (10.3):

39. drag coefficient (10.8):

40. streamline (10.5):

ANSWERS TO 1–21, IMPORTANT TERMS

1. fluid; 2. pascal; 3. gauge; 4. Archimedes' principle; 5. Reynolds number; 6. laminar; 7. capillary; 8. Poiseuille's; 9. Venturi meter; 10. buoyant; 11. terminal; 12. pressure; 13. Stokes's; 14. Pascal's; 15. viscosity; 16. continuity; 17. surface tension; 18. Bernoulli's; 19. streamlines; 20. Turbulent; 21. specific gravity.

SAMPLE QUESTIONS

QUESTION 10.1: You are floating on a rubber raft in a small swimming pool in which the water level has been carefully measured. You throw overboard some wooden blocks that had been on the raft and watch the blocks float on the water. What happens to the water level as measured on the edge of the pool? Would the water level behave differently if the blocks were concrete that sank to the bottom of the pool?

Answer:

The water level will be unchanged. A volume of water with a weight equal to the raft and contents has already been displaced. Spreading the contents out will not change the volume of displaced water. This is true even if the blocks were concrete and sank to the bottom.

QUESTION 10.2: A block of wood floats half submerged in a container of water. If the same container were in an earth-orbiting satellite, how would the block float? Explain your reasoning.

Answer:

On the earth the block sinks until it has displaced an equal weight of water. In orbit the water and block are in free fall, so there is no weight. If the water does not wet the block, the block would be repelled by the surface tension of the water. If the water wets the block, then the block will probably be completely inside the water glob.

SAMPLE SOLUTIONS

PROBLEM 10.1: A 200 kg motorized tricycle is supported by three identical tires inflated to a gauge pressure of 200 kPa. Ignoring any possible effects of tread thickness, calculate the area of contact between each tire and the road. Hint: Neglect any buoyant force due to the earth's atmosphere. (10.1)

Solution:

Strategy: The tricycle is in equilibrium so the upward and downward forces balance. Through a series of action-reaction pairs, the gravitational force is balanced by the force provided by the tires. This latter force is pressure times contact area.

Since the tricycle is in static equilibrium the normal force provided by the surface upon which the tricycle rests is equal to the mass times the acceleration of gravity as shown symbolically in Fig. 10.1. That is, if the mass is m and g is the acceleration of gravity, then

N = mg, and the reaction pair is the force provided by the tires. Hence

$$F = mg.$$

Fig. 10.1: Forces between tricycle and road.

The force, F, provided by the tires will be the pressure in the tires times the surface area of contact. This force must also be the weight of the tricycle, mg.

If we let A be the area of one tire and assume all tires are equal, then

$$F = 3 \times p \times A,$$

where the 3 comes from the three tires and p is the true pressure inside the tires. Equating these gives

$$A = \frac{mg}{3p}.$$

The true pressure is found from the gauge pressure by adding the pressure of one atmosphere

$$p \quad = p_{\text{gauge}} + p_{\text{atmosphere}} = 200 \text{ kPa} + 101.3 \text{ kPa}$$

$$= 301 \text{ kPa}.$$

Hence

$$A \quad = \frac{200 \text{ kg} \times 9.80 \text{ m/s}^2}{3 \times (3.01 \times 10^5 \text{ Pa})} = \frac{200 \times 9.8}{9.03 \times 10^5} \times \frac{\text{kg m/s}^2}{\text{Pa}}$$

$$= 2.17 \times 10^{-3} \times \frac{\text{kg m/s}^2}{\text{Pa}}$$

$$= 2.17 \times 10^{-3} \times \frac{\text{N}}{\text{N/m}^2} \quad = 2.17 \times 10^{-3} \text{ m}^2.$$

Each tire has a contact area of about 22 cm^2.

PROBLEM 10.2: A bicycle pump has a piston of diameter 2.54 cm (1 inch). What minimum force must be exerted on the piston to add air to a tire at a gauge pressure of 413 kPa (60 lb/in.^{2})? (10.1)

Solution:

> **Strategy:** The applied force plus the force of the atmosphere must be greater than the internal force applied by the tire.

Let F be the force we are supplying, P_a be the pressure of the atmosphere, and P be the internal pressure of the gas in the tire. The net force tending to compress the air in the tire as shown in Fig. 10.2b is

$$\text{net force} = F + P_aA - PA > 0.$$

Hence

$$F > (P - P_a) \times A = (\text{gauge pressure}) \times A$$

$$> 21 \text{ N (approx. 47 lb of force)}$$

Fig. 10.2: (a) Air pump attached to bicycle tire. (b) Forces on air pump piston.

PROBLEM 10.3: A wooden board 1.0 m × 15 cm × 2.0 cm has a density of 0.50 g/cm³. The board is floated in water and lead is placed on top of the board. How much lead can be placed on top of the board and have the top of board level with the water? Assume the experiment is done at 20°C. (10.3)

Solution:

Strategy: Use Archimedes' principle. Calculate the volume (and hence the mass) of water displaced by the board and set this equal to the mass of the board plus the mass of the lead.

This problem can be solved by an application of Archimedes' principle. The board plus lead will displace a mass of water equal to the combined mass of the board plus lead. Since the board floats level with the water, as seen in Fig. 10.3, the volume of water displaced is the same as the volume of the board.

Fig. 10.3: Lump of lead on a wooden board which is floating on water.

LET:

Symbol	Quantity	Value
m_{Pb}	mass of lead	
m_b	mass of board	
m_w	mass of displaced water	
V_b	volume of the board	
V_w	volume of displaced water	
ρ_b	density of wood	0.50 g/cm³
ρ_w	density of water	998 kg/m³

Then

$$m_w = m_b + m_{Pb} \quad \text{and} \quad V_w = V_b,$$

with

$$m_w = V_w \times \rho_w \quad \text{and} \quad m_b = V_b \times \rho_b.$$

Thus

$$m_{Pb} = m_w - m_b = V_w \times \rho_w - V_b \times \rho_b = V_b (\rho_w - \rho_b).$$

We can now enter the numbers:

$$V_b = 1.0 \text{ m} \times 0.15 \text{ m} \times 0.02 \text{ m} = 3.0 \times 10^{-3} \text{ m}^3,$$

and using the numbers in text Table 10.2 (convert the density of wood to kg/m³) we find that the lead has a mass of 1.5 kg.

PROBLEM 10.4: A container filled partway with water has a total mass of 2.50 kg (container plus water). A piece of lead with a mass of 1.13 kg is suspended from a spring scale. The scale is calibrated to read weight. If the scale is arranged such that the lead is completely submerged in the water, what is the reading on the spring scale? Assume the experiment is carried out at 20°C. (10.3)

Solution:

Strategy: Use Archimedes' principle. The water will exert an upwards force equal to the weight of the displaced water.

Fig. 10.4: Free body diagram for the lead held in water. The buoyant force is denoted by F_w while the tension acting on the spring is T.

Archimedes' principle tells us that the water exerts an upwards force equal to the weight of the displaced water. The spring scale should read the weight of the lead less this upward force. This is expected since the tension of the lower end of the spring scale is the sum of the gravitational force on the mass and the buoyant force of the water. That is, in terms of the free body diagram for the lead as shown in Fig. 10.4,

$$T = mg - F_w,$$

where m is the mass of the lead, g the acceleration of gravity, F_w is the buoyant force, and T is the tension applied to the scale and hence the weight read by the spring scale.

To calculate the weight of the displaced water we need to find the volume of lead:

$$\text{volume of lead} = \frac{\text{mass of lead}}{\text{density of lead}} = \frac{1.13 \text{ kg}}{1.13 \times 10^4 \text{ kg/m}^3} = 1.00 \times 10^{-4} \text{ m}^3.$$

This corresponds to a weight of water of

$$F_w = \text{density} \times \text{volume} \times \text{acceleration of gravity}$$

$$= 998 \text{ kg/m}^3 \times 1.00 \times 10^{-4} \text{ m}^3 \times 9.80 \text{ m/s}^2 = 0.978 \text{ N}.$$

The spring scale will read

$$T = 1.13 \text{ kg} \times 9.8 \text{ m/s}^2 - 0.978 \text{ N} = 10.1 \text{ N}.$$

The spring scale will read 10.1 N.

PROBLEM 10.5: Oil flows through 5.0 cm diameter pipe at a velocity of 0.75 m/s. The pipe diameter reduces to a 3.0 cm diameter for a distance of 1.0 m before enlarging to 4.0 cm. What is the velocity of the oil in the 4.0 cm pipe? Assume the size transitions are smooth and tapered. (10.5)

Solution:

Strategy: Apply the equation of continuity. Only the behavior in the beginning and ending tubes will be needed.

Text Eq. 10.7, which applies to incompressible flow, gives us the relation between the area, A, and the flow velocity, v. Since no oil is lost and viscosity is ignored, the constriction to 3.0 cm diameter will not affect the final answer. You can verify this.

Let the subscript "1" apply to the 5.0 cm pipe and the subscript "2" apply to the 4.0 cm pipe. Then

$$A_1 v_1 = A_2 v_2,$$

$$v_2 = \frac{A_1}{A_2} v_1 = \frac{R_1^2}{R_2^2} v_1 = 1.2 \text{ m/s}.$$

You should apply the equation of continuity to the intermediate pipe to verify that we did not need to know the velocity in that pipe.

PROBLEM 10.6: Water is traveling at 2.5 m/s from the Colorado River to LA in a pipe which is 1.5 m in diameter. As the pipe lowers 50 m over one of the many hills on its path it tapers to a diameter of 1.3 m. Assuming that the water pressure is 2.0 atmospheres at the top of the hill, what is the pressure at the bottom? Assume a water density of 1.0×10^3 kg/m^3 and neglect viscosity. (10.6)

Solution:

> **Strategy:** Use Bernoulli's equation. The equation of continuity can be used to find the water velocity at the bottom of the hill.

Fig. 10.5 shows the problem. Quantities at the top of the hill are denoted by the subscript "1", while the same quantities at the bottom of the hill are indicated by the subscript "2". Bernoulli's equation should relate these quantities at the top and bottom. We are given the initial pressure, height and velocity (P_1, h_1, and v_1). We know the final height (h_2) but do not know the final pressure (P_2) or final velocity (v_2). Since we know the pipe sizes(R_1, R_2), we can get the final velocity from the equation of continuity.

Pipe 1.5 m in dia carrying water at 2.0 atm at 2.5 m/s

50 m

Pipe 1.3 m in dia

FIG 10.5: LA water pipe.

In terms of our variables, Bernoulli's equation is

$$P_1 + \rho g h_1 + \frac{1}{2}\rho v_1^2 = P_2 + \rho g h_2 + \frac{1}{2}\rho v_2^2,$$

where g is the acceleration of gravity, while the equation of continuity is

$$4\pi R_1^2\, v_1 = 4\pi R_2^2\, v_2 .$$

Solving the latter equation first gives us $v_2 = 2.2$ m/s; then Bernoulli's equation gives

$$P_2 = P_1 + \rho g(h_1 - h_2) + \frac{1}{2}\rho(v_1^2 - v_2^2)$$

$$= 2.0 \text{ atm} \times 1.01 \times 10^5 \text{ N/m}^2 + 4.9 \times 10^5 \text{ N/m}^2 + 0.07 \times 10^4 \text{ N/m}^2$$

$$= 7.0 \times 10^5 \text{ N}\bullet\text{m}^2 = 6.9 \text{ atm}.$$

The water pressure will increase to 6.9 atmospheres.

PROBLEM 10.7: A 25.0 cm high by 20 cm diameter can, resting on a table, is filled to the brim with SAE 10 motor oil (density = 0.95 g/cm^3). 20.0 cm from the top of the can is a 10.0 cm long horizontal tube with a cross-section of 0.20 cm^2. How far from the base of the can will the oil hit the table? (10.7)

Solution:

> **Strategy:** Use Bernoulli's equation to calculate the pressure at the depth of the fluid. Use Poiseuille's law to calculate the flow rate and hence the exit velocity. Newton's laws will then determine the distance of travel while falling the extra 5.0 cm.

Since the diameter of the can is large when compared to the diameter of the tube, it is reasonable to neglect the motion of the oil except when it flows through the tube and leaves the can. We can then use Bernoulli's equation to calculate the pressure in the can at the level of the tube. We take the pressure at the surface of the oil to be zero — i.e., neglect air pressure. Then the oil pressure at the level of the tube is

$$\text{pressure} = p_i = \rho g h,$$

where p_i is the interior pressure at the level of the tube, ρ is the oil density, g is the acceleration of gravity and h is the 20 cm depth of the oil at the tube.

Since the pressure of the exit of the tube is nearly zero, we can use Poiseuille's law to find the flow rate through the tube; that is,

$$p_i - p_e = p_i = 8\frac{Q\eta L}{\pi R^4},$$

where p_e is the exit pressure, Q is the flow rate, η is the viscosity, L is the tube length, R is the tube radius, and SI units must be used. Solving this for the flow rate gives

$$Q = \frac{\pi R^4 p_i}{8\eta L}.$$

Fig. 10.6:   Oil flowing out of a tube in the side of a can.

We can convert this flow rate to an exit velocity, v_e by

$$v_e = \frac{Q}{\text{tube area}} = \frac{Q}{\pi R^2}.$$

Thus

$$v_e = \frac{p_i R^2}{8\eta L} = \frac{\rho g h R^2}{8\eta L}.$$

The vertical and horizontal motions of the fluid will be assumed independent, so a fluid element should hit the table in a time given by

$$h_o = \frac{1}{2} g \, t_f^2,$$

where h_o is the height above the table and t_f is the time it takes to hit the table. Then the horizontal distance traveled is

$$\text{distance} = v_e\, t_f = \frac{\rho g h\, R^2}{8\eta L} \sqrt{\frac{2h_0}{g}} = \frac{\rho h\, R^2 \sqrt{2gh_0}}{8\eta L}.$$

Using SI units, R^2 = area/π, and text Table 10.4 for the viscosity, we find

$$\text{distance} = 0.0075 \text{ m} = 0.75 \text{ cm}.$$

This is the horizontal distance from the end of the tube, so the total distance is 10.75 cm.

PROBLEM 10.8: A steel sphere of radius 0.42 cm falls through castor oil at 20°C. What is the maximum velocity of fall? (10.8)

Solution:

> **Strategy:** Since the object is small, we use the formula for terminal velocity.

The terminal velocity is given by

$$v_t = \frac{2r^2 g}{9\eta} (\rho - \rho')$$

where $r = 0.0042$ cm, g is the acceleration of gravity, $\eta = 0.986$ Pa·s from Table 10.4, $\rho = \rho_{steel} = 7.8 \times 10^3$ kg/m³ the density of steel and $\rho' =$ density of air $<< \rho_{steel}$.

Substituting in the numbers yields $v_t = 0.30$ m/s

PROBLEM 10.9: Water at 20°C flows through a pipe of varying diameter. The first part of the pipe has a radius of 3.0 cm. The pipe then contracts to 2.5 cm for a length of 1.0 meter before expanding to 2.7 cm as illustrated in Fig. 10.7. What is the maximum inlet velocity for which we are sure that the flow is smooth? (10.9)

Solution:

> **Strategy:** The Reynolds number for the flow determines if the flow is smooth or turbulent.

FIG. 10.7: Pipe of radius 3.0 cm with restriction to 2.5 cm followed by 2.75 cm radius portion. The portion marked "1" is 3.0 cm in diameter, the portion marked "2" is 2.5 cm in diameter and the portion marked "3" is 2.7 cm in diameter.

To solve this we need to apply the Reynolds rule that the Reynolds number is less than 2000 for smooth (laminar) flow and that it be greater than 3000 for turbulent flow. The Reynolds number is given by

$$\text{Re} = \frac{\rho \, v \, D}{\eta},$$

where ρ is the density, v is the velocity and D is the tube diameter. If we let "1" be the values when $D = 6.0$ cm, "2" the values when $D = 5.0$ cm, and "3" the values when $D = 5.4$ cm, then the equation of continuity requires

$$D_1{}^2 \, v_1 = D_2{}^2 \, v_2 = D_3{}^2 \, v_3,$$

so the highest speed occurs in region "2". Requiring that Re < 2000 in region "2" gives

$$v_2 < 2000 \, \frac{\eta}{\rho \, d} = 2000 \times \frac{1.0 \times 10^{-3} \text{ N} \cdot \text{s/m}^2}{1.0 \times 10^3 \text{ kg/m}^3 \times 5.0 \times 10^{-2} \text{ m}} = 4.0 \text{ cm/s}.$$

The inlet velocity is v_1, so

$$v_1 = \frac{D_2{}^2}{D_1{}^2} \, v_2 = 2.8 \text{ cm/s}.$$

The maximum inlet velocity to assure laminar flow is 2.8 cm/s.

PROBLEM 10.10: Water flows in a pipe of 4.0 cm diameter at a flow rate of 138 cm³/s. Is the flow turbulent or smooth? (10.9)

Solution:

Strategy: Calculate the Reynolds number for the flow. Use the flow rate to calculate the velocity.

Let D be the pipe diameter and v be the flow velocity. The flow rate is given in terms of the flow velocity and the diameter by

$$\text{flow rate} = \pi \frac{D^2}{4} \times v = 138 \text{ cm}^3/\text{s}.$$

So the velocity is 11 cm/s $= v$. Then

$$\text{Re} = \frac{\rho \, v \, D}{\eta} = \frac{1.0 \times 10^3 \text{ kg/m}^3 \times 0.11 \text{ m/s} \times 0.04 \text{ m}}{1.0 \times 10^{-3} \text{ N} \cdot \text{s/m}^2} = 4,400.$$

Since this is greater than 3000, the flow is turbulent.

SUGGESTED READINGS

Badeer, H. S., and C. E. Synolakis, "The Bernoulli-Poiseuille Equation." *The Physics Teacher,* November 1989, p. 598.

MacDonald, W. M., and S. Hanzely, "The Physics of the Drive in Golf." *American Journal of Physics,* March 1991, p. 213.

Dean, N., "The Brief, Swift Reign of the Clippers." *American Heritage of Invention & Technology,* Fall 1989, p. 48.

Flink, J. J., "The Path of Least Resistance." *American Heritage of Invention & Technology,* Fall 1989, p. 34. (The aerodynamics of automobiles.)

Frisch, U., and S. A. Orszog, "Turbulence: Challenges for Theory and Experiment." *Physics Today,* January 1990, p. 24.

Heavers, R. M., and M. G. Medeiros, "Laminar and Turbulent Flow in a Glass Tube." *The Physics Teacher,* May 1990, p. 297.

Heppenheimer, T. A., "The Closing of the Aviation Frontier." *American Heritage of Invention & Technology,* Winter 1991, p. 24.

Jakab, P. L., *Visions of a Flying Machine.* Washington: Smithsonian Institution Press, 1990. The story of the Wright brothers and the process of invention.

Jones, R. T., *Wing Theory.* Princeton, N. J.: Princeton University Press, 1990.

Kuethe, D. O., "Confusion about Pressure." *The Physics Teacher,* January 1991, p. 20.

Peterson, C. W., "High-Performance Parachutes." *Scientific American,* May 1990, p. 108.

Van Heuvelen, A., "Physics of the Circulatory System." *The Physics Teacher,* November 1989, p. 590.

Walker, J., "The Amateur Scientist: Why Are the First Few Puffs the Hardest When You Blow Up a Balloon?" *Scientific American,* December 1989, p. 136.

Weltner, K., "Aerodynamic Lifting Force." *The Physics Teacher,* February 1990, p. 78.

Weltner, K., "Bernoulli's Law and Aerodynamic Lifting Force." *The Physics Teacher,* February 1990, p. 84.

Witten, T. A., "Structured Fluids." *Physics Today,* July 1990, p. 21.

Zebrowski, E., Jr., "Superterminal Velocities." *The Physics Teacher,* November 1989, p. 618.

Chapter 11
Thermal Physics

IMPORTANT TERMS

Fill in the blanks with the appropriate word or words.

1. The _____ temperature scale has the steam point at 212°.

2. Two objects at the same temperature are said to be in thermal _____.

3. The _____ temperature scale divides the interval between the ice point and the steam point into 100 intervals and takes the ice point as its zero.

4. If a material undergoes a change of phase, the amount of heat absorbed or emitted per unit mass is called the _____.

5. The ratio of the thickness of a material to its thermal conductivity is called the material's _____ .

6. _____ is the minimum possible temperature.

7. The quantity which relates the size of an object at a given temperature to its size at another temperature is called the thermal _____.

8. An iron poker has one end in a fire. After a period of time, the other end also gets hot. The heat has been transferred by the process of _____.

9. The heat from the sun reaches the earth by _____.

10. The _____ of an object is a number assigned to indicate the heat contained in the object.

11. The basic division of all currently used temperature scales is called the _____ .

12. _____ are devices used to measure temperatures.

13. The _____ temperature scale uses absolute zero as its reference zero.

14. The heat required to raise the temperature of one gram of water 1°C is called the _____.

15. Hot air rises and carries off heat from a hot body. This process is called _____.

16. The ratio of the amount of heat required to raise the temperature of a body 1°C to the mass of the body is called the _____ of the body.

17. For small temperature changes, we assume that one-dimensional thermal behavior is _____ thermal expansion.

18. The _____ is the energy absorbed or emitted as a substance changes from a solid to a liquid or conversely.

19. The _____ is the energy absorbed or emitted as a substance changes from a liquid to a vapor or conversely.

20. The _____ is a measure of the ability of a substance to transfer heat by conduction from a boundary at one temperature to a boundary at another temperature.

Write the definitions to the following:

21. Kelvin temperature scale (11.2):

22. latent heat of transformation (latent heat) (11.6):

23. temperature (11.1):

24. thermometer (11.1):

25. heat capacity (11.5):

26. specific heat (11.5):

27. convection (11.7):

28. radiation (11.7):

29. conduction (11.7):

30. R value (11.7):

31. Fahrenheit temperature scale (11.2):

32. absolute zero (11.2):

33. linear thermal expansion (11.3):

34. thermal conductivity (11.7):

35. calorie (11.4):

36. thermal equilibrium (11.1):

37. thermal expansion coefficient (11.3):

38. Celsius temperature scale (11.2):

39. heat of fusion (11.6):

40. heat of vaporization (11.6):

ANSWERS TO 1–20, IMPORTANT TERMS

1. Fahrenheit; 2. equilibrium; 3. Celsius; 4. latent heat of transformation (latent heat); 5. R value; 6. absolute zero; 7. expansion coefficient; 8. conduction; 9. radiation; 10. temperature; 11. degree; 12. thermometers; 13. Kelvin; 14. calorie; 15. convection; 16. specific heat capacity (specific heat); 17. linear; 18. heat of fusion; 19. heat of vaporization; 20. thermal conductivity.

SAMPLE QUESTIONS

QUESTION 11.1: How does thermal expansion affect the precision of a pendulum clock?

Answer:

The length of the pendulum changes as its temperature changes. If the temperature increases, the pendulum will get longer. Since the period of a pendulum depends upon its length, the period will change.

QUESTION 11.2: What happens to the heat of transformation absorbed during a phase transition?

Answer:

The energy is stored internally in the "order" of the substance. When a liquid freezes, the energy available for transformation comes from the loss of random arrangement in the substance. Likewise when a substance melts, the released transformation comes from the energy released as the substance becomes disordered.

SAMPLE SOLUTIONS

PROBLEM 11.1: A concrete highway in Alaska consists of slabs 12 m in length. How wide must the expansion joints be to allow for thermal changes over the temperature range from $-30°C$ to $30°C$? Assume the concrete is of a special type which has a linear expansion coefficient of $9 \times 10^{-6}/°C$. (11.3)

Solution:

> **Strategy:** Calculate the linear expansion of each slab over the temperature range. The expansion joints must be larger than the maximum change. Investigate the effect of choosing reference length at different temperatures to determine if this choice is important (it should not be important).

The length of the slab as temperature changes should be given by Eq. 11.4 of the text:

$$L = L_0 [1 + \alpha(T - T_0)],$$

where L is the length at temperature T, L_0 is the length at temperature T_0, and α is the linear expansion coefficient.

We are given that $\alpha = 9 \times 10^{-6}/°C$ and we know that the length is about 12 m. What may not be clear from the problem statement is if we should take $L_0 = 12$ m or if L_0 is the length at some intermediate temperature. We will show that the choice of the temperature for L_0 has no effect on the answer. The problem is illustrated in Fig. 11.1.

Fig. 11.1: Concrete slabs with gaps for thermal expansion.

First Choice of L_0

Take T_0 to be 30°C and $L_0 = 12$ m. Then for $T = -30$°C, L is given by

$$L = 12 \text{ m } [1 + \frac{9 \times 10^{-6}}{°C}(-30°C - 30°C)]$$

$$= 12 \text{ m} \times [1 + 0.00054] .$$

So the change in length,

$$\Delta L = L - L_0,$$

is given by
$$\Delta L = 12 \text{ m} \times 0.00054 = 0.00648 \text{ m} = 6.5 \text{ mm}.$$

Second Choice of L_0

Take $\qquad T_0 = -30°C \qquad$ and $\qquad L_0 = 12$ m.

Calculate ΔL for 30°C. This gives the same number.

Third Choice for L_0

Take $\qquad T_0 = -30°C \qquad$ and $\qquad L$ at 30°C to be 12 m,

so $\qquad\qquad\qquad\qquad\qquad \Delta L = 12 \text{ m} - L_0.$

This gives us the following two equations:

$$L(-30°C) = L_0,$$

$$L(30°C) = 12 \text{ m} = L_0[1 + \alpha(30°C - \{-30°C\})] = 1.00054 \times L_0.$$

The latter equation gives us

$$L_0 = \frac{12 \text{ m}}{1.00054},$$

so

$$\Delta L = 12 \text{ m} \frac{1.00054 - 1}{1.00054} = \frac{6.5 \text{ mm}}{1.00054}.$$

Considering the significant figures known, this answer is the same as in the first two approaches. What this result suggests is that if it is not clear which length is L_0, then it probably does not matter which temperature is assigned the value of L_0 as long as you are consistent.

All three approaches give us essentially the same answer.

PROBLEM 11.2: What is the percent increase in volume of a glass sphere with a radius of 10 cm when it is heated from 20°C to 100°C? (11.3)

Solution:

Strategy: Use the relation for the volume change in the text. Since a percentage change is desired and since the volume change is proportional to the volume, we expect the result to be independent of the volume of the sphere.

This problem can be solved by using the formula for volume expansion (text Eq. 11.5) which is

$$\Delta V = \beta V_0 \Delta T$$

where

Symbol	Quantity
ΔV	change in volume
β	volume coefficient of expansion of glass = $27 \times 10^{-6}/°C$
V_0	initial volume of glass ($\frac{4}{3} \pi r^3$)
ΔT	change in temperature = 80°C

The percentage change will be

$$\frac{\Delta V}{V_0} \times 100 = 100\, \beta\, \Delta T = 0.22\%$$

so the volume does not need to be calculated and the percentage change is 0.22%.

PROBLEM 11.3: Three hundred grams of lead shot is placed in a 2.0 m long wooden tube which is capped on both ends. If the tube is in a vertical position and then quickly inverted, the shot will fall the length of the tube. If this is done 100 times in succession, what will be the increase of the temperature of the shot? (11.4)

Solution:

Strategy: The work done in inverting the tube is put into potential energy of the shot. The shot falls, converting the potential energy to kinetic energy. Upon impact this kinetic energy is converted to heat. Use energy conservation to calculate the kinetic energy which is then completely converted to heat.

The idea behind this problem is illustrated in Fig. 11.2. In Fig. 11.2a we have the tube at the beginning of an inversion. The tube is quickly inverted. By "quickly inverted," we mean to make the approximation that none of the shot has a chance to start to fall before the inversion process is completed. Inverting the tube requires work. Part of this work appears as potential energy in the shot. As the shot falls, this potential energy changes to kinetic energy. When the shot hits the bottom of the tube, the kinetic energy is converted into heat. Since the tube is wood, we will assume that none of this heat is lost to the tube.

(a) **(b)** **(c)**

Fig. 11.2: (a) Tube with shot before inversion. (b) Tube with shot just after inversion and with shot starting to fall. (c) Tube with shot after shot has fallen to bottom of the tube.

One additional assumption must be made. We will assume that all of the shot falls through a distance equal to the length of the tube.

GIVEN:

Quantity	Symbol	Value
length of tube	L	2.0 m
mass of shot	m	0.300 kg
number of inversions	n	100
potential energy added per inversion	PE	$m\,g\,L$
heat added to one shot per inversion	ΔQ	PE (in kcal)
specific heat of lead (Table 10.3)	c	0.0305 kcal/kg•°C

If total heat added is $n\,\Delta Q$, then the temperature change ΔT will be given by

$$n\Delta Q = c\,m\Delta T,$$

so

$$\Delta T = \frac{n\,\Delta Q}{mc}.$$

Before we can use the given numbers, we must convert the change in potential energy to kcal. The expression PE = mgL gives the PE in joules. From text Table 10.2 we have

$$1\text{ J} = 2.38 \times 10^{-4}\text{ kcal} \qquad \text{or} \qquad 1\text{ kcal} = 4187\text{ J}$$

Then

$$\Delta T = \frac{100 \times 0.30\text{ kg} \times 9.8\text{ m/s}^2 \times 2.0\text{ m} \times 2.38 \times 10^{-4}\text{ kcal/J}}{0.0305\text{ kcal/kg•°C} \times 0.30\text{ kg}}$$

$$= 15.29°\text{C} = 15°\text{C}.$$

One hundred inversions will raise the temperature of the shot 15°C. This result is independent of the mass of lead provided, as we can neglect the heat absorbed by the wood.

PROBLEM 11.4:   A cup of negligible heat capacity and heat conductivity holds 210 g of hot water at 90°C. A 200 g block of glass at 27°C is immersed in the water and comes to equilibrium with the water. What is the final temperature of the block of glass? (11.5)

SOLUTION:

> **Strategy:** Conserve energy. Let the heat gained by the block of glass equal the heat lost by the water.

Since the water is at a higher temperature and the glass at a lower temperature, we expect the final temperature to be between the two temperatures. Since all temperatures are below or above phase changes, we will not have to worry about heats of transformation. Our basic assumptions are that the heat lost by the water is the heat gained by the glass, as shown in Fig. 11.3c, and that, in equilibrium, both are at the same temperature. We can get the needed heat capacities (specific heats) from Table 11.3 of the text.

Fig. 11.3:   (a) System of water and glass before contact. (b) System of water and glass in contact. (c) Heat flow from water to glass.

GIVEN:

Quantity	Symbol	Value
heat capacity of water	c_W	1.00 cal/g•°C
heat capacity of glass	c_G	0.199 cal/g•°C
mass of water	m	210 g
mass of glass	M	200 g
initial temperature of water	T_W	90°C

Law to Use

$$\text{Heat loss of water } (-\Delta Q_W) \; = \; \text{heat gain of glass } (\Delta Q_G).$$

Note that we express the quantities ΔQ as the appropriate quantities for ΔQ positive if there is a heat gain. That is, a negative heat gain is a heat loss. Then,

$$\Delta Q_W \; = \; c_W \times m \times (T - T_W) \; = \; -c_W \times m \times (T_W - T),$$

and

$$\Delta Q_G \; = \; c_G \times M \times (T - T_G).$$

Equating gives

$$c_W \times m \times (T_W - T) \; = \; c_G \times M \times (T - T_G)$$

or

$$T \; = \; \frac{c_W \, m \, T_W + c_G \, M \, T_G}{c_W \, m + c_G \, M} \; = \; \frac{19974 \text{ cal}}{249.8 \text{ cal/}°C} \; = \; 79.96°C \; = \; 80.0°C$$

to 3 significant figures.

So the water and glass come to equilibrium at 80.0°C assuming no heat loss to the container.

PROBLEM 11.5: How many calories are required to change 800 grams of ice at $-20°C$ to steam at 135°C? (11.6)

Solution:

> **Strategy:** There are two transformations in this problem (ice to water and water to steam) in addition to the heat required to heat the different phases. Use the tabulated heat capacities to calculate the energy required for each step.

This process can be broken into 5 distinct physical steps:

1. Raise the temperature of the ice from $-20°C$ to 0°C.
2. Melt the ice (temperature remains constant).
3. Raise the temperature of the water from 0°C to 100°C.
4. Change the water to steam (temperature remains constant).
5. Raise the temperature of the steam from 100°C to 135°C.

We will let the symbols ΔQ_1, ΔQ_2, ΔQ_3, ΔQ_4, and ΔQ_5 represent the heat that must be added in each step as shown in Fig. 11.4.

For steps 1, 3, and 5, $\qquad \Delta Q_i = c \times m \times \Delta T \qquad$ (Eq. 11.6 of text)

where c is the appropriate specific heat, m is the mass, and ΔT is the temperature change.

For steps 2 and 4, $\qquad\qquad \Delta Q_i = L \times m \qquad$ (Eq. 11.7 of text)

where L is the appropriate latent heat of transformation and m is the mass.

GIVEN:

Quantity	Symbol	Value
specific heat of ice	c_I	0.50 cal/g•°C
specific heat of water	c_W	1.00 cal/g•°C
specific heat of steam	c_S	0.48 cal/g•°C
heat of fusion of water	L_I	80 cal/g
heat of vaporization of water	L_S	540 cal/g
mass (constant for all steps)	m	800 g

Then, for each step, we have:

$$\Delta Q_1 = m \times c_I \times (0°C - [-20°C]) = 8,000 \text{ cal},$$

$$\Delta Q_2 = m \times L_I = 64,000 \text{ cal},$$

$$\Delta Q_3 = m \times c_W \times (100°C - 0°C) = 80,000 \text{ cal},$$

$$\Delta Q_4 = m \times L_S = 432,000 \text{ cal},$$

$$\Delta Q_5 = m \times c_S \times (135°C - 100°C) = 13,440 \text{ cal}.$$

So the total heat needed, ΔQ, is

$$\Delta Q = 597,440 \text{ cal} = 6.0 \times 10^5 \text{ cal},$$

since some of the specific heats are given to only two figures.

Fig. 11.4: (a) The steps needed to change ice at −20°C to steam at 135°C.
(b) The heat needed per gram for each step.

As an alternative approach, one can add up the heat needed per gram for each step and then multiply by the mass. That is,

total heat per gram =
$$0.5 \text{ cal/g} \cdot °C \times 20°C + 80 \text{ cal/g} + 1 \text{ cal/g} \cdot °C \times 100°C$$
$$+ 540 \text{ cal/g} + 0.48 \text{ cal/g} \cdot °C \times 35°C = 747 \text{ cal/g}.$$

Then

$$\Delta Q = 747 \text{ cal/g} \times 800 \text{ g} = 6.0 \times 10^5 \text{ cal},$$

as before.

PROBLEM 11.6: A styrofoam cooler with a surface area of 0.75 m^2 and an average thickness of 2.5 cm is filled with 2.0 kg of ice at a temperature of 0°C and taken on a fishing trip. How long will it take the ice to melt if the thermal conductivity of styrofoam is 3.2×10^{-4} W/cm•°C and the outside temperature is 35°C? Neglect the specific heat of the styrofoam. (11.7)

SOLUTION:

Strategy: The energy that flows through the styrofoam changes the state of the ice to water. Assume there is no change in temperature until all the ice is melted.

Fig. 11.5:   (a) Cooler with interior at 0°C and exterior at 35°C. (b) Heat flow across wall. Positive direction chosen to make ΔQ positive.

During the entire time that the ice is in the cooler, heat is being conducted through the outside walls to the inside as shown in Fig. 11.5a. Since the ice starts at 0°C and we are considering only the time it takes the ice to melt, the interior temperature remains at 0°C. Thus the situation is as illustrated in Fig. 11.5b.

We are interested in finding how long it takes for enough heat to enter the container to melt 2.0 kg of ice. This amount of heat, ΔQ, is given by

$$\Delta Q = (\text{mass of ice}) \times (\text{heat of fusion of ice}) = 2000 \text{ g} \times 80 \text{ cal/g} = 160,000 \text{ cal}.$$

The heat flow equation is given by text Eq. 11.8 as

$$\frac{\Delta Q}{\Delta t} = -\mathbf{K} A \frac{\Delta T}{\Delta x} ,$$

where ΔQ is the heat that flows in the time interval Δt, $\mathbf{K}$ is the thermal conductivity, A is the surface area, and ΔT is the temperature change across thickness Δx. The unknown quantity is Δt, so

$$\Delta t = \frac{\Delta Q}{-\mathbf{K} A \dfrac{\Delta T}{\Delta x}} .$$

Before substituting in numbers, we should get all quantities in a consistent set of units. Alternatively, we could work out the units and convert the final answer. Usually the former procedure is simplest. Let us consider each separately:

$\Delta T = 0°C - 35°C = -35°C$ leave as is — notice sign follows from choice of direction in Fig. 11.6b.

$\Delta x = 2.5$ cm   leave as is

$A = 0.75$ m^2 = 7500 cm^2 use cm or convert Δx to meters

$\mathbf{K} = 3.2 \times 10^{-4}$ W/cm•°C   consistent with A and Δx in cm

$\Delta Q = 160{,}000$ cal   convert to W•s or change $\mathbf{K}$

$\qquad = 1.6 \times 10^5$ cal $\times$ 4.187 J/cal $= 6.6992 \times 10^5$ J.

So

$$\Delta Q = 6.6992 \times 10^5 \text{ W•s.}$$

Then

$$\Delta t = \frac{6.6992 \times 10^5 \text{ W•s.}}{-3.2 \times 10^{-4} \text{ W/cm•°C} \times 7500 \text{ cm}^2 \dfrac{-35°C}{2.5 \text{ cm}}} = 1.9 \times 10^4 \text{ s} \cong 5.3 \text{ hr.}$$

It will take the ice about $5\frac{1}{3}$ hr to melt.

GENERAL COMMENTS

In mechanics you were urged to draw motion plots or force diagrams. The analogous plots for thermal physics are heat flow diagrams. Large bodies or reservoirs of heat are indicated by their temperature. Heat flow and work can be indicated by arrows. An example is shown in Fig. 11.6 where heat Q_1 is extracted at temperature T_1, heat Q_2 is exhausted at temperature T_2 while work W is done. Figures such as 11.6 not only help your thinking, but they also make the meaning of the variables more obvious.

Fig. 11.6: An example of a heat flow diagram.

SUGGESTED READINGS

Bartels, R. A., "Do Darker Objects Really Cool Faster?" *American Journal of Physics*, March 1990, p. 244.

Brown, S. C., "Benjamin Thompson, Count Rumford." *The Physics Teacher*, May 1976, p. 263.

Donnely, R. J., "Leo Dana: Cryogenic Science and Technology." *Physics Today*, April 1987, p. 38.

Gibbons, J. H., P. D. Blair, and H. L. Gwin, "Strategies for Energy Use." *Scientific American*, September 1989, p. 136.

Jones, E. R., "Fahrenheit and Celsius, a History." *The Physics Teacher*, November 1980, p. 594.

Rosenfeld, A. H., and D. Hafemeister, "Energy-Efficient Buildings." *Scientific American*, April 1988, p. 78.

Ward, D. S. "Advances in Solar Heating and Cooling Systems." *The Physics Teacher*, April 1976, p. 199.

Wineland, D.J., and W.M. Itano, "Laser Cooling." *Physics Today*, June 1987, p. 34.

Chapter 12
Gas Laws and Kinetic Theory

IMPORTANT TERMS

Fill in the blanks with the appropriate word or words.

1. The fraction of molecules in a gas found as a function of temperature and velocity is given by the _____.

2. The relation between the volume and the temperature for an ideal gas at constant pressure is called _____ law.

3. The number of molecules in a mole of a substance is called _____ number.

4. The _____ of gases assumes that a gas consists of particles in motion.

5. The model of a gas in which the gas is treated as made up of many particles is the _____.

6. The five principal assumptions of the kinetic theory model of gases define what is called a(n) _____ gas.

7. _____ law states "The pressure exerted by a gas is inversely proportional to the volume of space in which it is enclosed."

8. The quantities that describe the condition or state of a system are called _____.

9. The constant appearing in the ideal gas law is called _____ constant.

10. Any equation linking pressure, temperature, and volume is called an equation _____.

11. The gas constant per molecule is called _____ constant.

12. The relation between the pressure, temperature, number of moles, and the volume of an ideal gas is called the _____.

13. The _____ is the amount of a substance of a system that contains as many elementary entities as there are atoms in 0.012 kg of carbon-12.

Write the definitions of the following:

14. Avogadro's number (12.4):

15. Boyle's Law (12.2):

16. law of Charles and Gay-Lussac (12.3):

17. Maxwell-Boltzmann distribution (12.7):

18. mole (12.4):

19. ideal gas law (12.4):

20. universal gas constant (12.4):

21. ideal gas (12.5):

22. Boltzmann constant (12.6):

23. kinetic theory of gases (12.5):

24. state variable (12.4):

25. equation of state (12.4):

ANSWERS TO 1–13, IMPORTANT TERMS

1. Maxwell-Boltzmann distribution; 2. Charles' and Gay-Lussac's; 3. Avogadro's; 4. kinetic theory; 5. kinetic theory model; 6. ideal; 7. Boyle's; 8. state variables; 9. universal gas; 10. of state; 11. Boltzmann; 12. ideal gas law; 13. mole.

SAMPLE QUESTIONS

QUESTION 12.1: A car owner's manual says to measure the tire pressure before driving. What difference would it make if the pressure were measured after driving several miles at highway speeds?

Answer:

After driving several miles, friction would have heated the air in the tires. This increased temperature would increase the pressure of the air in the tires.

QUESTION 12.2: Give an explanation for the difference between heat and temperature, based on the kinetic theory.

Answer:

For gases, temperature is a measure of the average kinetic energy of the gas molecules. Kinetic energy is one of the forms that energy can be stored in a substance. For example, energy can be stored in a phase change. Thus it is possible for heat to flow into or out of a body without the temperature changing. We should think of heat as a net energy flow and avoid terms such as the "total heat in the body."

SAMPLE SOLUTIONS

PROBLEM 12.1: What is the total downward atmospheric force on the top of a 2.0 m by 3.0 m table top? Why doesn't this force break the table or its legs? (12.1)

Pressure on table top

Pressure on table bottom

Fig. 12.1: Table with upward and downward atmospheric forces shown.

Solution:

> **Strategy:** Force equals pressure times area. If there is no motion all forces must add to zero.

Since force is pressure times area, the total downward force on the table top, F, is given by:

$$F = p \times A.$$

Atmospheric pressure is 1.01×10^5 Pa, so the force is:

$$
\begin{aligned}
F &= p \times A = 1.01 \times 10^5 \text{ Pa} \times (2.0 \text{ m} \times 3.0 \text{ m}) \\
&= 1.01 \times 10^5 \text{ N/m}^2 \times (2.0 \text{ m} \times 3.0 \text{ m}) = 6.1 \times 10^5 \text{ N}.
\end{aligned}
$$

This downward force does not break the table since it is balanced by an equal upward atmospheric force on the lower side of the table as shown in Fig. 12.1.

PROBLEM 12.2: An upright glass cylinder 75 cm tall and 3.0 cm in diameter is fitted with a light piston that is free to slide. The cylinder is closed at its lower end. First the piston is placed on the cylinder. Initially the piston sinks to 73 cm from the bottom of the cylinder. Water is poured into the cup-like cavity formed by the top of the piston and the cylinder until the cavity is full. At what fraction of the total height of the cylinder will the piston be? Assume the lower portion of the cylinder contains an ideal gas at constant temperature. (12.2)

Solution:

> **Strategy:** Since the temperature is assumed constant, this should be an application of Boyle's law. That is, pressure times volume is constant. The gravitational force on the water provides an added pressure on the gas trapped below the piston.

The physical arrangement before the piston is added is shown in Fig. 12.2a, before the water is added is shown in Fig. 12.2b, and after the water is added is shown in Fig. 12.2c.

Since the area of the piston is constant, we can expect the diameter of the cylinder not to appear in the answer.

We define the following variables [note that the area is π (diameter/2)^2].

GIVEN:

Quantity	Symbol	Value
area across cylinder	A	7.0686 cm^2
length of cylinder	L	75 cm = 0.75 m
initial height of piston above bottom of cylinder	h_0	73 cm = 0.73 m
final height of piston above bottom of cylinder	h	unknown
density of water	ρ	1000 kg/m^3
pressure in gas after piston is added	P_i	unknown
atmospheric pressure	p_a	1.01325×10^5 Pa

We assume that the pressure of the ideal gas in the bottom of the cylinder was initially at one atmosphere.

First Step

The gas was compressed by the weight of the piston until the gas pressure was able to support the piston. This occurred when the piston was a height h_0 above the bottom. The law which covers this behavior at constant temperature is Boyle's law, which states

$$PV = \text{constant}, \qquad \text{(text Eq. 12.1)}$$

where P is the gas pressure and V is the gas volume.

Fig. 12.2: (a) Cylinder before piston is added. (b) Piston in cylinder before the water is added. (c) Piston and cylinder after water is added.

Second Step

We now add water. The weight of the water will compress the gas in the cylinder. The compressed gas will have a higher pressure. When the gas is compressed enough, this pressure will balance force provided by the added water. Of course, more water will be needed than the amount that could originally fit into the "cup" since the piston is now lower in the cylinder. Finally, the piston will be in balance between the opposing forces of the trapped gas and the piston and water.

Before the water is added, Boyle's law gives

$$P_i V = P_i \times h_0 \times A = \text{constant.}$$

The weight of the added water is $(L - h) \times g \times \rho \times A$, so the pressure added by the water is

$$\frac{(L - h)\, g\, \rho\, A}{A} = \rho\, g\, (L - h).$$

Hence,

$$(P_i + \rho\, g\, [L - h]) \times h \times A = \text{constant} = P_i \times h_0 \times A.$$

At first inspection we seem to have one equation with two unknowns (h and P_i). This apparent difficulty can be resolved by realizing that before the piston was added, the pressure of the gas in the cylinder was atmospheric.

Let us use P_p to represent the pressure added to the gas by the piston. That is,

$$P_p = \frac{\text{weight of piston}}{\text{area of piston}},$$

then we have the following three equations with the same constant:

Before the piston is added, $\quad P_a \times (L \times A) = \text{constant.}$ (I)

where the choice of length $= L$ is made since this is the gas trapped by the piston.

After the piston is added, $\quad (P_a + P_p) \times (h_0 \times A) = \text{constant.}$ (II)

After the water is added, $\quad (P_a + P_p + \rho g [L - h]) \times (h \times A) = \text{constant.}$ (III)

Equation III follows from the weight of the water being $\rho g (L - h) A$ as is shown in Fig. 12.2a. Hence the pressure due to the water is $\rho g (L - h)$.

Equations I and II give $\quad P_a + P_p = P_a \dfrac{L}{h_0}.$

While Eqs. I and III give $\quad [P_a + P_p + \rho g (L - h)] \times h = P_a L.$

We can thus eliminate the effect of the piston and get

$$\left(\frac{P_a L}{h_0} + \rho g [L - h] \right) h = P_a L.$$

The solution to this quadratic equation is:

$$h = \frac{\left(\dfrac{P_a L}{h_0} + \rho g L\right) \pm \sqrt{\left(\dfrac{P_a L}{h_0} + \rho g L\right)^2 - 4\,\rho\, g\, L\, P_a}}{2\rho g} = Z \pm \sqrt{Z^2 - \frac{L P_a}{\rho g}},$$

where

$$Z = \frac{\left(\dfrac{P_aL}{h_0} + \rho g L\right)}{2\rho g} = \frac{P_aL}{2\rho g h_0} + \frac{L}{2}.$$

Now

$$\frac{P_aL}{2\rho g h_0} = 5.3 \, \frac{N \cdot s^2}{kg} = 5.3 \, m,$$

$$\frac{L}{2} = 0.37 \, m, \quad \text{and} \quad \frac{LP_a}{\rho g} = 7.6 \, m^2.$$

Thus

$$h = 5.7 \, m \pm \sqrt{(5.7 \, m)^2 - 7.6 \, m^2} = 11 \, m \text{ or } 0.71 \, m.$$

The 11 m answer is clearly not correct, so we conclude that when the water is poured into the top of the cylinder, the piston sinks to 70 cm from the bottom of the cylinder.

PROBLEM 12.3: A column of dry air was sealed from the atmosphere by closing one end of a glass tube and placing a drop of mercury in the tube 0.40 m from the end of the tube. This was done at a temperature of 30°C. The tube was then placed in a freezer whose temperature is unknown. After the tube reached the temperature of the freezer, the mercury slug was found to be 0.35 m from the closed end. What is the temperature of the freezer? (12.3)

Solution:

> **Strategy:** Since this is a constant pressure (isobaric) experiment, use the law of Charles and Gay-Lussac.

Fig. 12.3 shows the tube at 30°C in 12.3a and at the unknown temperature in 12.3b. Since in both cases, one end of the tube is open, the external pressure is atmospheric. We must assume the atmospheric pressure outside and inside the freezer is the same. The mercury slug is free to move, so the pressure of the gas trapped inside the tube is constant as the tube and gas cool.

Fig. 12.3: (a) Tube with mercury slug at 30°C. (b) Tube with mercury slug at unknown temperature.

Since the experiment is carried out at constant pressure, the applicable relation is the law of Charles and Gay-Lussac as given in text Eq. 12.3. Thus

$$\frac{V}{T} = \text{constant},$$

where V is the volume of the trapped gas and T is its temperature.

If we neglect the small decrease in the internal diameter of the glass tube as it cools, the volume of the gas is proportional to the length of the gas column, so

$$\frac{\text{length at } 30°C}{273 + 30} = \frac{\text{length at } T°C}{273 + T}.$$

Hence

$$T = \left((273 + 30)\frac{\text{length at } T°C}{\text{length at } 30°C} - 273\right)°C = \left(303\frac{0.35 \text{ m}}{0.40 \text{ m}} - 273\right)°C$$

$$= [265.125 - 273]°C = -7.875°C = -8.0°C.$$

The temperature of the freezer is $-8.0°C$.

PROBLEM 12.4: A cylinder closed at both ends has a piston in between which is free to slide. At 20°C the piston is exactly one third of the way from the left end when hydrogen gas is in the left side and helium is in the right side. What is the ratio of the number of hydrogen molecules to the number of helium molecules? (12.4)

Solution:

Strategy: Since we have different gases on each side of the piston, we will need to use the ideal gas law since the number of moles enters into that law's formulation. Equate temperatures and pressures to find the relative number of moles.

By the statement of the problem, both gases have the same temperature and the same pressure. They have different volumes as shown in Fig. 12.4.

Fig. 12.4: Cylinder containing hydrogen and helium separated by a piston.

We will assume that hydrogen and helium both obey the ideal gas law, that is, Eq. 12.4 of the text,

$$PV = nRT,$$

where P is the pressure, V the volume, n the number of moles, T the temperature, and R is a constant. Thus the number of moles we have for an ideal gas is

$$n = \frac{PV}{RT}.$$

If we let the subscript "hy" denote the hydrogen and the subscript "he" denote the helium, we have

$$n_{hy} = \frac{P_{hy}V_{hy}}{RT_{hy}}$$

and

$$n_{he} = \frac{P_{he}V_{he}}{RT_{he}}.$$

So

$$\frac{n_{hy}}{n_{he}} = \frac{V_{hy}}{V_{he}} = \frac{1}{2}.$$

The mole is a count of molecules, so the ratio of moles is the same as the ratio of the number of molecules; that is, there are twice as many helium molecules as hydrogen molecules. Note however that a hydrogen molecule has two atoms and helium is monoatomic, so in this case there are the same number of atoms.

PROBLEM 12.5: A common type of laboratory vacuum pump can reduce the pressure in a container to about 10^{-5} atm. How many molecules still remain in a volume of 0.20 m^3 that has been pumped to this "vacuum" pressure of 1.0×10^{-5} atm if the temperature is 300 K? (12.4)

Solution:

> **Strategy:** Use the ideal gas law.

The ideal gas law is

$$PV = nRT,$$

where P is the pressure, V is the volume, T is the Kelvin temperature, n is the number of moles, and R is the universal gas constant.

Expressing P in Pa , $P = 1.0 \times 10^{-5}$ atm $\times 1.0 \times 10^5$ Pa/atm $= 1.0$ Pa, then

$$n = \frac{PV}{RT} = \frac{1.0 \text{ Pa} \times 0.20 \text{ m}^3}{8.314 \text{ J/mol·K} \times 300 \text{ K}} = 8.0 \times 10^{-5} \text{ mol}.$$

Multiplying by Avogadro's number gives the number of molecules as 4.8×10^{19} in the "evacuated" volume.

PROBLEM 12.6: What is the average kinetic energy of a carbon dioxide molecule at 300 K? Assume carbon dioxide at this temperature is an ideal gas. What is the root mean square average speed of a molecule? (12.6)

Solution:

> **Strategy:** Use the results of the kinetic theory of gases, in particular text Eq. 12.8.

The kinetic theory of gases gives the relation between average kinetic energy (KE_{avg}) and the temperature (T) in text Eq. 12.8; then,

$$KE_{avg} = \frac{3}{2} k \, T.$$

Carbon dioxide has a molecular mass of 44 g/mol (see text Table 12.1). If we denote this by M_C, then the actual mass of a single molecule would be

$$m_C = M_C/N_A \text{ (expressed in grams)}$$

where N_A is Avogadro's number.

In terms of the root mean square speed, v_{rms}, the average kinetic energy is

$$KE_{avg} = \frac{1}{2} m_C \, v_{rms}^2,$$

or

$$v_{rms} = \sqrt{\frac{2 \, KE_{avg}}{m_C}}.$$

To summarize, we have the following information.

GIVEN:

Quantity	Symbol	Value
temperature	T	300 K
mass of a CO_2 molecule	m_C	$44/N_A = 7.31 \times 10^{-23}$ g 7.31×10^{-26} kg
Boltzmann constant	k	1.3807×10^{-23} J·K^{-1}molecule^{-1}

Then

$$KE_{avg} = \frac{3}{2} k \, T = \frac{3}{2} \times 1.3807 \times 10^{-23} \text{ J·K}^{-1}\text{molecule}^{-1} \times 300 \text{ K}$$

$$= 6.21315 \times 10^{-21} \text{ J·molecule}^{-1} = 6.2 \times 10^{-21} \text{ J·molecule}^{-1}$$

and

$$v_{rms} = \sqrt{\frac{2 \, KE_{avg}}{m_C}} = \sqrt{\frac{2 \times 6.21315 \times 10^{-21} \text{ J}}{7.31 \times 10^{-26} \text{ kg}}}$$

$$= 4.123 \times 10^2 \text{ km/s} = 4.1 \times 10^2 \text{ km/s}.$$

The average kinetic energy per molecule is 6.2×10^{-21} J, which corresponds to an rms speed of 4.1×10^2 km/s.

PROBLEM 12.7: Calculate the root-mean-square (rms) velocities in a gas at 300 K which is a mixture of diatomic molecules H_2 (M = 2.0×10^{-3} kg/mol) and diatomic deuterium molecules D_2 (M = 4.0×10^{-3} kg/mol). (12.6)

Solution:

Strategy: The two gases must be in equilibrium. Use the relation giving rms velocity in terms of mass and temperature.

The rms velocity is given by

$$v_{rms} = \sqrt{\overline{v^2}} = \sqrt{\frac{3kT}{m}}$$

where $k = 1.38 \times 10^{-23}$ J/K, T is the Kelvin temperature, and m is the molecular mass in kg.

We must first convert the molecular masses to kg. This conversion requires dividing the given numbers by Avogadro's number:

$$m_{H_2} = \frac{2.0 \times 10^{-3} \text{ kg/mol}}{6.0 \times 10^{23}/\text{mol}} = 3.3 \times 10^{-27} \text{ kg}$$

and

$$m_{D_2} = 6.6 \times 10^{-27} \text{ kg}.$$

Using these, we find

$$(v_{rms})_{H_2} = 1.9 \times 10^3 \text{ m/s} \quad \text{and} \quad (v_{rms})_{D_2} = 1.4 \times 10^3 \text{ m/s}.$$

These differ by enough to allow the two isotopes to be easily separated.

PROBLEM 12.8: Calculate the height at which the atmospheric pressure is one third (0.333) the sea level pressure, P_0, for air with average molecular mass of 28.9 and a uniform temperature of 295 K. (12.7)

Solution:

Strategy: Use the exponential law for the pressure with height as given in text Eq. 12.11.

The dependence of pressure, P, upon height is given in the text by Eq. 12.11 as

$$P(z) = P_0 e^{-mgz/kT},$$

where m is the mass of a molecule, g is the acceleration of gravity, z is the height, k is the Boltzmann constant and T is the temperature.

Calculate

$$\frac{1}{z_0} = \frac{mg}{kT} = \frac{4.80 \times 10^{-26} \text{ kg} \times 9.80 \text{ m/s}^2}{1.3807 \times 10^{-23} \text{ J·K}^{-1}\text{molecule}^{-1} \times 295 \text{ K}}$$

$$= 1.1547 \times 10^{-4} \text{ m}^{-1} = 1.2 \times 10^{-4} \text{ m}^{-1}.$$

GIVEN:

Quantity	Symbol	Value
height	z	unknown
sea level pressure	P_0	
pressure	$P(z)$	$0.333\, P_0$
temperature	T	295 K
acceleration of gravity	g	9.80 m/s^2
Boltzmann constant	k	1.3807×10^{-23} J·K^{-1}molecule^{-1}
molecular mass	M	28.9
mass of molecule	$m = M/N_A$	4.80×10^{-23} g $= 4.80 \times 10^{-26}$ kg.

Then

$$P(z) = 0.333\, P_0 = P_0\, e^{-z/z_0},$$

or

$$0.333 = e^{-z/z_0}.$$

The inverse of the operation e^x is denoted by ln. That is, if

$$y = e^x \qquad \text{then} \qquad ln(y) = x.$$

For example if $x = 2$, then $y = e^x = 7.3890561$, so $ln(7.3890561) = 2$.

Most calculators have a ln key (this key is not the same as the log key). Use your calculator to find $ln(3) = 1.0986123$ and $ln(12) = 2.4849066$. Your calculator may round off these to numbers with fewer figures.

Returning to our problem, use your calculator to find

$$ln(0.333) = -1.0996128,$$

so

$$-1.0996128 = -z/z_0 = -z \times 1.2 \times 10^{-4} \text{ m}^{-1},$$

or

$$z = 9.5 \times 10^3 \text{ m}.$$

With the molecular mass and temperature we assumed, the atmospheric pressure is 1/3 that of sea level at 9.5 km.

SUGGESTED READINGS

Baierline, R., "The Meaning of Temperature." *The Physics Teacher,* February 1990, p. 94.

Negret, J. P., "Boiling Water and the Height of Mountains." *The Physics Teacher,* May 1986, p. 290.

Romer, R. H., "Covering the U.S.A.: Exponential Growth of Railroad Tracks." *The Physics Teacher,* January 1990, p. 46.

Schneider, S. H., "Climate Modeling." *Scientific American,* May 1987, p. 72.

Wirth, D., *Ballooning, the Complete Guide to Riding the Winds.* New York: Random House, 1980. An interesting summary of the historical and technical facts about ballooning.

Chapter 13
Thermodynamics

IMPORTANT TERMS

Fill in the blanks with the appropriate word or words.

1. An increase of the _____ of a system is a measure of the increase of the disorder of the system.

2. A thermodynamic process in which the volume does not change is called a(n) _____ process.

3. A(n) _____ process is one in which no heat enters or leaves the system.

4. A(n) _____ system interacts with its environment via heat transfer and work.

5. The _____ law of thermodynamics states that "Two systems, each in thermal equilibrium with a third, are in thermal equilibrium with each other."

6. A system that requires work to move heat from a lower temperature region to a higher temperature region is called a(n) _____.

7. Thermodynamic properties of a system, such as volume, temperature and pressure, are called _____ variables.

8. The total kinetic and potential energy associated with the internal state of the atoms composing a system is called the _____ energy of the system.

9. The ratio of the work done to the heat input is called the _____.

10. _____ is concerned with the relationships between heat and work.

11. Conservation of energy is the basic content of the _____ law of thermodynamics.

12. The _____ cycle consists of two reversible isothermal and two reversible adiabatic processes.

13. The _____ law of thermodynamics tells us that heat cannot flow from a cooler body to a warmer body.

14. A process performed at constant pressure is called a(n) _____ process.

15. If the reversal of controlling factors causes a reversal of the energy transformation, the process is said to be _____.

16. The ratio of the heat delivered to the work supplied is called the _____.

17. A(n) _____ process is one in which the temperature of the system does not change.

18. A refrigerator that cools the outside and provides heat to the interior of a building is called a(n) _____.

Write the definitions of the following:

19. isochoric process (13.1):
20. isobaric process (13.1):
21. zeroth law of thermodynamics (13.1):
22. second law of thermodynamics (13.4):
23. adiabatic process (13.1):
24. thermal efficiency (13.2):
25. coefficient of performance (13.3):
26. entropy (13.5):
27. thermodynamic system (13.1):
28. state variable (13.1):
29. reversible process (13.1):
30. internal energy (13.1):
31. Carnot cycle (13.2):
32. heat pump (13.3):
33. thermodynamics (13.0):
34. first law of thermodynamics (13.1):
35. isothermal process (13.1):
36. refrigerator (13.3):

ANSWERS TO 1–18, IMPORTANT TERMS

1. entropy; 2. isochoric; 3. adiabatic; 4. thermodynamic; 5. zeroth; 6. refrigerator; 7. state; 8. internal; 9. thermal efficiency; 10. thermodynamics; 11. first; 12. Carnot; 13. second; 14. isobaric; 15. reversible; 16. coefficient of performance; 17. isothermal; 18. heat pump.

SAMPLE QUESTIONS

QUESTION 13.1: A weight is placed on top of a vertically held brass rod that has its lower end on a fixed support. Discuss what happens from an energy standpoint as the rod is heated.

Answer:

As the rod is heated, we would expect the rod to expand and become longer. This means that work is done to raise the height of the weight. Some of the energy provided to heat the rod must go into this work. The rod would thus expand slower than expected.

QUESTION 13.2: Can you cool a kitchen by leaving the refrigerator door open? What will happen?

Answer:

No! This will heat the room more than it can cool the room. The refrigerator transfers energy from inside the cool box to the room. As thermodynamics predicts, this process cannot be 100% efficient. More energy is released into the room than is removed from the cool box. If the refrigerator door is left open, the pump will have to run more. Of course, standing in front of the open door can feel good.

SAMPLE SOLUTIONS

PROBLEM 13.1: Consider a system whose pressure versus volume behavior follows the one shown in Fig. 13.1a. The system undergoes the transformation $A \Rightarrow B \Rightarrow C$. The total heat taken in is 500 J and the system does 200 J of work. What is the change in the internal energy of the system? What would the change be if the path taken had been $A \Rightarrow D \Rightarrow C$? (13.2)

Solution:

> **Strategy:** Conserve energy. Apply the first law to calculate the change in internal energy.

This problem can be simply answered using the first law. This tells us:

$$\text{change in internal energy } = \text{ heat flow } - \text{ work done.}$$

Fig. 13.1: (a) Pressure versus volume diagram for the fluid. (b) Heat and work flow for the system under consideration.

Thus for the path $A \Rightarrow B \Rightarrow C$,

$$\text{change in internal energy } = \text{ } + 500 \text{ J} - 200 \text{ J } = \text{ } 300 \text{ J.}$$

Since the internal energy of a system does not depend upon how it got to its present state, we know for the path $A \Rightarrow D \Rightarrow C$ that the change in internal energy is the same as previously calculated. The individual values for heat flow and work done may differ, but their difference will have the same value.

Before leaving this problem, let us examine what is happening in the two situations when we also consider the second law. The second law tells us that the system does not just absorb heat and do work; it must also eject heat at a lower temperature. This is shown in Fig. 13.1b. Some heat, Q_1, flows into the system and some heat, Q_2, is ejected from the system while work, W, is done. The values of Q_1, Q_2, and W that we have will depend upon the particular path taken from A to B; that is, if we let

$$B \text{ denote the path } A \Rightarrow B \Rightarrow C$$

and

$$D \text{ denote the path } A \Rightarrow D \Rightarrow C,$$

then, in general,

$$Q_{1B} \neq Q_{1D}, \quad Q_{2B} \neq Q_{2D}, \text{ and } W_B \neq W_D$$

but

$$\Delta U = Q_{1B} - Q_{2B} - W_B = Q_{1D} - Q_{2D} - W_D.$$

That is, the net change in the internal energy of the system, ΔU, is independent of the path used to reach the final state.

PROBLEM 13.2: Problem 11.22 of the text expresses energy in "jelly donut" units of 1 "jelly donut" = 1 MJ. Assuming that jelly donuts cost \$1 each, is it worth the effort to pick up a penny? That is, compare the amount of energy expended by picking up a penny (\$0.01) with the amount of energy that could be purchased by that penny spent on jelly donuts. *Hint:* To do this problem you must make some reasonable assumptions about the process of picking up the penny and about your mass. (13.2)

Solution:

Strategy: Make a simple model of the process of bending over to pick up the penny. Neglect the mass of the penny and consider only the work you must do to raise and lower your center of mass. Assume you have a weight of 150 lbs and a squat involves moving your center of mass 0.50 m up and down. Calculate the energy expended in picking up one penny.

(a)　　　　　(b)　　　　　(c)

Fig. 13.2: The process of picking up a penny.

The simplest process of picking up a penny is shown in Fig. 13.2 We will assume the picking up process is done by a squat. A squat moves the center of mass straight up and down. We will make the approximation that it takes no work to smoothly squat down and assume that all the work is done in standing back up. If d is the distance the center of mass moves in the squat and m is your mass, then to to pick up a penny we have

$$\text{work done} = m\,g\,h,$$

where g is the acceleration of gravity.

Step 1
Determine the work done in joules when picking up the penny. Assuming a weight of 150 pounds and a squat distance of about 0.50 m,

$$\text{work done} = \frac{150 \text{ lb (weight)}}{32 \text{ ft/s}^2} \times \frac{0.454 \text{ kg}}{\text{lb (mass)}} \times 9.8 \text{ m/s}^2 \times \frac{1}{2} \text{ m} = 10.4278 \text{ J} = 10 \text{ J}.$$

Neglecting any internal energy losses of the human body, it takes 10 J of work to stand after a squat.

Step 2
Determine the energy in joules which could be purchased by the penny spent on jelly donuts.

$$1 \text{ penny} \Rightarrow \$0.01 \times \frac{1 \text{ "jelly donut"}}{\$1} \times \frac{1 \text{ MJ}}{1 \text{ "jelly donut"}} \times \frac{10^6 \text{ J}}{\text{MJ}} = 1 \times 10^4 \text{ J}$$

Step 3
Comparing the results we see a gain of 1000:1; therefore, it is "worth" the energy expended to pick up the penny.

Or alternatively, argue that since $\dfrac{1 \text{ MJ}}{10 \text{ J}} = 10^5$,

you could pick up 100,000 pennies or $1,000 in pennies on the energy of 1 "jelly donut." This is a 1000% return on your effort.

PROBLEM 13.3: An inventor claims to have developed a steam turbine which is 74% efficient. Steam enters the turbine at a temperature of 300°C and is exhausted as water at a temperature of 30°C. Is such a turbine theoretically possible and is the claimed efficiency correct? (13.3)

Solution:

> **Strategy:** Calculate the efficiency of a Carnot cycle and compare with the given 74%. Carnot efficiency is less, so analyze what factors affecting the turbine have been neglected. The heat flow and work done are shown in Fig. 13.3.

The maximum thermal efficiency is that of a Carnot engine. If a Carnot engine operates between a hot reservoir at temperature T_H and a cool reservoir of temperature T_C, its efficiency is given by

thermal efficiency$_{maximu}$

$$= 1 - \frac{T_C}{T_H}$$

$$= 1 - \frac{(273+30)K}{(273+300)K} = 0.47.$$

Since 47% is less than 74%, we know the 74% efficiency claim is wrong.

What might the inventor have done incorrectly to get his 74% figure?

He has probably used text Eq. 13.3 incorrectly. This equation gives

thermal efficiency $= \dfrac{Q_H - Q_C}{Q_H}$.

Fig. 13.3: Heat flow and work for turbine.

He probably identified $Q_H - Q_C$ as the heat gained when 1 kg of steam at 300°C is changed to 1 kg of water at 30°C and computed Q_H by adding the heat that could be extracted by cooling 30°C water to 0 K ice.

Using Tables 13.3 and 13.4 of the text we have:

Heat which is given off by cooling 1 kg of 300°C steam to 30°C water

cool 300°C steam to 100°C	$\Rightarrow 0.48 \times (300 - 100)$ kcal $\Rightarrow$	96 kcal
convert steam to 100°C water		540 kcal
cool water to 30°C	$\Rightarrow 1.00 \times (100 - 30)$ kcal	<u>70</u> kcal
		706 kcal

The heat which can be extracted from 30°C water is:

cool water to 0°C	$1.00 \times (30 - 0)$ kcal	30 kcal
freeze water		80 kcal
cool water to −273.16°C	$0.50 \times (0 - 273.16)$ kcal	<u>137</u> kcal
		247 kcal

He then erroneously identified

$$Q_H - Q_C = 706 \text{ kcal} \qquad \text{and} \qquad Q_H = 951 \text{ kcal}.$$

and calculated

$$\text{thermal efficiency} = \frac{W}{Q_H} = \frac{706 \text{ kcal}}{951 \text{ kcal}} = 0.74.$$

There are several mistakes he has made; among them are:

1. The heat into and out of the turbine cannot be calculated as he has done. The heat flow will not be the amount of heat released by a kilogram of steam converted to water multiplied by the number of kilograms used. There will be additional heat flowing from the hot reservoir to keep the hot side of the turbine at its temperature and additional heat flowing to the environment from the turbine to keep its outer jacket at the surrounding temperature.

2. No mention is made of possible changes in the volume and pressure of the working fluid. The steam will be at greater than one atmosphere pressure in order to get it to

flow through the turbine. The implication is that the outlet water is at one atmosphere pressure. Probably some heat flow from the high temperature reservoir will be needed to equalize these pressures.

The proper way to find Q_H is to measure the heat released by the fuel needed to produce the steam that keeps the turbine running. Q_C is extremely difficult to calculate since there are many energy losses in practice. Experience permits an engineer to make a good estimate of thermal efficiency before a turbine is built.

PROBLEM 13.4: A refrigerator maintains inside temperature of 5.0°C when the exterior temperature is 25°C. It maintains the same interior temperature when the room temperature is 30°C. How big is the percentage change in the coefficient of performance for the two cases? (13.4)

Solution:

> **Strategy:** Calculate the coefficient of performance for the two cases using text Eq. 13.5.

As shown in Fig. 13.4a and b, a refrigerator extracts heat from a low temperature reservoir and deposits it into a high temperature reservoir. The second law of thermodynamics tells us that this is only possible if work is done on the refrigerator.

(a) (b)

Fig. 13.4 (a) Refrigerator operating between 5°C and 25°C. (b) Refrigerator operating between 5°C and 30°C.

The coefficient of performance of a refrigerator is given by text Eq. 13.5 as:

$$\text{c.p.(refrigerator)} = \frac{T_C}{T_H - T_C}$$

where T_H is the hot temperature and T_C is the cold temperature.

For the first case $\text{c.p. (25°C)} = \dfrac{273 + 5}{(273+25)-(273+5)} = 13.9 = 14,$

while in the second case

$$\text{c.p. (30°C)} = \frac{273 + 5}{(273+30)-(273+5)} = 11.12 = 11.$$

The percent change in coefficient of performance from the first to the second case is

$$100 \times \frac{14 - 11}{14} = 21.14\% = 21\%.$$

Assuming the cost of operation is proportional to the c.p., we see that it will cost nearly 1/5 more to operate the refrigerator at 30°C than at 25°C.

PROBLEM 13.5: Determine the change in entropy that occurs when 30 grams of cream at 5°C is stirred into 400 grams of coffee at 85°C. Assume the specific heat of both the cream and the coffee is essentially the same as that of water and that the stirring process introduces 1,000 J of energy. (13.6)

Solution:

Strategy: Distribute the stirring energy to each substance in a way proportional to each mass and assume no energy losses to the environment. Use the methods of the previous chapter to find the final energy. Then calculate the entropy by text Eq. 13.7.

Neglect all heat losses. There is a heat flow into the cream and a heat flow out of the coffee. We will assume the energy of stirring is distributed proportional to the masses of each substance.

The heat flows must be equal, so we can use the methods of Chapter 10 to calculate the final temperature. We then find the average temperature for each substance by

$$T_{avg} = \frac{T_{initial} + T_{final}}{2}$$

and use text Eq. 13.7 to calculate each entropy change.

GIVEN:

Quantity	Symbol	Value
mass of cream	m	30 g
mass of coffee	M	400 g
initial temperature of cream	T_C	5°C = 278 K
initial temperature of coffee	T_H	85°C = 358 K
final temperature of mix	T	unknown
energy of stirring added to cream	ΔQ_1	$\frac{30 \times 1000}{400+30}$ J
energy of stirring added to coffee	ΔQ_2	$\frac{400 \times 1000}{400+30}$ J
specific heats of both substances	c	1 cal/g·K.

Note that knowing the temperature in °C to 2 significant figures is the same as knowing the temperatures to 3 significant figures in K.

The heat <u>gained</u> by the cream is $\quad\quad\quad\quad \Delta Q_{\text{cream}} = \Delta Q_1 + m\,c\,(T - T_C),$

while the heat <u>lost</u> by the coffee is $\quad\quad\quad \Delta Q_{\text{coffee}} = -\Delta Q_2 + M\,c\,(T_H - T).$

Notice the minus sign put in for ΔQ_2 since the energy of stirring is added to the coffee. It is probably most convenient to use calories, grams and Kelvin as our units, since that makes c unity and saves converting temperatures at the end. Expressing ΔQ_1 and ΔQ_2 in cal by the conversion $1\ \text{J} = 0.2388$ cal, we get

$$\Delta Q_{\text{cream}} = 16.7\ \text{cal} + 30 \times (T - 278\ \text{K})\,\frac{\text{cal}}{\text{K}}$$

$$\Delta Q_{\text{coffee}} = -222.1\ \text{cal} + 400 \times (358\ \text{K} - T)\,\frac{\text{cal}}{\text{K}}.$$

Equating these we get

$$(M+m)cT = c(M\,T_H + m\,T_C) - \Delta Q_1 - \Delta Q_2$$

or

$$T = \frac{M\,T_H + m\,T_C}{M + m} - \frac{\Delta Q_1 + \Delta Q_2}{c(M + m)} = (352.4 - 0.5)\ \text{K} = 352\ \text{K}.$$

Thus the final temperature is $352\ \text{K} = 79°\text{C}.$

The second term is the effect of stirring, which is negligible for this example. Remember that K to 3 significant figures is °C to 2 significant figures.

The average temperatures are now

$$T_{\text{avg-cream}} = 316\ \text{K}$$

and

$$T_{\text{avg-coffee}} = 356\ \text{K}.$$

Before we can calculate the entropy change, we must get the (equal) heat flows for each substance. Using the final temperature we get

$$\Delta Q_{\text{cream}} = (16.7 + 2220)\ \text{cal} = 2.2 \times 10^3\ \text{cal},$$

and

$$\Delta Q_{\text{coffee}} = (-222.1 + 2400)\ \text{cal} = 2.2 \times 10^3\ \text{cal} = \Delta Q_{\text{cream}}.$$

These two are equal as expected, since ΔQ_{cream} is <u>heat added</u> to the cream and ΔQ_{coffee} is the <u>heat lost</u> by the coffee.

Before we calculate the entropy change we must be careful and remember that ΔQ_{cream} is energy added and ΔQ_{coffee} is energy lost, so the total entropy change, ΔS, is

$$\Delta S = \frac{\Delta Q}{T} = \frac{2.2 \times 10^3\ \text{cal}}{316\ \text{K}} - \frac{2.2 \times 10^3\ \text{cal}}{356\ \text{K}} = (6.96 - 6.18)\,\frac{\text{cal}}{\text{K}} = 0.78\,\frac{\text{cal}}{\text{K}}.$$

The entropy change is positive as required.

PROBLEM 13.6: What is the efficiency and thermal pollution of a power plant which produces 500 MW of work per unit time by operating between two heat reservoirs, one at 1500 K and the other at 400 K? (13.7)

Solution:

> **Strategy:** Use the definition of thermal efficiency to find the efficiency and the ratio of heat in to heat out. This combined with the total work done will give us the thermal pollution produced.

If Q_H is the heat input per unit time at temperature T_H, Q_C the heat output per unit time at temperature T_C, and W the work possible for the plant, then

$$\text{thermal efficiency} = 1 - \frac{T_C}{T_H} = 1 - \frac{Q_C}{Q_H} = \frac{W}{Q_H} = 1 - \frac{400 \text{ K}}{1500 \text{ K}} = 0.733,$$

so
$$Q_H = \frac{W}{\text{thermal efficiency}} = 6.82 \times 10^8 \text{ W}$$

and
$$Q_C = Q_H \frac{T_C}{T_H} = 1.82 \times 10^8 \text{ W}.$$

The power plant has an efficiency of 73.3% and produces 1.82×10^2 MW of thermal pollution.

MODELING

Extensive use of models is made in thermodynamics. We usually do not need to look at the internal structure of the thermodynamic system. Thus thermodynamics is independent of the kinetic theory of gases. Thermodynamics makes the following assumptions: The system has a temperature and an internal energy and is able to do work. Heat is transferred into and out of the system by bringing the system into contact with other bodies whose internal structure is often not specified.

Often these external bodies are assumed to be what are called constant temperature reservoirs. Such reservoirs are able to give up heat or absorb heat without changing their temperatures. Such ideal reservoirs do not exist in nature. However, they can be approximated in nature by making the constant temperature body sufficiently large that the internal energy of the reservoir is very much greater than the changes of the internal energy of the system under study. We will make extensive use of this model in this and the next chapter.

SUGGESTED READINGS

Issue on Energy for Planet Earth. *Scientific American,* September 1990.

Jones, P. D., and T. M. L. Wigley, "Global Warming Trends." *Scientific American,* August 1990, p. 84.

DiLavore, P., *Energy: Insights from Physics.* New York: John Wiley & Sons, 1984. The physics of energy sources and their uses.

Chapter 14
Periodic Motion

IMPORTANT TERMS

Fill in the blanks with the appropriate word or words.

1. A physical phenomenon that repeats itself regularly is termed _____.

2. Another name for the natural frequency of a system is the _____ frequency. This term is used because the amplitude of the system increases rapidly at that frequency.

3. The angle by which a harmonic function is offset at the time chosen as zero is called the _____ angle.

4. When two waves are in phase the amplitude builds up. This is called _____.

5. The force that a displaced stretched spring exerts to return to its undistorted position is called a(n) _____ force.

6. If an oscillator also has frictional forces, we refer to these forces as _____ forces.

7. Motion whose acceleration is proportional to displacement is called _____ motion.

8. The period of a freely oscillating system is called the _____ period.

9. The _____ is the frequency of a freely oscillating system.

10. The proportionality constant in Hooke's law is called a(n) _____.

11. A(n) _____ consists of a mass swinging from the end of a light string of constant length.

12. The maximum value of displacement of a wave is called the _____ of the wave.

13. Any object whose motion obeys Hooke's law is called a simple _____ oscillator.

Write the definitions of the following:

14. amplitude (14.2):

15. resonance (14.11):

16. spring constant (14.1):

17. restoring force (14.1):

18. resonant frequency (14.7):

19. periodic (14.0):

20. damping (14.6):

21. natural frequency (14.4):

22. phase angle (14.2):

23. simple harmonic motion (14.2):

24. natural period (14.4):

25. simple pendulum (14.5):

26. simple harmonic oscillator (14.2):

27. Hooke's law (14.1):

ANSWERS TO 1–13, IMPORTANT TERMS

1. periodic; 2. resonant; 3. phase; 4. resonance; 5. restoring; 6. damping; 7. simple harmonic; 8. natural; 9. natural frequency; 10. spring constant; 11. period; 12. amplitude; 13. harmonic.

SAMPLE QUESTIONS

QUESTION 14.1: The pendulum on the clock that strikes the hours on the bell, Big Ben, in London was particularly noted for its accuracy. Formerly the period was regulated by placing a penny on top of the bob of the 3.98-m long pendulum. Will placing a penny there make the clock run faster or slower? Explain.

Answer:

The period of a pendulum is independent of the mass; but the period depends upon the length of the pendulum. Neglecting the mass of the pendulum rod, this length is determined by the location of the center of mass of the bob. Laying a penny on top of the bob raises the center of mass relative to the pivot point. This makes the length shorter and thus the period smaller. A smaller period will make the clock run faster.

QUESTION 14.2: If the outer case of an air conditioner rattles when the unit runs, what can you do that will probably make it stop?

Answer:

You could turn the air conditioner off, but that solution will make life less pleasant. A better solution is to place a brick or two on top of the outer case. This will change the resonance frequency of the case so that it no longer is close to the frequencies of the air conditioner.

SAMPLE SOLUTIONS

PROBLEM 14.1: A mass of 0.140 kg is hung on the end of a spring as shown in Figure 14.1a. At the new equilibrium point the spring is stretched 12 .0 cm from its initial length. The mass is now pulled down a further 15.0 cm and released at time $t = 0$. Answer the following:
  (a) What is the spring constant? (14.1)
  (b) What is the period of oscillations? (14.4)
  (c) What is the general form of the motion after release? (14.2)
  (d) What are the displacement, velocity and acceleration at
    $t = t_1 = 0.100$ s, $t = t_2 = 0.300$ s, and $t = t_3 = 0.520$ s?   (14.2)

Solution:

Strategy: This problem involves calculating the spring constant using Hooke's law, then using that to find the period and hence the displacement.

In order to express our ideas, we will make the drawing shown in Fig. 14.1a. The following variables will be used:

Quantity	Symbol	Value
spring constant	k	unknown
mass added	m	0.140 kg
change in spring length	d	12.0 cm
further initial displacement	x_0	15.0 cm
later displacement at time t	$x(t)$	unknown
velocity at time t	$v(t)$	unknown
acceleration at time t	$a(t)$	unknown
period of oscillation	τ	unknown
acceleration of gravity	g	9.80 m/s^2

When we hang the mass on the spring, the spring stretches until the Hooke's law force plus the gravitational force is zero. That is

$$F = -k\,d, \qquad \text{(Hooke's law)} \qquad \text{(text Eq.14.1)}$$

and

$$F + mg = 0 \qquad \text{(condition for equilibrium)}$$

Thus $k = \dfrac{mg}{d} = 0.140 \text{ kg} \times \dfrac{9.80 \text{ m/s}^2}{0.12 \text{ m}} = 11.4333 \text{ N/m.} = 11.4 \text{ N/m.}$

Considering the significant figures, the spring constant $k = 11.4$ N/m.

Since the gravitational force is constant, this initial stretching of the spring will always take care of the gravitational force. So the subsequent motion of the mass is not affected by the gravitational force, provided we measure the displacement from the new equilibrium position. This is the choice we have made for $x(t)$ in Fig. 14.1.

The period of oscillation is given by text Eq. 14.8 as

$$\tau = 2\pi \sqrt{\frac{m}{k}} = 2\pi \sqrt{\frac{m}{mg/d}} = 2\pi \sqrt{\frac{d}{g}} = 0.69528 \text{ s,}$$

or, considering significant figures, $\tau = 0.695$ s.

Note that for this problem, as in the case of the simple pendulum, the motion is independent of the mass when measured relative to the equilibrium point.

(a)   **(b)**

Fig. 14.1: (a) Mass of 1.0 kg hanging on a spring. (b) Time displacement of the mass following the initial displacement.

The general displacement, velocity, and acceleration will be given by text Eqs. 14.3 to 14.6:

$$x(t) = x_0 \cos\left(2\pi\,\frac{t}{\tau}\right) = 0.15 \cos(9.04 \times t) \times \text{m}$$

$$v(t) = -\sqrt{\frac{k}{m}}\,x_0 \sin\left(2\pi\,\frac{t}{\tau}\right) = -1.35 \sin(9.04 \times t) \times \text{m/s}$$

$$a(t) = -\frac{k}{m}\,x_0 \cos\left(2\pi\,\frac{t}{\tau}\right) = -12.2 \cos(9.04 \times t) \times \text{m/s}^2.$$

Fig. 14.1b shows the general displacement curve and the time values of the critical points.

We can now answer the last question.

Time	Position	Velocity	Acceleration
0.100 s	0.0928 m	− 1.06 m/s	− 7.54 m/s^2
0.300 s	− 0.136 m	− 0.562 m/s	1.23 m/s^2
0.520 s	− 0.00174 m	1.35 m/s	0.142 m/s^2

Since we have chosen x positive downwards, positive velocity and positive acceleration are also downwards. So when 0.100 second has passed, we are still below equilibrium but moving upwards with upwards acceleration. After 0.300 seconds the mass has almost reached its maximum upward displacement so it is moving upwards with downward acceleration. And finally, after 0.520 seconds the body still has an upward displacement but it is moving downward, with a downwards acceleration. Each of these answers agrees with what we expect when we examine Fig. 14.1b.

PROBLEM 14.2: For the harmonic oscillator in Problem 14.1, what are the maximum potential and kinetic energies of oscillation? What are the first two times the maximum of each occurs, and where is the mass at that time? (14.3)

Solution:

> **Strategy:** Use energy conservation and the expression for the potential energy of a stretched spring.

We are asked for potential energy relative to the equilibrium position. That is, we again do not need to include the gravitational potential energy of the mass provided the position is measured relative to the equilibrium position. With this in mind, we see that the maximum potential energy, due only to the spring force, occurs first when we start. The maximum potential energy is given by

$$PE_{max} = \frac{1}{2} k \, x_0^2.$$

This energy will be a maximum when the spring has maximum extension or maximum compression (since the potential is not changed by $x_0 \rightarrow -x_0$). When the potential energy is at maximum, the mass is at rest, so the kinetic energy is zero.

Whenever the mass passes through the equilibrium point the potential energy is zero and hence, since the total energy is conserved, the kinetic energy is a maximum. Thus

$$KE_{max} = PE_{max}.$$

It will take one quarter cycle for the motion to reach the original equilibrium position and that position will be reached every half cycle (see Fig. 14.1b), so

$$KE_{max} = PE_{max} = \frac{1}{2} \times 11.4 \text{ N/m} \times (0.15 \text{ m})^2 = 0.128 \text{ J}$$

and

Elapsed time (periods)	Energy max	Time (seconds)
0.0	PE	0.0
0.25	KE	0.174
0.50	PE	0.348
0.75	KE	0.521

PROBLEM 14.3: A 66 pound child (mass = 30 kg) is playing on a swing of negligible mass and of length 3.6 m. The child is being pushed by her father each time she swings, and is now swinging so high that her mass reaches a distance of 1.5 m above her starting position. How fast is she moving when she passes through the bottom of her swing? How often must her father push her? (14.5)

Solution:

> **Strategy:** Consider the motion as having two different intervals. During the first interval the child is swinging freely and energy is conserved. During the second interval the child is being pushed by her father and energy is not conserved. First use conservation of energy to find the kinetic energy at the bottom of the swing in terms of her maximum height. Calculate the period from the small angle approximation for the simple pendulum.

We can answer the first question without approximation by using conservation of energy. At any point on her swing her potential energy is given by

$$PE = mgh,$$

where m is her mass, g is the acceleration of gravity and h is her height above the lowest point as shown in Fig. 14.2a. At the highest point, all the energy is potential energy since her motion is undergoing reversal. This maximum potential energy must be numerically equal to her kinetic energy at the bottom of her swing.

Fig. 14.2:   (a) Girl on swing.   (b) Evaluation of the small angle approximation.

GIVEN:

Quantity	Symbol	Value
mass of girl	m	30 kg
length of swing	L	3.6 m
height of swinging	h	1.5 m
acceleration of gravity	g	9.8 m/s^2
maximum velocity	v_{max}	?

$$KE_{max} = \frac{1}{2} m (v_{max})^2$$

$$= PE_{max} = mgh,$$

so

$$v_{max} = \sqrt{2gh}$$

$$= \sqrt{2 \times 9.8 \text{ m/s}^2 \times 1.5 \text{ m}} = 5.4 \text{ m/s}.$$

She is moving 5.4 m/s at the bottom of her swing.

If we assume that the simple pendulum approximation is close enough, then the period of the swing is given by

$$\tau = 2\pi\sqrt{\frac{L}{g}} = 2\pi \times \sqrt{\frac{3.6 \text{ m}}{9.8 \text{ m/s}^2}} = 3.8 \text{ s}.$$

Her father must push her every 3.8 s and she is moving 5.4 m/s as she passes her lowest point.

The approximation we made is that the angle θ in Fig. 14.2b is small. To check this note that

$$\cos\theta = \frac{L-h}{L} = \frac{2.1}{3.6} = 0.58333,$$

which gives

$$\theta = 54°.$$

This θ is not small so the assumption of small swing angle is not very good, but it is the best that can be done at this level of mathematics.

PROBLEM 14.4: A mass on the end of a spring is found to have a period of 2.00 s when displaced vertically. When the same mass and spring are placed horizontally on a fairly smooth surface, the amplitude is observed to decrease by 50% after 20 oscillations and that this takes a total time of 41.0 s. What is the damping constant for the second situation? What is the percentage change in the period caused by the damping? (14.6)

Solution:

> **Strategy:** Assume the motion is described by text Eq. 14.12 to find the damping constant. The observed time and number of periods gives us the new period.

The two cases are shown in Fig. 14.3. Fig. 14.3a shows the vertical oscillations and Fig. 14.3b shows the second case with the oscillator horizontal.

(a)   **(b)**

Fig. 14.3: (a) Oscillator is suspended vertically. (b) Oscillator is displaced while horizontal on a smooth surface.

Since the surface is smooth and 20 cycles are needed for a 50% decrease in amplitude, the damping force is small, so we can use text Eq. (14.12):

$$y = y_0\, e^{-\gamma t}\cos(2\pi f t).$$

where γ is the damping constant, y is the displacement, t is the time, and f is the frequency of the damped oscillator. Since the damping is small, we expect that f is nearly the natural frequency. However, we can find the damping constant without approximating the frequency, since we are given that the amplitude decreases by 50.0% when $t = 41.0$ s. That is,

$$\frac{y}{y_0} = 0.500 = e^{-\gamma \times 41.0 \text{ s}}.$$

If we take the natural logarithm of both sides we get

$$\ln(e^{-\gamma \times 41.0 \text{ s}}) = -41.0 \text{ s} \times \gamma = \ln(0.50) = -0.6931$$

or

$$\gamma = \frac{-0.6931}{-41.0 \text{ s}} = 0.0169/\text{s}.$$

The damping constant is 0.0169/s.

To find the percentage change in the period, we note that 20 cycles with damping takes 41.0 s. So the damped period is

$$\tau_{\text{damped}} = \frac{41.0 \text{ s}}{20} = 2.05 \text{ s},$$

while

$$\tau_{\text{undamped}} = 2.00 \text{ s}.$$

Thus the percentage change is

$$100 \times \frac{\tau_{\text{undamped}} - \tau_{\text{damped}}}{\tau_{\text{undamped}}} = 2.5\%.$$

The period changes $2\frac{1}{2}$ % due to the damping.

PROBLEM 14.5: The fan unit for a building's heating and air-conditioning system is mounted on six support springs. Assuming that the fan shaft turns at 2100 rpm (revolutions per minute) and the mass of the unit is 610 kg, what is the proper stiffness constant for the springs used to support the fan unit? (14.7)

Solution:

Strategy: This is a problem in resonance. As indicated in the text, the natural resonance of the fan system should be 5 times the resonant frequency of the support system. Find the fan frequency, then divide by 5 to find the desired support frequency. The support is equivalent to a mass on a spring so we can find the needed spring constant. Since there are six support springs, each spring must have a constant 1/6 that of the desired frequency.

The natural frequency of the system is

$$2100 \text{ rpm} \times \frac{1 \text{ min}}{60 \text{ s}} = 35 \text{ Hz},$$

so the desired resonance frequency of the spring support system is $\frac{1}{5} \times 35$ Hz = 7.0 Hz.

Considering the support as a spring supporting a mass, we can find the needed spring constant as

$$k = m\,(2\pi f)^2 = 610 \text{ kg} \times (2\pi \times 7.0 \text{ Hz})^2 = 1.18 \times 10^6 \text{ N/m}.$$

Since each of the six springs provided 1/6 the support, each spring must have a spring constant, k_s, which is

$$k_s = \frac{1}{6} \times 1.18 \times 10^6 \text{ N/m} = 1.97 \times 10^5 \text{ N/m}.$$

Each of the six springs must have a spring constant of 1.97×10^5 N/m.

PROBLEM 14.6: One of the spring supports of the fan unit in the preceding problem (Problem 14.5) is broken. Will the vibration of the fan be worse than before? (14.7)

Solution:

> **Strategy:** The text criterion was that the support resonant frequency should be at least a factor of 3 away from the fan resonant frequency. We must determine how close the resonant frequency is with the broken spring.

The support in the preceding problem was designed to be a factor of 5 away from resonance. We must determine if the loss of one spring will shift us into the very noisy region.

With only five springs the support system has a spring constant of 5/6 to the design constant. The new resonant frequency of the support is now

$$f_{new} = \sqrt{\frac{k_{new}}{2\pi m}} = \sqrt{\frac{\frac{5}{6}k_{old}}{2\pi m}} = \sqrt{\frac{5}{6}f_{old}} = 0.913 \times f_{old} = 0.913 \times 7 \text{ Hz} = 6.4 \text{ Hz.}$$

The new frequency is further away from resonance so there will be no noise increase, that is, provided the five remaining springs can still support the fan unit without being squeezed so much that the spring coils come into contact and the spring ceases to be a spring. If this happens, then the noise will get worse. Design engineers usually pick springs so that one can be broken without the unit failing completely.

SUGGESTED READINGS

Carlson, J. E., "The Pendulum Clock." *The Physics Teacher,* January 1991, p. 8.

Crane, H. R., "The Foucault Pendulum as a Murder Weapon and a Physicist's Delight." *The Physics Teacher,* May 1990, p. 264.

Crummett, B., "Measurements of Acceleration Due to Gravity." *The Physics Teacher*, May 1990, p. 291.

Egan, D. M., *Concepts in Architectural Acoustics*. New York: McGraw-Hill, 1972. A handy reference for solving a broad range of acoustical and vibrational problems.

Jones, G.E., and J. L. Ferguson, "Easy Displacement Versus Time Graphs for a Vibrating String: Tuning a Guitar by Television." *American Journal of Physics*, Vol. 48, May 1980, p. 362.

Rossing, T. D., "The Physics of Kettledrums." *Scientific American*, November 1982, p. 172.

Rugar, D., and P. Hansma, "Atomic Force Microscope." *Physics Today*, October 1990, p. 23.

Walker, J., " The Amateur Scientist: Strange Things Happen When Two Pendulums Interact Through a Variety of Interactions." *Scientific American*, October 1985, p.176.

Weinreich, G., "The Coupled Motions of Piano Strings," *Scientific American*, January 1979, p. 118.

Yin, T. P., "The Control of Vibration and Noise." *Scientific American*, January 1969, p. 98.

Chapter 15
Wave Motion

IMPORTANT TERMS

Fill in the blanks with the appropriate word or words.

1. For a standing wave on a string, the point on the string that vibrates with the greatest amplitude is called a(n) _____ .

2. The _____ frequency of a string is its lowest resonant frequency.

3. Waves in which the propagating medium moves back and forth along the direction of propagation are called _____ waves.

4. A(n) _____ wave pulse is one for which the displacement of the pulse is perpendicular to the direction the pulse travels.

5. If a source is moving through a medium at a speed greater than the wave velocity for the medium, the source produces a single conical wave front known as a(n) _____ .

6. Waves in which the propagating medium moves back and forth perpendicular to the direction of propagation are called _____ waves.

7. When two or more waves combine such as to cancel each other's amplitude, we have _____ interference.

8. For a standing wave on a string, the point on the string that vibrates with the least amplitude is called a(n) _____ .

9. A(n) _____ is a disturbance that transfers energy through a medium without imparting net motion to the medium.

10. If wave fronts are nearly flat, we call the wave a(n) _____ wave.

11. If waves combine in phase so their amplitudes reinforce, we have _____ .

12. A wave generated by simple harmonic motion is called a(n) _____ wave.

13. The _____ frequencies of a string are integer multiples of the fundamental frequency of that string.

14. The _____ of a wave is defined as the power flowing per unit area.

15. At any instant the resultant of a number of waves is the algebraic sum of all the component waves. This is a statement of the _____ .

16. The common unit of sound measurement on the logarithmic scale is the _____ .

17. The compressions (or the rarefactions) produced by a sound wave are referred to as the _____.

18. The _____ is the change in frequency due to relative motion of source and receiver.

19. The separation between successive wave crests or wave fronts is referred to as the _____ of the wave.

20. Relative _____ level is usually measured in decibels.

21. A resonant frequency above the fundamental frequency is called a(n) _____.

22. Two sound waves of nearly the same frequency produce _____.

23. For a(n) _____ wave on a string there exist fixed points on the string where the amplitude is zero and other fixed points where the amplitude is maximum.

24. A(n) _____ wave is a physical effect that travels from one location to another.

Write the definitions of the following:

25. transverse wave pulse (15.1):

26. beats (15.11):

27. harmonic wave (15.2):

28. Doppler effect (15.6):

29. intensity (15.3):

30. wavefront (15.4):

31. plane wave (15.4):

32. shock wave (15.7):

33. harmonic wave (15.2):

34. principle of superposition (15.9):

35. transverse wave (15.4):

36. traveling wave (15.1):

37. constructive interference (15.9):

38. decibel (15.5):

39. fundamental frequency (15.9):

40. standing wave (15.9):

41. nodes (15.9):

42. antinode (15.9):

43. overtone (15.9):

44. intensity level (15.5):

45. harmonic frequency (15.9):

46. longitudinal wave (15.4):

ANSWERS TO 1–24, IMPORTANT TERMS

1. antinode; 2. fundamental; 3. longitudinal; 4. transverse; 5. shock wave; 6. transverse; 7. destructive; 8. node; 9. wave; 10. plane; 11. constructive interference; 12. harmonic; 13. harmonic; 14. intensity; 15. principle of superposition; 16. decibel; 17. wavefronts; 18. Doppler shift; 19. wavelength; 20. intensity; 21. overtone; 22. beats; 23. standing; 24. traveling.

SAMPLE QUESTIONS

QUESTION 15.1: You can estimate the distance to a lightning flash by counting the seconds that elapse between the time you see the flash and when you hear the accompanying thunder. (a) The approximate distance in miles is obtained by dividing the time in seconds by 5. Explain why this is so. (b) Estimate the distance to a lightning flash that you see 10 s before you hear the thunder.

Answer:

Light travels 3.0×10^8 m/s (186,000 mi/s) while sound travels at about 340 m/s (1100 ft/s). So it takes light a very short time to travel a mile, while sound takes about 5 seconds to travel the same mile. If a lightning flash is one mile from us, we will hear the sound of the lightning flash (thunder) about 5 seconds after we see it. If we hear the thunder 10 seconds after we see the flash, the flash occurred about 2 miles from us.

QUESTION 15.2: How could you get a stretched string to vibrate in only its second harmonic mode?

Answer:

The fundamental (first harmonic) mode of a stretched string has maximum amplitude in the middle of the string. If you hold the middle of the string fixed, all odd number modes will be suppressed. If you then pluck the string one-fourth of the way from one end, the second mode will be the strongest of the excited even harmonics.

SAMPLE SOLUTIONS

PROBLEM 15.1: A 3000 Hz sound wave passes from air into water. What is the frequency in the water? What is the ratio of the wavelength in air to that of the wave in the water? Assume the temperature is 15°C. (15.2)

Solution:

> **Strategy:** The number of wavefronts entering a material must be the same as the number just inside the material. Use the equation relating wavelength, frequency and velocity along with simple conservation of the number of compressions across a surface.

If we stand at the air/water interface, we see that the frequency is unchanged as the sound wave passes from the air to the water or we would sometimes see a compression of air corresponding to a rarefaction of water or conversely. Thus the frequency in the water is the same as the frequency in the air as shown in Fig. 15.1.

Fig. 15.1: Sound wave passing from air to water.

This conclusion about the frequencies cannot hold for the wavelengths since the frequency, f, wavelength, λ, and the velocity, v, are related by

$$v = \lambda f, \qquad \text{or} \qquad \lambda = \frac{v}{f}. \qquad \text{(text Eq. 15.2)}$$

We can use text Table 15.1 to get the velocities, giving us

$$\frac{\lambda_{air}}{\lambda_{water}} = \frac{\dfrac{v_{air}}{f_{air}}}{\dfrac{v_{water}}{f_{water}}} = \frac{v_{air}}{v_{water}} = \frac{340 \text{ m/s}}{1500 \text{ m/s}} = 0.227 .$$

The wavelength in the air is 0.227 times the wavelength in water.

PROBLEM 15.2: When sounded, two identical car horns produce sound intensities at your ear in the ratio of 10 : 1. If the nearer car is 6.5 m away from you, how far away is the other car? (15.4)

Solution:

> **Strategy:** Since the car horns are identical, we assume the difference in intensity is due to a difference in distance from us. Since we are not given any other information we must assume that there are no reflections so the wavefronts are essentially spherical. We can then apply the inverse square law.

Let I_1 be the intensity of the sound produced by the closer horn at a distance of r_1 and I_2 be the sound intensity of the further horn at a distance r_2. We are given

$$\frac{I_1}{I_2} = 10 \qquad \text{and} \qquad r_1 = 6.5 \text{ m}.$$

The intensity should be given by the inverse square law, so we can write

$$I_1 = \frac{k}{r_1^2} \qquad \text{and} \qquad I_2 = \frac{k}{r_2^2},$$

where k is the sound intensity of either horn if it were 1.0 m away.

Rather than find the value of k by using the known values of I_1 and r_1, we take the ratio of the two values to get

$$\frac{I_1}{I_2} = \frac{\dfrac{k}{r_1^2}}{\dfrac{k}{r_2^2}} = \frac{r_2^2}{r_1^2} = 10.$$

Thus

$$r_2^2 = 10 \times r_1^2 = 10 \times (6.5 \text{ m})^2.$$

This gives

$$r_2 = 21 \text{ m}.$$

The second car is located 21 m away from us.

PROBLEM 15.3: A given sound level is quoted as –2.0 dB above the reference. What would the sound level be if the intensity were increased by a factor of 8.0? What is the original intensity in W/m^2 if the reference intensity is 10^{-12} W/m^2? (15.5)

Solution:

Strategy: Use the definition of intensity in dB. The first method is to calculate the initial intensity in terms of the reference intensity and then get the new intensity level for a factor of 8. The second method involves using the properties of the logarithm.

First Method

The intensity level, IL, in dB is given in the text by

$$IL = 10 \log_{10}\left(\frac{I}{I_0}\right),$$

where I is the intensity in watts and I_0 is the reference intensity.

We are given that

$$-2.0 = 10 \times \log_{10}\left(\frac{I}{I_0}\right)$$

(1)

and asked to find

$$IL = 10 \times \log_{10}\left(\frac{8I}{I_0}\right).$$

(2)

We can solve Eq. 1 for I as

$$I = I_0 \times 10^{-0.2}$$

and substitute this into Eq. 2 to get

$$IL = 10 \times \log_{10}\left(\frac{8 \times 10^{-0.2}I_0}{I_0}\right) = 10 \times \log_{10}(8 \times 10^{-0.2})$$

$$= 10 \times \log_{10}8 + 10 \times \log_{10}10^{-0.2} = 9.031 + (-2.00) = 7.0.$$

The new power is 7.0 dB.

Second Method
Use the *log* property that

$$\log(a \times b) = \log a + \log b.$$

Subtract Eq. 2 from Eq. 1 to get

$$IL - (-2.0) = 10 \times \log_{10}\left(\frac{8.0\ I}{I_0}\right) - 10 \times \log_{10}\left(\frac{I}{I_0}\right)$$

$$= 10 \times \log_{10}8.0 + 10 \times \log_{10}\left(\frac{I}{I_0}\right) - 10 \times \log_{10}\left(\frac{I}{I_0}\right)$$

$$= 10 \times \log_{10}8 = 9.0.$$

Again $IL = -2.0 + 9.0 = 7.0$ dB.

Since I_0 is given as 10^{-12} W/m^2, we can use the first equation to find that

$$I = I_0 \times 10^{-0.2} = 6.3 \times 10^{-13}\ \text{W/m}^2.$$

Whenever the intensity is less than the reference level, the relative intensity in dB will be negative.

PROBLEM 15.4: A train is approaching an observer at a speed of 60 m/s. A horn of 5,000 Hz on the train sounds as if it has a frequency which is offset 1000 Hz from the original 5,000 Hz. What is the speed of sound at the time of this observation? (15.6)

Solution:

> **Strategy:** Since the train is approaching we expect the sound to be Doppler shifted to a shorter wavelength. The emitted wavelength is longer than the wavelength we hear. Use the Doppler shift formula for an approaching source to determine the velocity of approach.

If we let f be the emitted frequency, f' the observed frequency, v the approach velocity, and v_0 the velocity of sound, then

$$f' = \frac{f}{1 - \dfrac{v}{v_0}},$$

where the observed frequency must be larger than the emitted frequency.

Let Δf be the difference between the emitted and the observed frequency; then

$$f' = = f + \Delta f = \frac{f}{1 - \dfrac{v}{v_0}}.$$

So

$$v = v_0 \times \left(1 - \frac{f}{f + \Delta f}\right) = \frac{v_0 \, \Delta f}{f + \Delta f},$$

and we get

$$v_0 = v \times \frac{f + \Delta f}{\Delta f}.$$

Entering the given numbers we get a velocity of sound of 360 m/s.

PROBLEM 15.5: A train approaching at a speed of 40 m/s sounds as if it has a horn frequency of 1500 Hz. What is its true (rest) frequency? What frequency do you hear as the train recedes? Assume the temperature is 15°C. (15.6)

Solution:

> **Strategy:** Since the train is approaching, the sound we hear has a shorter wavelength than the emitted sound, so we expect the emitted wave to be longer than the observed sound. Apply the Doppler effect rule for a moving source that is approaching to determine the true wavelength. We can then repeat this process for a receding source.

The two situations are shown in Fig. 15.2a and b. Text Eqs. 15.5 apply. If we let f be the true frequency, f' be the frequency we hear, v the relative velocity, and v_0 the velocity of sound in air, then as the train approaches

$$f'_{app} = \frac{f}{(1 - v/v_0)}$$

and as it recedes

$$f'_{rec} = \frac{f}{(1 + v/v_0)} .$$

GIVEN:

Quantity	Symbol	Value
frequency heard at approach	f'_{app}	1500 Hz
true frequency	f	unknown
frequency heard at recession	f'_{rec}	unknown
relative speed	v	40 m/s
speed of sound at 15°C	v_0	340 m/s

Fig. 15.2: (a) Approaching train sounding its horn. (b) Receding train sounding its horn.

The first step is to find f using the first Doppler effect relationship and then use this f to find f'_{rec}:

$$f = f'_{app}(1 - v/v_0) = 1500 \text{ Hz} \left(1 - \frac{40 \text{ m/s}}{340 \text{ m/s}}\right),$$
$$= 1.3235294 \times 10^3 \text{ Hz} = 1.3 \times 10^3 \text{ Hz}.$$

The rest frame or true frequency is 1.3×10^3 Hz.

Knowing the true frequency, we can calculate the frequency heard when the train is receding:

$$f'_{rec} = \frac{1323.5294}{1.1176471} = 1.184 \times 10^3 \text{ Hz} = 1.2 \times 10^3 \text{ Hz}.$$

We conclude that the horn is emitting a frequency of about 1300 Hz and that we hear a 1200 Hz signal as the train recedes.

PROBLEM 15.6: You hear a sonic boom from the plane exactly 3.50 s after it passes directly overhead in level flight. At the time you hear the boom, you see the plane at an angle of 25.0° above the horizon. How fast is the plane traveling and what was its height as it passed over you? Assume that the speed of sound at the altitude of the plane is 325 m/s. (15.7)

Solution:

Strategy: Since we hear a sonic boom, it is reasonable to assume the Mach number is greater than 1. Since we know the angle to the plane when we hear the sonic boom, we can calculate the Mach number and hence the velocity of the plane. Knowing the velocity of the plane and the time delay until we hear the sonic boom will allow us to determine the height of the plane as it passes over us.

Assume the air is still. We let v denote the velocity of sound (given as 325 m/s) and v_S the velocity of the airplane relative to the ground as shown in Fig 15.3.

Fig. 15.3   Flight path of plane relative to observer. The elapsed time Δt = 3.5 s.

Since we hear the sonic boom when the airplane appears to be 25.0° above the horizon, the angle of the shock wave is also 25.0°. Text Eq. 15.8 gives the relation between the shock angle, θ, the speed of sound, v, and the speed of the source, v_S:

$$\sin \theta = \frac{v}{v_S} = \frac{1}{\text{Mach number}}.$$

This gives a Mach number = $1/(\sin \theta)$ = 2.37, and a plane velocity of 769 m/s.

If the plane is moving 769 m/s horizontally, then it travels 2.69×10^3 m in 3.50 s. This is the distance $v_S \Delta t$ in Fig. 15.3. As we can see from that figure:

$$\sin \theta = \frac{h_0}{v_S \, \Delta t}, \quad \text{which gives} \quad h_0 = 1.14 \text{ km.}$$

The plane is traveling at 2.69 km/s and passes overhead at a height of 1.14 km.

PROBLEM 15.7: A wire is stretched between two supports 1.100 m apart. If the speed of sound in the wire is 2040 m/s, what is the frequency of the fundamental vibration and the first overtone? If two identical such strings are plucked with the first producing the fundamental and the second the first overtone only, what beat frequencies will be heard? (15.11)

Solution:

Strategy: The frequencies of a string are given in the text by Eq. 15.9. Use that to calculate the frequencies and then the beat frequency. To answer the final question we need to consider the beats produced by these frequencies.

The fundamental frequency of a string is given by $f_{\text{fundamental}} = \dfrac{v}{2L}$,

where v is the sound velocity in the string and L is the length, while the first overtone is given by $f_{\text{first overtone}} = \dfrac{v}{L}$.

For this string these frequencies are 927.3 Hz and 1.855×10^3 Hz.

The beat frequency is given by the difference in these two frequencies, or by

$$f_{beat} = |f_{fundamental} - f_{first\ overtone}| = 927.7\ \text{Hz} = f_{fundamental},$$

where the last equality follows from the two equations without using the given numerical values. That is, we are caught in significant figures. Since we are subtracting two numbers we lose accuracy and must consider 927.3 and 927.7 the same number.

But this means there is no beat effect since the beat frequency is one of the frequencies with which we began. If we have only one string with the fundamental and any number of overtones excited, no beats will be heard. Beats only are heard when the difference in frequencies is not one of the frequencies present.

COMMENTS ON MODELS

Up to this point we have considered the properties of waves in several media — air, water, springs. From these considerations we can extract the abstract idea of a wave. We will return to this abstract idea of a wave in later chapters when we consider electromagnetic waves and quantum matter waves.

As is stated in the text, there are two types of waves: transverse waves and longitudinal waves. Ripples on the surface of a pond are a good example of a transverse wave, while a low intensity sound wave provides an example of a longitudinal wave. In both of these examples we restrict our consideration to small disturbances. That is, if you splash the pond hard enough, water will travel horizontally as well as vertically. If you sneeze, there is a net motion of air towards the listener as well as a compression and rarefaction.

In a transverse wave, the material that is carrying the wave moves perpendicular to the direction the wave travels. The surface of a pond moves up and down as a ripple travels across the pond in a horizontal direction. In a sound wave, on the other hand, the air moves back and forth in the same direction as the sound propagates.

SUGGESTED READINGS

Fletcher, N. H., and S. Thwaites, "The Physics of Organ Pipes." *Scientific American,* January 1983, p. 94.

Jeffries, A. D., P. R. Saulson, R. E. Spero, and M. E. Zuker, "Gravitational Wave Observatories." *Scientific American,* June 1987, p. 50.

Kottick, E. L., K. D. Marshall, and T. J. Hendrickson, "The Acoustics of the Harpsichord." *Scientific American*, February 1991, p. 110.

Luzader, S., "PV and PVC Acoustics." *The Physics Teacher,* October 1990, p. 459.

Rossing, T. D., *The Science of Sound.* Reading, Mass.: Addison-Wesley, 1982. This book covers a broad range of topics from musical instruments to high-fidelity electronics to the acoustical properties of auditoriums. It contains much material not previously available in one book. The mathematical level of Rossing's book is about the same as in this text.

Chapter 16
Electric Charge and Electric Field

IMPORTANT TERMS

Fill in the blanks with the appropriate word or words.

1. The negative movable charges often found in solids are called _____.

2. The principle that allows us to treat the total electrical force on an object as the sum of the individual forces is called the law of _____.

3. The nuclei of atoms have a positive _____ charge.

4. The law giving the force exerted by one point charge on another is called _____ law.

5. The _____ is the electric force per unit charge felt by a small test charge.

6. Two equal and opposite charges separated by a small distance are called a(n) _____ .

7. In Coulomb's law $\left(\dfrac{1}{4\pi\varepsilon_0} \dfrac{q_1 q_2}{r^2} \right)$, ε_0 is called the _____ constant.

8. A(n) _____ is a material through which charge can easily flow.

9. The _____ through a small element of surface is the product of the magnitude of the electric field and the projection of the surface perpendicular to the field.

10. Electrons have a negative electric _____.

11. The strength of an electric dipole can be given by either the magnitude of one of the dipole charges involved and the separation of the charges or by the product of these, which is called the dipole _____.

12. A(n) _____ does not permit charges to move freely.

13. The technique of holding a metal object with an insulating handle, bringing it near a charged object and then touching the metal with a finger, removing the finger and then removing the globe is the process of charging _____ .

14. _____ law is that the net electric flux through any closed surface is directly proportional to the net electric charge enclosed within that surface.

15. Single charges can neither be created nor destroyed is a statement of the _____ .

16. The SI unit of electric charge is the _____.

17. A(n) _____ surface is an imaginary surface used for Gauss's law.

18. The study of electrical charges at rest is called _____.

Write the definitions of the following:

19. dipole moment (16.8):

20. permittivity constant (and value) (16.2):

21. charging by induction (16.1):

22. conductor (16.1):

23. insulator (16.1):

24. electron (16.2):

25. coulomb (16.4):

26. electric charge (16.2):

27. Gaussian surface (16.6):

28. Gauss's law (16.6):

29. conservation of charge (16.1):

30. Coulomb's law (16.2):

31. electrostatics (16.0):

32. electric field (16.4):

33. electric flux (16.6):

34. electric dipole (16.8):

ANSWERS TO 1–18, IMPORTANT TERMS

1. electrons; 2. superposition; 3. electric; 4. Coulomb's; 5. electric field; 6. electric dipole; 7. permittivity; 8. conductor; 9. electric flux; 10. charge; 11. moment; 12. insulator; 13. by induction; 14. Gauss's; 15. conservation of charge; 16. coulomb; 17. Gaussian; 18. electrostatics.

SAMPLE QUESTIONS

QUESTION 16.1: It is common to observe a static spark when you touch the door knob after shuffling across a rug in the winter. Why is it uncommon to get a similar spark in the summer?

Answer:

In the winter the humidity inside a house is often very low. The ability of air to hold water increases as the temperature of the air increases. As a furnace of a house heats the air, the relative humidity decreases. In summer, just the opposite occurs.

QUESTION 16.2: Discuss the similarities and differences between the gravitational force between masses and the electric force between charges.

Answer:

Both the gravitational and the electric forces are $1/r^2$ forces. The gravitational force is much weaker than the electric force. There are two kinds of electric charges (positive and negative) but apparently only positive masses.

SAMPLE SOLUTIONS

PROBLEM 16.1: Two charges exert an attractive force of 6.2×10^{-4} N on each other. The larger charge is twice the magnitude of the smaller. The smaller charge is 2.0×10^{-6} C. What is the larger charge and the separation of the two charges? (16.2)

Solution:

> **Strategy:** Use Coulomb's law to calculate the forces and solve for the unknowns.

Since the forces are attractive the charges must have opposite sign. If we let q represent the 2.0×10^{-6} C charge, then the other charge must be $-2q$ or -4.0×10^{-6} C. We let d be the separation and F be the attractive force of 6.2×10^{-4} N as shown in Fig. 16.1.

Fig. 16.1: Charges q and $-2q$ separated by a distance d.

Write Coulomb's law as

$$F = \frac{1}{4\pi\varepsilon_0} \frac{(-2q)(q)}{d^2} = \frac{1}{4\pi\varepsilon_0} \frac{2q^2}{d^2}.$$

We know all the values except for d.

GIVEN:

Quantity	Symbol	Value
charge of smaller	q	2.0×10^{-6} C
attractive force	F	6.2×10^{-4} N
separation	d	?
constant	$\dfrac{1}{4\pi\varepsilon_0}$	9.0×10^9 N·m²/C²

So
$$F = \frac{1}{4\pi\varepsilon_o} \frac{2q^2}{d^2} = \frac{9.0 \times 10^9 \text{ N·m}^2/\text{C}^2 \times 2 \times (2.0 \times 10^{-6} \text{ C})^2}{d^2} = 6.2 \times 10^{-4} \text{ N}.$$
Thus
$$d^2 = \frac{9.0 \times 10^9 \text{ N·m}^2/\text{C}^2 \times 2 \times (2.0 \times 10^{-6} \text{ C})^2}{6.2 \times 10^{-4} \text{ N}} = 11.6 \text{ m}^2.$$

So the other charge is -4.0×10^{-6} C and the two charges are separated by 3.4 meters.

PROBLEM 16.2: Two charges of charge $+3.5 \times 10^{-8}$ C and mass 0.10 g are held in position at a separation of 20 cm. Midway between them a charge of -2.0×10^{-8} C and mass 0.20 g is also held. What are the forces needed on each charge to maintain the position of the charges? Assume the experiment was done in a space station so that no gravitational forces are involved. If one of the two outside charges is released, what is its initial acceleration? (16.3)

Solution:

Strategy: An equilibrium problem. Assume the sum of the mechanical plus electrical forces on each particle is zero. Use Coulomb's law to determine the electric forces.

We first need to find the electrostatic forces on each charge. Let us call the left hand object "1", the middle object "2", and the right hand object "3" as shown in Fig. 16.2a. Since the middle object is equidistant for both end bodies, we let this separation be $d = 10$ cm. In our figure distance is measured positively to the right.

Electrical forces will tend to separate or attract the bodies. Therefore, there must be other forces needed to keep the separations constant. We term these other forces mechanical forces and denote them by F_M, while the electrical forces are denoted by F_E.

GIVEN:

Quantity	Symbol	Value
mass of object "1"	m_1	0.10 g
mass of object "3"	$m_3 = m_1$	0.10 g $= 1.0 \times 10^{-4}$ kg
mass of object "2"	m_2	0.20 g $= 2.0 \times 10^{-4}$ kg
separation 1-2, 2-3	d	10 cm $= 1.0 \times 10^{-1}$ m
charge of object "1"	q_1	3.5×10^{-8} C
charge of object "2"	q_2	-2.0×10^{-8} C
charge of object "3"	$q_3 = q_1$	3.5×10^{-8} C
electrical force on "1"	F_{E1}	unknown
electrical force on "2"	F_{E2}	unknown
electrical force on "3"	F_{E3}	unknown
mechanical force on "1"	F_{M1}	unknown
mechanical force on "2"	F_{M2}	unknown
mechanical force on "3"	F_{M3}	unknown
constant in Coulomb's law	k	9.0×10^9 N·m^2/C^2

We will have to be careful with sign convention when we apply Coulomb's law. A positive sign means a repulsive force and a negative sign is an attractive force. Consider charges "1" and "2" only. Since these charges are of the opposite sign, the force between them is attractive. This means that the force on charge "1" due to charge "2" is directed to the right and is hence along the +x direction and therefore a positive force in the vector sense. On the other hand, the force on charge "2" due to charge "1" is directed in the –x direction and is therefore a negative force in the vector sense.

Fig. 16.2: (a) The three bodies and the total electric force on each shown. The electric forces were arbitrarily shown to the right. (b) The electric forces on "1" and "2" due to each other. (c) The electric forces on "1" and "3" due to each other. (d) The total electric forces on "1". (e) The total electric forces on "2". (f) The total forces needed to keep the system in equilibrium. Electric forces are now shown in their proper orientation.

There are three pairs of forces needed:

$$\text{The force between "1" and "2"} = F_{1\text{-}2} = k\frac{q_1 q_2}{d^2} = -6.3 \times 10^{-4} \text{ N.}$$

Since this force has a negative sign, it is an attractive force. Since particle "1" is to the left of particle "2", the force on "1" is to the right and the force on two is towards the left as shown in Fig. 16.2b.

$$\text{The force between "1" and "3"} = F_{1\text{-}3} = k\frac{q_1 q_3}{(2d)^2} = 2.8 \times 10^{-4}\ \text{N}.$$

Since the sign of this force is positive, it is a repulsive force. Hence the force on "1" is to the left and the force on "2" is towards the right as shown in Fig. 16.2c.

We can now vectorially add the forces on "1" to get

$$F_{E1} = 6.3 \times 10^{-4}\ \text{N} - 2.8 \times 10^{-4}\ \text{N} = 3.5 \times 10^{-4}\ \text{N}.$$

This force is positive since it is directed towards the right (in a positive x direction).

A repeat of our analysis for "3" must, by symmetry, give us the same numerical value but the opposite direction, so, as shown in Fig. 16.2d,

$$F_{E3} = -6.3 \times 10^{-4}\ \text{N} + 2.8 \times 10^{-4}\ \text{N} = -3.5 \times 10^{-4}\ \text{N}.$$

Finally $F_{2\text{-}1}$ has the same magnitude as $F_{2\text{-}3}$ but is in the opposite direction as shown in Fig. 16.2e:

$$F_{E2} = -6.3 \times 10^{-4}\ \text{N} + 6.3 \times 10^{-4}\ \text{N} = 0.$$

These three forces are shown in Fig. 16.2f, pointing in the proper directions.
The mechanical forces needed to maintain the configuration are also shown in Fig. 16.2f. They have the same magnitude but are in the opposite direction to the electric forces.

$$F_{M1} = -F_{E1}; \qquad F_{M2} = -F_{E2} = 0; \qquad F_{M3} = -F_{E3}.$$

If the left outside charge were released, the initial acceleration would be given by

$$\text{acceleration} = \frac{\text{Force}}{\text{mass}} = \frac{F_{E1}}{m_1} = 3.5\ \text{m/s}^2.$$

The acceleration is positive, indicating that particle "1" is accelerated in the positive x direction.

PROBLEM 16.3: A negative charge of 2.7×10^{-7} C is held in a uniform electric field of 1.0×10^3 V/m. Another negative charge of 7.2×10^{-7} C is brought to a distance d from the first charge. What is the value of d such that that the force of either charge on the other is equal in magnitude to the force on the first charge due to the uniform field? (16.5)

Solution:

Strategy: Use Coulomb's law to calculate the electric field due to the second negative charge and use the definition of the force due to an electric field.

For the purposes of a drawing, we imagine the electric field is directed to the left as shown in Fig. 16.3. We will let q_1 be the first charge and q_2 be the second. To stress the nature of the question, the two particles have not been placed along the direction of the field.

$$E = 10^3 \text{ V/m}$$

Fig. 16.3: Two charges a distance d apart in a uniform electric field E.

GIVEN:

Quantity	Symbol	Value
uniform electric field	E	1.0×10^3 V/m
first charge	q_1	2.7×10^{-7} C
second charge	q_2	-7.2×10^{-7} C
separation of charges	d	unknown
constant in Coulomb's law	k	9.0×10^9 N·m²/C²

We are asked to choose d so that the magnitude of the force on "1" due to the field is the same as the magnitude of the force on "1" due to "2".

The force on "1" due to the field is

$$F_E = q_1 E = -2.7 \times 10^{-7} \text{ C} \times 1.0 \times 10^3 \text{ V/m} = -2.7 \times 10^{-4} \text{ N},$$

where the minus sign indicates the force is in a direction opposite to E (i.e., to the right).

The force on "1" due to "2" is

$$F_C = k \frac{q_1 q_2}{d^2} = \frac{1.7 \times 10^{-3} \text{ N·m}^2}{d^2},$$

where the positive sign indicates that the force is repulsive.

Equating the magnitudes of these two forces gives

$$2.7 \times 10^{-4} \text{ N} = \frac{1.7 \times 10^{-3} \text{ N·m}^2}{d^2},$$

or

$$d = \sqrt{6.3 \text{ m}^2} = 2.5 \text{ m}.$$

The second charge must be placed 2.5 m from the first charge.

PROBLEM 16.4: Gauss's law I. A spherical shell has an inner radius of 0.25 m and an outer radius of 0.30 m. If a total charge of 2.0×10^{-3} C is placed on the shell, give the electric field as a function of distance from the center of the shell if the shell is a perfect conductor. (16.7)

Solution:

> **Strategy:** Since we have an extended charge distribution with high symmetry, we expect this problem is most easily solved by use of Gauss's law. The electric field crossing a sphere centered on the center of symmetry times the area of that sphere must be the total charge contained within the sphere divided by ε_0. Choose the spherical surface to be inside the shell, within the shell and outside the shell.

Since this is a problem with spherical symmetry, we expect the electric field will depend only upon the distance from the center of the sphere and that any electric field present must be radially directed. Since we are not dealing with point charges, the only tool that seems reasonable to use is Gauss's law (text section 15.7). This law states that the flux through a surface is proportional to the net charge, Q, contained within the surface, i.e.,

$$\Phi_E = \frac{Q}{\varepsilon_0}.$$

The simplest choice to use for the Gaussian surface is to take a sphere of radius r, centered on the same center used for the spherical shell. Other choices can be made, but the mathematics are above the level of this course, and the results are the same.

With this choice we have

$$\Phi_E(r) = E(r) \times (\text{area of sphere of radius } r) = E(r) \times 4\pi r^2.$$

Gauss's law becomes

$$E(r) = \frac{Q}{4\pi \varepsilon_0 r^2},$$

where Q is the net charge contained within r, and

$$\frac{1}{4\pi \varepsilon_0} = 9.0 \times 10^9 \text{ N·m}^2/\text{C}^2.$$

For a perfect conductor, the entire charge must lie on the outer surface of the shell (see text section 16.6), so

$$E(r) = \begin{cases} 0 & r < 0.30 \text{ m} \\ \dfrac{1.8 \times 10^7 \text{ N·m}^2/\text{C}}{r^2} & r \geq 0.30 \text{ m} \end{cases}.$$

That is, the electric field is zero within the outer surface of the shell and falls off as the square of the distance from the center for distances outside the shell. This behavior is shown in Fig. 16.4c.

Fig. 16.4: (a) Charge on the surface of a conducting spherical shell. (b) Charge distributed uniformly in a spherical shell. (c) The electric field for Problem 16.4. (d) The electric field for Problem 16.5.

PROBLEM 16.5: Gauss's law II. A spherical shell has an inner radius of 0.25 m and an outer radius of 0.30 m. If a total charge of 2.0×10^{-3} C is placed on the shell, give the electric field as a function of distance from the center of the shell if the shell is an insulator and the charge is distributed uniformly throughout the shell. (16.7)

Solution:

Strategy: Since we have an extended charge distribution with high symmetry, we expect this problem is most easily solved by use of Gauss's law. The electric field crossing a sphere centered on the center of symmetry times the area of that sphere must be the total charge contained within the sphere divided by ε_0. Choose the spherical surface to be inside the shell, within the shell and outside the shell.

Since this is a problem with spherical symmetry, we expect the electric field will depend only upon the the distance from the center of the sphere and that any electric field present must be radially directed. Since we are not dealing with point charges, the only tool that seems reasonable to use is Gauss's law (text section 15.7). This law states that the flux through a surface is proportional to the net charge, Q, contained within the surface, i.e.,

$$\Phi_E = \frac{Q}{\varepsilon_0}.$$

The simplest choice to use for the Gaussian surface is to take a sphere of radius r, centered on the same center used for the spherical shell. Other choices can be made, but the mathematics are above the level of this course, and the results are the same.

With this choice we have

$$\Phi_E(r) = E(r) \times (\text{area of sphere of radius r}) = E(r) \times 4\pi \, r^2,$$

so Gauss's law becomes

$$E(r) = \frac{Q}{4\pi \, \varepsilon_0 \, r^2},$$

where Q is the net charge contained within r, and

$$\frac{1}{4\pi \, \varepsilon_0} = 9.0 \times 10^9 \text{ N·m}^2/\text{C}^2.$$

For this problem the total charge contained inside r is zero for $r < 0.25$ m and equal to 2.0×10^{-3} C for $r \geq 0.30$ m.

The charge for 0.25 m $\leq r \leq 0.30$ m will vary with r.

For $r < 0.25$ m

$$E(r) = \frac{Q}{4\pi \, \varepsilon_0 \, r^2} = \frac{0.0 \text{ C}}{4\pi \, \varepsilon_0 \, r^2} = 0,$$

while for $r > 0.30$ m

$$E(r) = \frac{Q}{4\pi \, \varepsilon_0 \, r^2} = \frac{2.0 \times 10^{-3} \text{ C}}{4\pi \, \varepsilon_0 \, r^2} = \frac{1.8 \times 10^7 \text{ N·m}^2/\text{C}}{r^2}.$$

We now examine the behavior for 0.25 m $< r < 0.30$ m. What we need is the total amount of charge contained within r. Since the charge is distributed uniformly within the shell, the total amount of charge within r will be proportional to the volume of the shell within r. The total volume between 0.25 m $= r_{in}$ and 0.30 m $= r_{out}$ can be calculated by subtracting the volume of a sphere of radius r_{in} from the volume of a sphere of radius r_{out}, giving:

$$\text{total volume} = \frac{4\pi}{3} \, r_{out}^3 - \frac{4\pi}{3} \, r_{in}^3 = \frac{4\pi}{3} \, (r_{out}^3 - r_{in}^3),$$

while the volume between 0.25 m $= r_{in}$ and some $r \leq r_{out}$ is

$$\text{volume at r} = \frac{4\pi}{3} r^3 - \frac{4\pi}{3} r_{in}^3 = \frac{4\pi}{3} (r^3 - r_{in}^3).$$

We can find the charge inside r by a proportion:

$$\frac{\text{charge within } r}{\text{total charge}} = \frac{\text{volume (inside } r) \text{ containing charge}}{\text{total volume containing charge}}$$

or

$$\frac{Q(r)}{Q} = \frac{Q(r)}{2.0 \times 10^{-3} \text{ C}} = \frac{\text{volume at } r}{\text{total volume}} = \frac{\frac{4\pi}{3}(r^3 - r_{in}^3)}{\frac{4\pi}{3}(r_{out}^3 - r_{in}^3)} = \frac{r^3 - r_{in}^3}{r_{out}^3 - r_{in}^3}.$$

This gives for $0.25 \text{ m} \le r \le 0.30 \text{ m}$

$$Q(r) = 2.0 \times 10^{-3} \text{ C } \frac{r^3 - (0.25 \text{ m})^3}{(0.30 \text{ m})^3 - (0.25 \text{ m})^3} = 0.18 \frac{\text{C}}{\text{m}^3} \times [(0.30 \text{ m})^3 - (0.25 \text{ m})^3].$$

Hence for all r

$$E(r) = \begin{cases} 0 & r < 0.25 \text{ m} \\ 1.6 \times 10^9 \frac{\text{N}}{\text{C·m}}\left(r - \frac{(0.25 \text{ m})^3}{r^2}\right) & 0.25 \text{ m} \le r \le 0.30 \text{ m} \\ \frac{1.8 \times 10^7 \text{ N·m}^2/\text{C}}{r^2} & r > 0.30 \text{ m} \end{cases}$$

This field is shown in Fig. 16.4d. At the boundaries you must change the formulas used to calculate $E(r)$. You can easily verify that each formula gives the same value at the boundaries. Note that the electric field, if it is not trivially zero, must depend upon the distance r from the center of symmetry.

PROBLEM 16.6: Gauss's law III. A point charge of 2.0×10^{-8} C is located at the center of an uncharged conducting spherical shell of inner radius 0.15 m and outer radius 0.20 m (Fig. 16.5a). What is the electric field inside the shell, within the surfaces of the shell, and outside the shell? What charges are induced on the surfaces of the shell? (16.7)

Solution:

Strategy: Since we have an extended charge distribution with high symmetry, we expect this problem is most easily solved by use of Gauss's law. The electric field crossing a sphere centered on the center of symmetry times the area of that sphere must be the total charge contained within the sphere divided by ε_0. Choose the spherical surface to be inside the shell, within the shell and outside the shell.

Reread the first parts of Problems 16.4 and 16.5. For spherical symmetry the electric field, $E(r)$, is given by

$$E(r) = \frac{Q}{4\pi \varepsilon_0 r^2},$$

where Q is the net charge contained within r, and

$$\frac{1}{4\pi \varepsilon_0} = 9.0 \times 10^9 \text{ N·m}^2/\text{C}^2.$$

There are once again three regions to analyze: $0 < r < 0.15$ m, 0.15 m $< r < 0.20$ m, and $r > 0.20$ m. We also know that the presence of the point charge may cause charge to be induced on the inner and outer surface of the conducting shell. We will denote these charges by Q_{in} and Q_{out} respectively; and note that $Q_{in} + Q_{out} = 0$ since there in no net charge on the sphere. We can now make a table of $Q(r)$ for the various regions:

Radius	Total charge contained
$0 < r < 0.15$ m	2.0×10^{-8} C
0.15 m $< r < 0.20$ m	2.0×10^{-8} C $+ Q_{in}$
0.20 m $< r$	2.0×10^{-8} C $+ Q_{in} + Q_{out}$

Point Charge of 2.0×10^{-8} C

Conducting shell
Inner radius 0.15 m
Outer radius 0.20 m

(a)

$E(r)$

0   r

(b)

Fig. 16.5:   (a) A point charge is located inside a hollow spherical conducting sphere.
(b) The resulting electric field.

Since $Q_{in} + Q_{out} = 0$, we know $Q(r)$ for $r > 0.20$ m, and hence the electric fields for $r < 0.15$ m and for $r > 0.20$ m are known and are the same form in each region since Q is constant in each region. Thus

$$E(r) = \frac{Q}{4\pi \, \varepsilon_0 \, r^2} = \frac{1.8 \times 10^2 \text{ N·m}^2}{r^2} \quad \begin{cases} r < 0.15 \text{ m} \\ r > 0.20 \text{ m} \end{cases}$$

It remains then to determine Q_{in} and the electric field for 0.15 m $< r < 0.20$ m. But this range of r is inside the conductor. We know that no electric field exists inside a conductor so $E(r)$ for 0.15 m $< r < 0.20$ m must be zero; and hence the induced charge $Q_{in} = -2.0 \times 10^{-8}$ C on the inner surface and $Q_{out} = 2.0 \times 10^{-8}$ C on the outer surface:

$$E(r) = 0 \quad 0.15 \text{ m} < r < 0.20 \text{ m}.$$

Note that in this problem the electric field is not continuous across the surfaces (Fig. 16.5b). This should be expected whenever there is a surface charge distribution present.

PROBLEM 16.7: A molecule of HCl, which has an electric dipole moment $p = 3.4 \times 10^{-33}$ C·m and moment of inertia about its center of mass of $I = 2.65 \times 10^{-47}$ kg·m^2, is initially at rest in a uniform electric field, E, of 1.0×10^3 V/m and oriented at an angle, θ, of 30° to the field direction. What is the initial acceleration that the molecule undergoes? (16.8)

Solution:

Strategy: Apply the electric torque law to get the torque. Then use Newton's law relating applied torque and resulting angular acceleration.

We can visualize the HCl molecule as an ideal dipole of strength p located in the field as shown in Fig. 16.6. The torque on the dipole is given by

$$\tau = pE \sin(\theta),$$

where θ is the angle between P and E as shown in Fig. 16.6.

Fig. 16.6:   HCl molecule in electric field.

The law relating angular acceleration and torque was given in Chapter 9 of the text as

$$\tau = I\,\alpha,$$

where I is the moment of inertia and α is the angular acceleration. Thus

$$\alpha = \frac{pE \sin(\theta)}{I}.$$

GIVEN:

Quantity	Symbol	Value
dipole moment	p	3.4×10^{-33} C·m
moment of inertia	I	2.65×10^{-47} kg·m^2
electric field strength	E	1.0×10^3 V/m
angle between p and E	θ	30°
angular acceleration	α	unknown

So the angular acceleration will be 6.4×10^{16} rad/s^2 in a direction tending to reduce θ (counterclockwise in the figure).

A WORD ON MODELS

We have no pure charges. That is, all charged bodies have mass; however, not all bodies with mass have net charge. So we only need add a new property, the charge, to our idealized particles. We idealize an electron to have a mass ($m_e = 9.11 \times 10^{-31}$ kg) and a charge ($e = -1.60 \times 10^{-19}$ C) among other properties, but not to have extension. We will later see that in everyday experience all charges come in multiples of the electron charge. (The exception is quarks in nuclei that may have $-e/3$ or $+2e/3$ as their charge.) The charge on an electron is very small so we can assume that we have test particles with as small a charge as desired. These test particles can be used to investigate the electric fields produced by other charges. In general a test charge is similar to the test particle used for the gravitational field. That is, a test charge is assumed to have a small enough charge that it does not disturb the positions of the charges we are investigating. Likewise a point particle is one whose size is very small compared with other lengths in the problem under investigation.

All of our models in this chapter do not explain why the negative charges are usually the movable ones. For now we still assume a solid has no structure other than its shape, mass and electrical conductivity. Solids will be taken as either perfect metals or perfect insulators. Real solids have more complicated properties.

SIGN CONVENTION

Coulomb's law has the sign convention that an attractive force is negative and a repulsive force is positive. This tells you the direction of the force with respect to the line joining the two charges. This does not tell you the direction of the force with respect to an external coordinate system. You will need to draw a picture of the physical situation to determine whether the forces point in a positive or a negative vector sense. Sample Solution 16.2 gives you a good example of this problem and how it should be handled.

SUGGESTED READINGS

Few, A. A., "Thunder." *Scientific American,* July 1975, p. 80.

Frank, N. H., "Faraday and the Field Concepts in Physics." *The Physics Teacher,* May 1968, p. 210.

Golembeski, D. J., "Struggling to Become an Inventor." *American Heritage of Invention & Technology,* Winter 1989, p. 8. The invention of xerography.

Heilbron, J. L., "Franklin's Physics." *Physics Today,* July 1976, p. 32.

Lewis, H.W., "Ball Lightning." *Scientific American,* March 1963, p. 106.

Singer, S., *The Nature of Ball Lightning.* New York: Plenum, 1971.

Steyn-Ross, A., and A. Riddell, "Standing Waves in a Microwave Oven." *The Physics Teacher,* October 1990, p. 474.

Chapter 17
Electric Potential and Capacitance

IMPORTANT TERMS

Fill in the blanks with the appropriate word or words.

1. The _____ of a material is the ratio of the capacitance of a capacitor with this material between the plates to the capacitance of the same capacitor with a vacuum between the plates.

2. The ratio of the change in electric potential to the distance we move is called the _____ in that direction.

3. The maximum potential per unit thickness that a material can withstand without breakdown is called the _____ of the material.

4. The _____ between two points is defined as the ratio of the electric potential energy change that occurs when a test charge is moved between the two points to the value of the charge.

5. The _____ of a capacitor is the ratio of the charge stored to the potential difference.

6. _____ accelerators are machines used to accelerate charged particles to very high kinetic energies.

7. The _____ is the SI unit of electric potential.

8. Materials that do not permit the conduction of electric charge are called insulators or _____.

9. A surface of constant potential is called a(n) _____.

10. A(n) _____ is a device to store charge.

11. The unit of capacitance, defined as 1 coulomb per volt, is the _____.

12. _____ energy is the ability to do work stored in the electric field.

Write the definitions of the following:

13. electric potential (17.1):

14. equipotential surface (17.3):

15. volt (17.1):

16. dielectric strength (17.8):

17. Van de Graaff accelerator (17.2):

18. dielectric (17.8):

19. potential gradient (17.4):

20. capacitor (17.5):
21. capacitance (17.5):
22. farad (17.5):
23. electric potential difference (17.1):
24. dielectric constant (17.8):

ANSWERS TO 1–12, IMPORTANT TERMS

1. dielectric constant; 2. potential gradient; 3. dielectric strength; 4. electric potential;
5. capacitance; 6. Van de Graaff; 7. coulomb; 8. dielectrics; 9. equipotential surface;
10. capacitor; 11. farad; 12. electric potential.

MOTION IN THE ELECTRIC FIELD

The electric field is defined to go from high potential to low potential. This direction is defined by the way a positive charge will move if it is released in the field. This charge is a test charge — that is, a test charge has a charge sufficiently small that we can neglect the effects of the test charge on the electric fields present. Remember that positively charged particles feel a force in the same direction that the field lines point, while negatively charged particles feel a force in the opposite direction.

SAMPLE QUESTIONS

QUESTION 17.1: Why is it that the passengers in an automobile are not electrocuted if the automobile is struck by lightning?

Answer:

To be electrocuted a current must pass through your body. The car body provides a cage to conduct the current. This current flows over the car body and to the ground over the wet tires.

QUESTION 17.2: Work is required to pull apart two charged capacitor plates. What happens to the energy expended in pulling them apart?

Answer:

The energy required to separate the two plates is stored in the electric potential energy field. If the plates are released, they will accelerate towards each other. This potential energy will then be converted to kinetic energy.

SAMPLE SOLUTIONS

PROBLEM 17.1: Two charges of charge $+3.5 \times 10^{-8}$ C and mass 0.10 g are held in position at a separation of 20 cm. Midway between them a charge of -2.0×10^{-8} C and mass 0.20 g is also held. What is the electric field seen by each charge? What is the electric potential at a point midway between the first and second charge? (17.1)

Solution:

Strategy: Use the definition of electric field and vectorially sum the individual fields. (See Fig. 17.2 of the text for the definitions of the terms used.)

This is Problem 16.2 of the last chapter, so we can use the the figure for and the results of that problem. In terms of Fig. 16.2, the force on the left hand charge is

$$F_{E1} = 3.5 \times 10^{-4} \text{ N},$$

the force on the right hand charge is

$$F_{E3} = -3.5 \times 10^{-4} \text{ N},$$

and the central charge feels a force

$$F_{E2} = 0 \text{ N}.$$

The sign convention is positive towards the right.

The electric field is defined as
$$E = \frac{F}{q},$$

where the sign of the charge, q, must be included. That is, if the charge is negative the electric field will point in the opposite direction to the force.

So the electric fields are

$$E_{E1} = \frac{F_{E1}}{q_1} = 1.0 \times 10^5 \text{ N/C} = 1.0 \times 10^5 \text{ V/m},$$

$$E_{E2} = 0,$$

and

$$E_{E3} = -1.0 \times 10^5 \text{ V/m}.$$

The potential at any point along the line joining the charges is given by

$$V = \frac{q_1}{4\pi\varepsilon_0 \, r_1} + \frac{q_2}{4\pi\varepsilon_0 \, r_2} + \frac{q_3}{4\pi\varepsilon_0 \, r_3},$$

where r_i is the distance from the point in question to the ith charge as is shown in Fig. 17.1. That is, r_1 is the distance between the chosen point and the charge named "1", r_2 is the distance between the chosen point and the charge marked "2", and r_3 is the distance between the chosen point and the charge named "3".

This expression for force is correct only when all the charges are in a line and then only for points along that line.

Fig. 17.1: The field point located by an "×" relative to the three charges. This figure is drawn for a field point to the right of all three charges. The question has the field point midway between charges "1" and "2".

GIVEN:

Quantity	Symbol	Value
charge of object "1"	q_1	3.5×10^{-8} C
charge of object "2"	q_2	-2.0×10^{-8} C
charge of object "3"	$q_3 = q_1$	3.5×10^{-8} C
distance from q_1	r_1	5.0 cm = 0.050 m
distance from q_2	r_2	5.0 cm = 0.050 m
distance from q_3	r_3	15 cm = 0.15 m
constant in Coulomb's law	$1/(4\pi\varepsilon_0)$	9.0×10^9 N·m²/C²

So, at the point in question,

$$V = \frac{3.5 \times 10^{-8} \text{ C} \times 9.0 \times 10^9 \text{ N·m}^2/\text{C}^2}{0.050 \text{ m}}$$

$$+ \frac{-2.0 \times 10^{-8} \text{ C} \times 9.0 \times 10^9 \text{ N·m}^2/\text{C}^2}{0.050 \text{ m}}$$

$$+ \frac{3.5 \times 10^{-8} \text{ C} \times 9.0 \times 10^9 \text{ N·m}^2/\text{C}^2}{0.150 \text{ m}} \quad ,$$

giving,

$$V = 4.8 \times 10^3 \text{ N·m/C} = 4.8 \times 10^3 \text{ V}.$$

So midway between the two charges the potential is about 4.8×10^3 volts with respect to the potential at a large distance from the charges.

PROBLEM 17.2 Estimate the energy in electron volts necessary to completely separate an electron and a proton bound in a hydrogen atom. Take the separation of the proton and electron in a hydrogen atom as 1.0×10^{-8} cm. (17.3)

Solution:

> **Strategy:** Use conservation of energy: the work done is equal to the change in potential energy.

The work done, W, is given by

$$W = -e\,[V(\infty) - V(r_0)],$$

where e is the magnitude of the electronic charge, $V(\infty) = 0$ is the potential energy when the charges are at a very large separation, and $V(r_0)$ is the potential when the proton and electron are at a separation $r_0 = 1.0 \times 10^{-8}$ cm.

Thus $\quad W = \dfrac{e^2}{4\pi\varepsilon_0 r_0} = \dfrac{(1.6 \times 10^{-19}\text{ C})^2}{4\pi(8.85 \times 10^{-12}\text{ C}^2\text{N}^{-1}\cdot\text{m}^{-2})(10^{-10}\text{ m})} = 2.3 \times 10^{-18}\text{ J.}$

We convert this to electron volts by dividing by the value of an electron volt in joules, so the work done is

$$\frac{2.3 \times 10^{-18}\text{ J}}{1.602 \times 10^{-19}\text{ J}} = 14\text{ eV.}$$

That is, an electron accelerated through a potential of 14 V will have the same energy as the energy needed to completely separate (ionize) the electron and proton in a hydrogen atom.

Problem 17.3: A conducting sphere of radius $r = 10$ cm has a total charge of -1.0×10^{-7} C on its surface. How much work is needed to move a charge of -7.0×10^{-18} C in a complete circle of radius 30 cm centered on the center of the conducting sphere? Neglect any distortion of the charge on the conducting sphere by the external charge. (17.4)

Solution:

> **Strategy:** Both the surface of the conducting sphere and the path of the charge are equipotential surfaces (lines). Consider the work done along an equipotential surface.

Since the charge is moving on an equipotential surface, there would be no work done to move the charge.

PROBLEM 17.4: An air gap parallel plate capacitor is constructed of two plates of surface area 100 cm^2 and separation 0.0500 cm. The capacitor is oriented with the plates horizontal in the laboratory, which is on earth. The capacitor is charged to a potential of 700.0 V with the upper plate at the positive potential. A physically small charge of mass 0.300 g and charge -1.00×10^{-8} C is placed in the middle of the capacitor and midway between the plates. Describe the motion of the charge. Assume the presence of the charge between the plates does not affect the distribution of charges on the capacitor plates. (17.7)

Solution:

> **Strategy:** Calculate the electric field inside the capacitor and the resulting electric force on the charge. Vectorially add this force to the gravitational force and determine the resulting motion.

The problem is shown edge on in Fig. 17.2a. We show the upper plate being at a potential of +700 V with respect to the bottom plate and measure position positively upward from the bottom plate. The charge feels two forces, the electric force and the gravitational force, as shown in Fig. 17.2b.

To solve this problem we need to calculate the electric field between the plates and thus find the electric force on the charge. Since we are given the electric potential, we do not need to find the value of the capacitance; that is, the electric field depends only upon the potential difference and the separation of the plates.

GIVEN:

Quantity	Symbol	Value
potential difference	V	700 V
plate separation	d	5.00×10^{-4} m
charge of object	q	-1.00×10^{-8} C
mass of object	M	3.00×10^{-4} kg
acceleration of gravity	g	-9.80 m/s^2
electric field	$E = -\Delta V/\Delta x$	unknown
initial position of charge	x_0	2.5×10^{-4} m
position of charge	$x(t)$	unknown

Fig. 17.2: (a) The physical arrangement of a charge between the charged plates of a capacitor. (b) The forces acting on the charge.

We first find the electric field by the relation

$$E = -\frac{V}{d} = -\frac{700 \text{ V}}{5.00 \times 10^{-4} \text{ m}} = -1.40 \times 10^6 \text{ V/m}.$$

The minus sign tells us that the electric field points in the negative x-direction — i.e., downwards. If we let F_e be the electric force and F_g be the gravitational force, then

and
$$F_e = qE = -1.00 \times 10^{-8} \text{ C} \times (-1.40 \times 10^6 \text{ V/m}) = 1.40 \times 10^{-2} \text{ N},$$

$$F_g = Mg = 3.00 \times 10^{-4} \text{ kg} \times (-9.80 \text{ m/s}^2) = -2.94 \times 10^{-3} \text{ N}.$$

The acceleration, a is given by Newton's second law,

$$a = \frac{\text{sum of forces}}{M} = \frac{1.40 \times 10^{-2} \text{ N} + (-2.94 \times 10^{-3} \text{ N})}{3.00 \times 10^{-4} \text{ kg}}$$

$$= 36.9 \text{ N/kg} = 36.9 \text{ m/s}^2.$$

The force and hence the acceleration is positive (that is, upward).

The position is given by the relation
$$x(t) = x_0 + v_0 t + \frac{1}{2} a t^2,$$

where x_0 is the position at $t = 0$ and v_0 is the velocity at $t = 0$. We are measuring the position positive upwards from the bottom plate. The position is then

$$x(t) = 2.5 \times 10^{-4} \text{ m} + 18.5 \text{ m/s}^2 \times t^2.$$

This upward motion can continue until $x(t_f) = 5.00 \times 10^{-4}$ m, when the charge hits the plate. This defines $t_f = 3.67 \times 10^{-3}$ s.

The charge will move upward, accelerating at 18.5 m/s^2 until it hits the upper plate in 3.67 $\times 10^{-3}$ seconds. When the charge gets near the upper plate it will cause a redistribution of the charges on the plate, and hence the electric field will be slightly different than the field assumed for an ideal parallel plate capacitor, so our answer is not exact.

PROBLEM 17.5: Two large parallel plates in a vacuum chamber are separated by 1.00 mm. They are connected to a power supply of potential difference 2.5×10^3 V. A doubly charged oxygen ion (mass $= 2.8 \times 10^{-26}$ kg) starts from rest on the surface of one plate and is accelerated to the other. Neglecting gravity, what is the kinetic energy of the ion when it hits the other surface? Justify neglecting the force of gravity.

Solution:

Strategy: This problem can be solved in at least two ways. The easiest approach is to use conservation of energy, equating the initial potential energy to the final kinetic energy. The other approach is to use Newton's laws directly to find the acceleration and then use the kinematic laws to get the kinetic energy.

The problem is illustrated in Fig. 17.3. We assume the positive plate is at the bottom and that the ion is missing two electrons, so it is repelled by the lower plate and attracted to the upper plate.

Oxygen ion on lower plate

Fig. 17.3: Parallel plate capacitor with a positive ion located on the positive plate.

First Method

We will use conservation of energy:

(potential energy + kinetic energy)$_{initial}$ = (potential energy + kinetic energy)$_{final}$.

The initial kinetic energy is zero by assumption, while the change in the potential energy is given by the product of the charge of the particle with the potential difference. If the particle charge is $2e$ where e is the electron charge and the potential difference is V, then

$$(\text{kinetic energy})_{final} = 2e\,V$$
$$= 2 \times 1.6 \times 10^{-19}\,\text{C} \times 2.5 \times 10^{3}\,\text{V} = 8.0 \times 10^{-16}\,\text{J}.$$

The final kinetic energy is 8.0×10^{-16} J.

We can justify neglecting the gravitational force by calculating the change in gravitational potential energy for the mass, m, moving a height $d = 1.00$ mm. The gravitational potential energy change is given by

$$m\,g\,d = 2.8 \times 10^{-26}\,\text{kg} \times 9.8\,\text{m/s}^2 \times 10^{-3}\,\text{m} = 2.7 \times 10^{-28}\,\text{J}.$$

This is clearly very much less than the significant figures for the kinetic energy produced by the electrical force, so we are justified in neglecting the effects of gravity.

Second Method

Use the force laws to calculate the acceleration. Newton's third law gives

$$F = m\,a_y = q\,E_y$$

where a_y is the acceleration in the y-direction and E_y is the electric field. The electric field is given in terms of the potential difference by

$$E = \frac{V}{d}$$

so

$$a_y = \frac{2e}{m}\frac{V}{d}.$$

Since we don't know the time of travel between the plates, the obvious kinematic law to use is

$$v^2 = v_0^2 + 2\,a\,y.$$

Since the initial velocity is zero, this gives the velocity at the top plate as

$$v^2 = 2\,a\,d = \frac{4\,e}{m}\,V\,.$$

Hence the final kinetic energy is given by

$$(\text{kinetic energy})_{\text{final}} = \frac{1}{2}\,m\,v^2 = \frac{1}{2}\,m\,\frac{4\,e}{m}\,V = 2\,e\,V\,,$$

which is the same result as before.

PROBLEM 17.6: A pair of parallel plates is separated by 1.672 mm of Mylar. There is a charge of 2.52×10^{-8} C/m^2 on each plate (positive on the upper in Fig. 17.4 and negative on the lower). What is the electric field between the plates, neglecting edge effects? (17.8)

Solution:

> **Strategy:** Use the definition of capacitance for a parallel plate capacitor with a dielectric and the fundamental definition of capacitance. The problem is illustrated in Fig. 17.2. We are given the separation of the plates, but not their area. The other information is the charge density on the plates and that the plates are separated by Mylar. We are asked to find the electric field.

Since we know the plate separation, we could find the field if we knew the potential across the plates. This is given by text Eq. 17.7. First, determine what is needed and what is given.

GIVEN:

Quantity	Symbol	Value
area of plates	A	unknown
separation of plates	d	1.672 mm
dielectric constant (Table 17.2)	κ	3.1
relation for capacitance	C	$\kappa \varepsilon_0 \dfrac{A}{d}$
charge density	ρ	2.53×10^{-8} C/m^2
total charge	q	ρA (unknown)
potential across plates	V	unknown
electric field between plates	E	$-\dfrac{V}{d}$ (unknown)

The capacitance, charge and potential are related by text Eqs. 17.9 and 17.10,

$$V = \frac{q}{C} = \frac{\rho A}{\kappa \varepsilon_0 \frac{A}{d}} = \frac{\rho d}{\kappa \varepsilon_0},$$

so the electric field is given by

$$E = -\frac{V}{d} = -\frac{\rho}{\kappa \varepsilon_0}$$

$$= -\frac{2.53 \times 10^{-8} \text{ C/m}^2}{3.1 \times 8.85 \times 10^{-12} \text{ C}^2/\text{N}\cdot\text{m}^2} = -9.2 \times 10^2 \text{ V/m}.$$

Note that the electric field is independent of the plate separation for a given charge on the plates. The minus sign tells us that the electric field points in the opposite direction to the change in the potential.

1.672 mm

Mylar

Electric field = 2.52×10^{-8} C/m^2

Fig 17.4: A parallel plate capacitor with a Mylar dielectric.

PROBLEM 17.7: How much more charge can a given parallel plate capacitor hold without breakdown if the dielectric is Teflon rather than paper? (17.8)

Solution:

Strategy: The capacitance will change if we change dielectrics. We need to rewrite the relations to eliminate the voltage by using the electric field.

We are asked to compare two parallel plate capacitors which are identical except for the dielectric. The two capacitors are shown in Fig. 17.5. We let A be the area of the plates and d the plate separation. The breakdown strength of each substance will be denoted by $_{max}E_{material}$. We will also need the dielectric constants. Both are given in text Table 17.2.

Paper

Teflon

(a)   (b)

Fig. 17.5: (a) Paper filled capacitor. (b) Teflon filled capacitor.

GIVEN:

Quantity	Symbol	Value
dielectric constant of paper	κ_{paper}	3.5
dielectric constant of Teflon	κ_{Teflon}	2.1
breakdown strength of paper	$_{max}E_{paper}$	14×10^6 V/m
breakdown strength of Teflon	$_{max}E_{Teflon}$	60×10^6 V/m

For a capacitor the following equations hold:

$$q = CV \qquad C = \kappa\varepsilon_0 A/d,$$

where C is the capacitance, q the charge on either plate, and V the voltage. We want the maximum charge, given by

$$q_{max} = CV_{max},$$

where the maximum voltage before breakdown can be expressed in terms of the breakdown strength by

$$V_{max} = {}_{max}E\, d.$$

Putting these relations together gives

$$q_{max} = \kappa\varepsilon_0 \frac{A}{d} \times {}_{max}E\, d$$

$$= {}_{max}E\, \kappa\varepsilon_0 A.$$

So the maximum charge possible is directly proportional to the dielectric constant and the breakdown strength.

Finally, we get,

$$\frac{\text{maximum charge with Teflon}}{\text{maximum charge with paper}} = \frac{{}_{max}E_{Teflon}\, \kappa_{Teflon}}{{}_{max}E_{paper}\, \kappa_{paper}} = 2.6.$$

We can put over 2.5 times as much charge on the Teflon filled capacitor as on the paper filled capacitor.

PROBLEM 17.8: A pair of parallel plates, each of area 250 cm^2, is separated by an air gap of 2.00 mm. An electric field of 5.60×10^4 V/m exists between the plates. How much work must be done to separate the plates so the air gap is 2.50 mm? Neglect the effect of the edges of the plates on the electric field. (17.9)

Solution:

Strategy: We are asked to find the work; this suggests consideration of conservation of energy. That idea gives "work done = final energy − initial energy." We need the energy of the capacitor, which is given by text Eq. 17.2 as

$$\text{energy} = \frac{1}{2}CV^2.$$

So we need both initial and final C (capacitance) and V (voltage) to get the energy. How do we get C? If we use text Eq. 17.12, we can clearly calculate C. How do we get V? We are given the electric field and separation, so we can find the initial V. The final V is not obviously given, so we must calculate it. How do we do this? What is conserved during displacement? Charge on plates is conserved. So we can use text Eq. 17.7 to get the relation among charge, capacitance, and voltage. Use the initial voltage and capacitance to get the charge and then find the final voltage.

The problem is illustrated in Fig. 17.6.

We will solve this problem using conservation of energy. That is, the initial energy stored in the capacitor plus the work done is the final energy in the capacitor. The energy stored in a capacitor, with capacitance C charged to V volts, is given by text Eq. 17.13 as

$$\text{energy stored} = \frac{1}{2}C\,V^2,$$

while the capacitance of a parallel plate capacitor of area A and plate separation d is given by text Eq. 17.8 as

$$C = \varepsilon_0 \frac{A}{d}.$$

Since we know the initial electric field, we can calculate the initial potential as

$$V_{\text{init}} = E_{\text{init}} \times d_{\text{init}}$$

$$= 5.6 \times 10^4 \text{ V/m} \times 2.00 \times 10^{-3} \text{ m} = 112 \text{ V}.$$

The initial capacitance is

$$C_{\text{init}} = \varepsilon_0 \frac{A_{\text{init}}}{d_{\text{init}}}$$

$$= 8.85 \times 10^{-12} \text{ C}^2/\text{N·m} \times \frac{2.50 \times 10^{-2} \text{ m}^2}{2.00 \times 10^{-3}\text{m}}$$

$$= 1.11 \times 10^{-10} \text{ C}^2/\text{N} = 1.11 \times 10^{-10} \text{ F},$$

so the initial electrical energy is

$$\text{Energy}_{\text{init}} = \frac{1}{2}C_{\text{init}}V_{\text{init}}^2$$

$$= 6.96 \times 10^{-7} \text{ N·m}^2 = 6.96 \times 10^{-7} \text{ J}.$$

In order to calculate the work done, we will need the final energy of the capacitor, which means we need the potential after the plates are separated. To calculate this we will need to find the total charge on the plates using the definition of capacitance in text Eq. 17.7, which is

$$q = CV.$$

Since no charge is added

$$q_{final} = q_{init} = C_{init} V_{init} = 1.24 \times 10^{-8} \text{ C.}$$

The final capacitance is

$$C_{final} = \varepsilon_0 \frac{A_{final}}{d_{final}} = C_{init} \times \frac{d_{init}}{d_{final}} = 8.85 \times 10^{-11} \text{ F.}$$

Thus the final voltage is given by

$$V_{final} = \frac{q_{final}}{C_{final}} = 140 \text{ V,}$$

and the final electrical energy is

$$\text{energy}_{final} = \frac{1}{2} C_{final} V_{final}^2 = 8.67 \times 10^{-7} \text{ J.}$$

The work done, W, is

$$W = \text{energy}_{final} - \text{energy}_{init} = 1.71 \times 10^{-7} \text{ J.}$$

It will take 1.71×10^{-7} J of work to separate the plates an additional 0.50 mm.

Fig 17.6: (a) The capacitor with initial plate separation. (b) The capacitor with final plate separation.

PROBLEM 17.9: A parallel plate capacitor is hung by the center of one plate using a nonconducting wire, and another nonconducting wire is attached to the bottom plate. The dielectric between the plates is sticky and is just able to hold the capacitor together, along with the wires, if the capacitor is uncharged. The capacitor is now charged to a potential of 500 V, and the charging source removed. What is the maximum mass that can be hung on the lower wire without pulling the capacitor apart? *Hint:* Consider the potential energy gain due to the mass of the plate as the plate separation increases, and neglect the kinetic energy as the separation first occurs. (17.9)

Solution:

> **Strategy:** As the plates separate we must do work. Gravitational energy will be gained by the plate moving downward. If the new total energy is less than the original energy, the plate will fall.

It takes work to separate the plates of a charged capacitor. Work is done by a mass moving down in a gravitational field. This process will occur if the work provided by moving the mass, m, a distance $\Delta x = \Delta d$ in the gravitational field is greater than the work needed to separate the plates by a distance Δd.

The energy stored in the capacitor is given by text Eq. 17.13 as

$$W = \frac{1}{2} C V^2 ,$$

where C is the capacitance and V the voltage.

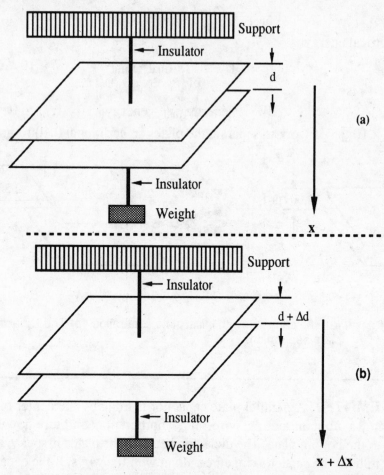

Fig 17.7: (a) Initial arrangement with plates a distance d apart and the mass added to weight. (b) After the plates have moved an extra distance Δd apart and the mass has moved to $x + \Delta x$.

Although the capacitor is initially charged to 500 V, this potential will not be constant if the plates separate. What is constant? The charge on the plates is constant, so text Eq. 17.7 should be used to relate the capacitance C and the voltage V. That is,

$$W = \frac{1}{2}CV^2 = \frac{1}{2}C\frac{q^2}{C^2}$$

$$= \frac{q^2}{2C} = \frac{q^2 d}{2\kappa\varepsilon_0 A} ,$$

where q is the initial charge on the plates, d the plate separation, A the plate area and κ is the dielectric constant. The change in energy, ΔW, if the plates separate by Δd, is then

$$\Delta W = \frac{q^2 \Delta d}{2\kappa\varepsilon_0 A}.$$

The problem is illustrated in Fig. 17.7. Figure 17.7a shows the initial plate separation and position of the mass, and Fig. 17.7b shows what would happen if the plate were to separate an amount Δd.

The energy gained by the mass moving Δx is

$$\Delta E = -mg\Delta x = -mg\Delta d.$$

The total energy change is given by

$$\Delta U = \Delta W + \Delta E$$

$$= \left(\frac{q^2}{2\kappa\varepsilon_0 A} - m g \right).$$

This change is positive provided

$$m < \frac{q^2}{2\kappa\varepsilon_0 A g}.$$

If the mass is larger than the critical value, $\dfrac{q^2}{2\kappa\varepsilon_0 A g}$, the plates will separate.

PROBLEM 17.10: A 500 μF capacitor is charged to a potential of 100 V. How much work is done by this charged capacitor if it is discharged? (17.9)

Solution:

Strategy: Use the equivalence between work and energy change and the value of the energy stored in the capacitor.

The work that can be done by a capacitor charged to 500 V is the same as its total energy, which is given by

$$\text{energy} = \frac{1}{2} C V^2$$

where CV is the capacitance and V is the voltage of the capacitor. Thus

$$\text{energy} = \frac{1}{2} C V^2 = \frac{1}{2} \times (500 \times 10^{-6} \text{ C}) \times (10^2 \text{ V})^2 = 2.50 \text{ J}.$$

There will be 2.50 J released if the capacitor is discharged.

SUGGESTED READINGS

Bassett, C. A. L., "Electrical Effects in Bone." *Scientific American,* October 1965, p. 18.

Berg, R. E., "Van de Graaff Generators: Theory, Maintenance, and Belt Fabrication." *The Physics Teacher,* May 1990, p. 281.

Jefimenko, O., and D. K. Walker, "Electrostatic Motors." *The Physics Teacher,* March 1971, p. 121.

Kristjansson, L., "On the Drawing of Lines of Force and Equipotentials." *The Physics Teacher,* April 1985, p. 202.

O'Neill, G. K., "The Spark Chamber." *Scientific American,* August 1962, p. 36.

Price, R. H., and R. J. Crowley, "The Lightning-Rod Fallacy." *American Journal of Physics,* September 1985, p. 843.

Regan, D., "Electrical Responses Evoked from the Human Brain." *Scientific American,* December 1979, p. 150.

Scott, B. I. H., "Electricity in Plants." *Scientific American,* October 1962, p. 107.

Smith, B. M., "The Polygraph." *Scientific American,* Janurary 1967, p. 25

Walker, J., "The Amateur Scientist: How to Map Electrically Charged Patches with Parsley, Sage, Rosemary and Thyme." *Scientific American,* April 1988, p. 114.

Chapter 18
Electric Current and Resistance

IMPORTANT TERMS

Fill in the blanks with the appropriate word or words.

1. The SI system of current is the _____ which is a current flow of 1 coulomb per second.

2. If the ratio of the applied voltage to the current flow through a circuit element is a constant, we say the circuit element obeys _____ law.

3. A(n) _____ is a source of emf.

4. The _____ of a battery is the potential that appears between its terminals when no current is flowing.

5. The _____ of a substance is the quantity which relates the electrical resistance of a given object to its physical dimensions.

6. When an unexpectedly large current flows in a circuit, we say that we have a(n) _____ circuit.

7. The electric _____ is the rate at which charge passes through a conductor.

8. The SI unit of resistance is the _____.

9. _____ relates the electric power consumed to the circuit's voltage and current.

10. If the magnitude of the current in a circuit is a constant, we have a(n) _____ current.

11. If the conductors in a circuit are broken so that no current can flow, we say that we have a(n) _____ circuit.

12. The electrical potential that a circuit has with respect to the earth is called the _____.

13. Electrical circuit elements which obey Ohm's law are called _____.

14. _____ rules govern the potential drops and currents in a circuit.

15. The ratio of the applied voltage to the resulting current is termed the _____ of the circuit.

Write the definitions of the following:
16. battery (18.1):
17. ampere (18.1):

18. electromotive force (emf) (18.1):
19. Kirchhoff's laws (18.6):
20. electric current (18.1):
21. Ohm's law (18.2):
22. resistivity (18.3):
23. Joule's law (18.4):
24. ohm (18.2):
25. direct current (dc) (18.1):
26. resistor (18.2):
27. short circuit (18.5):
28. open circuit (18.5):
29. resistance (18.2):
30. ground potential (18.5):

ANSWERS TO 1–15, IMPORTANT TERMS

1. ampere; 2. Ohm's; 3. battery; 4. electromotive force (or emf); 5. resistivity; 6. short; 7. current; 8. ohm; 9. Joule's law; 10. direct; 11. open; 12. ground potential; 13. resistors; 14. Kirchhoff's; 15. resistance.

SAMPLE QUESTIONS

QUESTION 18.1: Under what conditions would you arrange cells in series? In parallel?

Answer:

Cells are arranged in series when you need a larger voltage. If you need to draw a large current, then put the cells in parallel.

QUESTION 18.2: Explain why it is dangerous to reach into the back of a TV set even when it has been turned off and unplugged.

Answer:

When a TV set is operating, it needs a high voltage to run the picture tube. There are capacitors that are charged to this high voltage. These capacitors can remain charged for a long time after the TV is turned off.

SAMPLE SOLUTIONS

PROBLEM 18.1: We are given a "black box" with two terminals. When we apply a known voltage to the terminals the current which flows is measured. This data is given in Fig. 18.1. Graphically determine how the resistance of the "black box" depends upon voltage and plot it. Also plot the power dissipated in the box as a function of applied voltage. *Note:* The term "black box" means an object which we must analyze without opening or taking it apart. (18.2)

Solution:

> **Strategy:** Use the definition of resistance to make the graph. Use the definition of power in terms of voltage and current and make the plot.

Fig. 18.1:   A plot of voltage versus current. You should draw a smooth curve through the data.

The resistance is defined as the ratio of applied voltage to the current that flows. Read the current for each value of voltage and calculate the ratio. Check your results with those given in Table 18.1. Note that the resistance is defined as the ratio, not as the slope of the smooth curve.

The power used by the circuit at each voltage is given by power = voltage × current.

These values are given in Table 18.1 and the requested plots are Figs. 18.2 and 18.3.

The resistance at 0 V cannot be calculated as a ratio, but it can be calculated by determining the limit suggested by the values at $V = 5$ V, 10 V, and 15 V. We expect the correct value of this limit would be 10 Ω.

Table 18.1

applied voltage (Volts)	observed current (Amperes)	calculated resistance (Ohms)	Power (Watts)
0.0	0.0	0	0
5.0	0.5	10	2.5
10.0	1.0	10	10
15.0	1.5	10	23
20.0	2.0	10	40
25.0	2.4	10	60
30.0	2.8	11	84
35.0	3.2	11	1.1×10^2
40.0	3.4	12	1.4×10^2
45.0	3.6	12	1.6×10^2
50.0	3.8	13	1.9×10^2
55.0	3.9	14	2.0×10^2
60.0	4.0	15	2.4×10^2
65.0	4.1	16	2.6×10^2

Fig. 18.2: Resistance versus applied voltage.

Fig. 18.3: Power versus applied voltage.

PROBLEM 18.2: It is desired to run an electric drill at the end of a 50 foot extension cord. The cord is made of 18 gauge copper wire (18 gauge wire has a cross-sectional area of 8.231×10^{-7} m^2). Under these conditions the drill is observed to draw 5.0 amperes. Assume the supply voltage is 100 V. What is the voltage at the drill? What would the voltage drop be if a saw drawing 25 amperes was used on this cord? (18.3)

Solution:

Strategy: Use the formula relating resistance, resistivity, length and cross-sectional area. Calculate the voltage drop in the cord for the given currents.

The voltage at the drill will be less than 100 V since there will be a voltage drop in the extension cord. Since we are given the current when the extension cord is used, we do not need to worry about the voltage-current behavior of the drill. The cord and drill are shown schematically in Fig. 18.4.

Fig. 18.4: Electric drill with internal resistance R_{drill} and power cord with total resistance R_W.

Text Table 18.1 can be used to get the resistivity of copper. The resistance of a length of copper is given by text Eq. 18.4. Since the current flows down the wire and back, the total length of wire across which the voltage drop occurs is 100 ft:

$$100 \text{ ft} = 100 \text{ ft} \times \frac{12 \text{ in.}}{1 \text{ ft}} \times \frac{0.0254 \text{ m}}{1 \text{ in.}} = 30.48 \text{ m.}$$

So the total resistance in the wire, R_W (both directions), is given by

$$R_W = \frac{\rho l}{A} = \frac{1.72 \times 10^{-8} \ \Omega \cdot \text{m} \times 30.48 \text{ m}}{8.231 \times 10^{-7} \text{ m}^2} = 0.637 \ \Omega ,$$

where ρ is the resistivity of copper, l is the length of wire, and A is the cross-sectional area of the wire.

The voltage drop when the drill is used is then given by

$$V = I R_{drill} = 5.0 \text{ A} \times 0.637 \ \Omega = 3.2 \text{ V.}$$

When a 25 A saw is run the voltage drop will be 5 times larger or

$$V = 16 \text{ V.}$$

This is over 15% of the applied voltage, so the saw will be running at a much lower voltage than was probably built into the design of the saw by the manufacturer. This suggests the use of a heaver gauge power cord for the saw.

PROBLEM 18.3: How much does it cost to run the saw of Problem 18.2 for 5 hours? What is the cost of the energy wasted in heating the extension cord? Assume the cost of electricity is $0.09 per kilowatt-hour. (18.4)

Solution:

> **Strategy:** The total cost is given by: cost = power × time(in hours) × cost per hour.

Since we know the current and the applied voltage, we can find the power, so

$$\text{cost} = 100 \text{ V} \times 25 \text{ A} \times \frac{kW}{1000 \text{ W}} \times 5 \text{ hours} \times \$0.09 \text{ kW/hr}$$

$$= \$1.13.$$

The power loss in the extension cord is given by the same formula, where the voltage drop is 16 V. So if we divide \$1.13 by 100 and multiply by 16, we should get the correct answer:

$$\text{cost of wasted power} = \frac{\text{cost} \times 16 \text{ V}}{100 \text{ V}} = \$0.18.$$

In terms of percentages, the loss is given by

$$\text{loss} = \frac{\text{waste}}{\text{comsumption}} \times 100 = \frac{\$0.18}{\$1.13} \times 100 = 16\%.$$

About 16% of the power purchased is wasted in heating the extension cord.

PROBLEM 18.4: A voltmeter of resistance 1000 Ω, placed in the circuit of Fig. 18.5 as shown, reads a voltage of 40.0 V. Find the resistance R and the currents $I, I_1,$ and I_2. What is the potential at point A when the voltmeter is not in the circuit? (18.7)

Solution:

> **Strategy:** First use Ohm's law to calculate the currents I and I_2. Then use Kirchhoff's laws to find I_1 and hence R. When we know R, we can calculate the voltage drop without the meter.

For this problem we only need to know that a voltmeter has a known resistance for a given measured voltage. Since the voltage drop across the voltmeter is 70 V, the voltage at A is

$$V_A = 120.0 \text{ V} - 40.0 \text{ V} = 80.0 \text{ V}.$$

So there is an 80.0 V potential across the 500 Ω resistor, and hence

$$I = \frac{V_A}{500 \text{ Ω}} = 0.160 \text{ A}.$$

Likewise,

$$I_2 = \frac{\text{voltmeter reading}}{\text{voltmeter resistance}} = \frac{40.0 \text{ V}}{1000 \text{ Ω}} = 0.040 \text{ A}.$$

Fig. 18.5: Circuit with unknown resistance R and voltage measured by a voltmeter of internal resistance 1000 Ω.

Kirchhoff's laws require that

$$I = I_1 + I_2,$$

so

$$I_1 = 0.160 \text{ A} - 0.040 \text{ A} = 0.120 \text{ A},$$

and

$$R = \frac{\text{voltmeter reading}}{I_1} = \frac{40.0 \text{ V}}{0.120 \text{ A}} = 333 \ \Omega.$$

The unknown resistance is 333 Ω, and the currents I, I_1, and I_2 are 0.160 A, 0.120 A, and 0.040 A, respectively.

If the voltmeter were removed from the circuit, the total resistance of the circuit would be R + 500 Ω, so the current that flows would be

$$I = \frac{120.0 \text{ V}}{R + 500 \ \Omega} = 0.144 \text{ A},$$

and

$$V_A = I \times 500 \ \Omega = 72.0 \text{ V}.$$

So the true circuit voltage should be 72.0 V rather than the 80 V deduced with the voltmeter. That is to be expected since the voltmeter is in parallel and hence reduces the effective resistance between the source and point A, so a larger current flows and hence a larger voltage drop occurs across the 500 Ω resistor.

The voltmeter has too low a resistance to accurately measure this circuit. A better voltmeter will have a much higher internal resistance. Such a meter will read a voltage which is closer to the voltage that would be present if the meter were not attached.

PROBLEM 18.5: What is the equivalent resistance of the combination of resistors shown in Fig. 18.6? (18.7)

Solution:

Strategy: Use the rules for combining series and parallel resistors separately on parts of the circuit to simplify, repeating the process until you are left with only one resistor.

(a)

(b)

(c)

(d)

$$R_1 = R_2 = R_4 = R_5 = 5.0 \ \Omega \quad R_3 = 7.0 \ \Omega \quad R_6 = R_7 = 20.0 \ \Omega$$

Fig. 18.6: (a) A combination of resistors. (b) The equivalent combination after the first reduction. (c) The equivalent combination after the second reduction. (d) The equivalent combination after the third reduction.

The first choice in simplifying resistor combinations is in deciding where to start. This choice is completely arbitrary; however, if you approach the combination systematically, you can minimize your work. The following procedure is best:

1. Combine any resistors which are in series.
2. Now combine resistors that are in parallel.
3. Reexamine the combination, repeating steps 1 and 2 until only one resistor is left.

Consider the combination of seven resistors shown in Fig. 18.6a. R_1 and R_2 are in parallel. R_4 and R_5 are in series, and then they are in parallel with R_6 and R_7. These two combinations are in series with each other and with R_3.

The only resistors which are simply in series with each other are R_4 and R_5. Following rule 1, we combine these two resistors to get the equivalent resistance of

$$R_a = R_4 + R_5 = 5.0 \ \Omega + 5.0 \ \Omega = 10.0 \ \Omega$$

as shown in Fig. 18.6b.

We now should apply rule 2 to either of the two parallel combinations (right side or left side in Fig. 18.6b). Since we started on the right side, the following will continue working on that set of resistors; however, you could proceed to reduce the left side first without changing the answer.

Applying rule 2, the right side parallel combination is reduced to

$$\frac{1}{R_b} = \frac{1}{10 \ \Omega} + \frac{1}{20 \ \Omega} + \frac{1}{20 \ \Omega} = \frac{4}{20 \ \Omega} = \frac{1}{5.0 \ \Omega},$$

which reduces the combination to that shown in Fig. 18.6c.

We can still apply rule 2 to the combination of resistors on the left hand side of the combination, giving

$$\frac{1}{R_c} = \frac{1}{R_1} + \frac{1}{R_2} = \frac{2}{5.0\ \Omega} \quad \text{or} \quad R_c = 2.5\ \Omega$$

as shown in Fig. 18.6d.

Finally, apply rule 1 again, giving us the equivalent resistance of $R_{eq} = 14.5\ \Omega$.

PROBLEM 18.6: Use Kirchhoff's rules to find the current flowing through resistor R_5 in Problem 18.5 if the combination of resistors is connected across a 9.0 V battery. Neglect any internal resistance in the battery. (18.7)

Solution:

> **Strategy:** Apply Kirchhoff's voltage rule to get the total current flowing and then apply the current rules and the voltage rules to get the current through R5.

Since we have analyzed this combination of resistors in Problem 18.5 the majority of the work is done for us.

In general the following needs be done:

1. Reduce the combination to the equivalent resistance.
2. Use the voltage rule to find the total current.
3. Use the voltage rule to find the voltage drops across major portions of the combination.
4. Use the current rule to calculate the current through a major subsection.
5. Repeat steps 3 to 4 as needed.

In particular we want the current through R_5, which by the current rule is the current through R_4 also. If we could determine the voltage drop across the sum of R_4 and R_5 we could calculate the current. This voltage drop is just the voltage drop across R_b in Fig. 18.6c. With this in mind we apply the rules just given.

The equivalent resistance of the circuit was found in Problem 18.5 to be 14.5 Ω, so

$$i_{total} = \frac{V_{total}}{R_{eq}} = \frac{9.0\ \text{V}}{14.5\ \Omega} = 0.62\ \text{A}\ .$$

Since R_b is in series, this must be the current through R_b. This allows us to calculate the voltage drop across R_b, which we denote by V_b:

$$V_b = i_{total}\,R_b = 3.1\ \text{V}.$$

R_4 and R_5 together were denoted by the symbol R_a. The current through R_a, and hence the current through R_5, is given by

$$i = \frac{V_b}{R_a} = \frac{3.1\ \text{V}}{10\ \Omega} = 0.31\ \text{A}.$$

The current through R_5 is 0.31 A.

PROBLEM 18.7: Find the equivalent capacitance of the combination shown in Fig. 18.7. (18.8)

Solution:

Strategy: Follow the rules described in Problem 18.5.

$$C_1 = C_5 = 2.0 \ \mu F \quad C_2 = C_3 = C_4 = 3.0 \ \mu F \quad C_6 = C_7 = 4.0 \ \mu F$$

Fig. 18.7: (a) The combination of capacitors. (b) The combination after the first reduction. (c) The combination after the second reduction. (d) The combination after the third reduction.

Examine the combination and note that the right hand side is a combination of capacitors in series and parallel. Since C_2, C_3, and C_4 are in series we combine them to get

$$\frac{1}{C_a} = \frac{1}{C_2} + \frac{1}{C_3} + \frac{1}{C_4} = \frac{1}{1.0 \ \mu F}$$

as shown in Fig. 18.7b .

Likewise, we combine C_5 and C_6 to get $C_b = 1.3 \ \mu F$ as shown in Fig. 18.7c.

Now, apply rule 2 to reduce the parallel combination to get $C_c = 6.3 \ \mu F$ as shown in Fig. 18.7d.

Finally combine the two series capacitors to get $C_{equivalent} = 1.6 \ \mu F$.

PROBLEM 18.8: An automobile battery has a terminal voltage of 11.9 V when a current of 2.00 amperes is drawn. When the battery is used to start the automobile a current of 75.0 A is drawn by the starting motor and the terminal voltage of the battery drops to 8.00 V. What is the emf and the internal resistance of the battery? (18.9)

Solution:

> **Strategy:** Use the concept of internal resistance of a battery discussed in text section 18.9.

A battery can be considered as an ideal voltage source with a resistor in series as shown in Fig. 18.8. The terminal voltage, V_T, is given by

$$V_T = R_L \times i_L = emf - R_I \times i_L,$$

where R_L is the external load resistance, R_I is the internal resistance of the battery, i_L is the current that flows from the battery when a resistance of R_L is attached across the battery, and *emf* is the true emf of the battery.

Fig. 18.8: Battery (dotted box) with resistor as load.

We are given the terminal voltage for two different current loads, so we should be able to find the emf and the internal resistance from

$$V_{terminal} = emf - I_{load} R_{internal} .$$

That is,

$$11.9 \text{ V} = emf - 2.00 \text{ A} \times R_{internal}$$

and

$$8.00 \text{ V} = emf - 75.0 \text{ A} \times R_{internal}.$$

To solve these, multiply the first equation by 37.5 and subtract the second equation (see text Appendix: Solving Simultaneous Equations), giving

$$(446.25 - 8.00) \text{ V} = (37.5 - 1)\text{V} \times emf - (75 + 75)\text{A} \times R_{internal},$$

which is

$$438.25 \text{ V} = 36.5 \text{ A} \times emf$$

or

$$emf = 12.0 \text{ V}.$$

Using this value for *emf* in both equations gives 0.050 Ω and 0.053 Ω, respectively.

The process of subtracting the two 3-significant-figure numbers will, in this case, produce a result that is known to only 1 significant figure, so we have

$$R_{internal} = 0.05 \ \Omega.$$

The battery is an emf of 12.0 V with an internal resistance of 0.05 Ω.

PROBLEM 18.9: A household circuit is wired for 115 V with a 20 A circuit breaker. Which of the following items can be operated on that circuit at the same time? A 1800 W toaster, a 2200 W griddle, a 1000 W iron, a 1500 W mixer, lights totaling 400 W. Assume the rules for direct current and power apply in this case. (18.10)

Solution:

> **Strategy:** Use the definition of power in terms of voltage and current to determine the maximum wattage that the circuit breaker can support. Then add wattages to determine what combination of lights and appliances consumes less power than the maximum allowed.

The easiest way to answer this question is to determine the maximum wattage the circuit can carry. Since power equals voltage times current, we have the information to calculate the total wattage the circuit can carry.

Let $V = 110$ V be the line voltage and $I_m = 20$ A be the maximum current. Then, neglecting any losses in the lines, the maximum power, P_m, is given by

$$P_m = V \times I_m = 115 \ V \times 20 \ A = 2300 \ W.$$

The total power consumption will be the sum of the individual power consumptions. First of all, we see that if the griddle is running, nothing else, not even the lights, can be running. For the other combinations, try adding wattages together to see if the sum is less than or equal to 2300 W. Any individual appliance and any appliance plus the lights may be on, but any two appliances consume more power together than the circuit breaker will take.

PROBLEM 18.10: What is the power consumption of an electric toaster if its resistance is 10.6 Ω and it operates on a household circuit? (18.10)

Solution:

> **Strategy:** Apply Ohm's law.

This problem is similar to text Example 18.5. Here we are given the resistance (10.6 Ω) and the voltage (assume an effective household voltage of 120 V) and are asked to find the power. Using voltage and resistance we can easily find the current:

$$I = \frac{V}{R} = \frac{120 \text{ V}}{10.6 \, \Omega} = 11.3 \text{ A.}$$

We can then combine current and voltage to get the power.

$$P = IV = (11.3 \text{ A}) (120 \text{ V}) = 1360 \text{ W.}$$

The power consumption is 1360 W.

PROBLEM 18.11: Three identical light bulbs rated at 12 V are connected as shown in Figure 18.9 across a 12 V battery. Which bulb will be the brightest and why? Assume a bulb is brighter if more current flows through it. (18.10)

Solution:

Strategy: You can solve this by applying Kirchhoff's rules or more simply by analysis of the voltages and currents that must result from this arrangement.

Fig. 18.9   Three light bulbs denoted by L_1, L_2, and L_3 connected to a 12 V source.

By inspection of Fig 18.9, you should see that the same voltage (12 V) is applied to the bulb L_1 as to the series pair of $L_2 + L_3$. Since the current that flows is the voltage divided by the resistance, we expect that a larger current will flow through L_1 than through the series set $L_2 + L_3$. That is, the resistance of L_1 is less than the resistance of $L_2 + L_3$.

GENERAL COMMENTS

For problems in this and successive chapters, we model real circuits with some simplified properties. For example, unless otherwise specified, we assume the wires connecting circuit elements are without resistance, capacitance, etc. This is an approximation to the real case where the wires have a much lower resistance than the circuit elements. We assume that voltage sources, such as a battery, are a constant voltage source independent of load unless we specifically indicate that the concepts given in Section 17.9 are to be used. Only then do we consider the internal resistance of a battery. We also neglect any edge effects in capacitors.

In this chapter we do not consider the causes of the resistivity of a metal and the insulation properties of nonmetals. In Chapter 31 the authors discuss the fundamental physics that gives rise to resistivity. For this chapter we take the numbers in text Table 18.1 as experimentally measured numbers.

SUGGESTED READINGS

Barthold, L. O., and H. G. Pfeiffer, "High-Voltage Transmission." *Scientific American,* May 1964, p. 38.

Bassett, C. A. L., "Electrical Effects in Bone." *Scientific American,* October 1965, p. 18.

Chalmers, B., "The Photovoltaic Generation of Electricity." *Scientific American,* October 1976, p. 34.

de Santillana, G., "Alessandro Volta." *Scientific American,* January 1965, p. 82.

Ehrenstein, G., "Ion Channels in Nerve Membranes." *Physics Today,* October 1976, p. 33.

Hazen, R. M., *The Breakthrough.* New York: Simon & Schuster, 1988. A fascinating account of one of the most exciting technological developments of the century, the discovery of high-temperature superconductors.

McKnight, J. L., "Laboratory Notebooks of G. S. Ohm: A Case Study in Experimental Method." *American Journal of Physics,* February 1967, p. 110.

Pollard, W. G., "Prospects for Solar Electricity." *The Physics Teacher,* October 1976, p. 391.

Schagrin, M. L., "Resistance to Ohm's Law." *American Journal of Physics,* July 1963, p. 536.

Smith, F. A., and J. D. Wilson, "Electrical Circuits and Water Analogies." *The Physics Teacher,* October 1974, p. 396.

Wolsky, A. M., R. F. Giese, and E. J. Daniels, "The New Superconductors: Prospects for Applications." *Scientific American,* February 1989, p. 60.

Chapter 19
Magnetism

Important Terms

Fill in the blanks with the appropriate word or words. The three right hand rules are included as well as the important terms.

1. The magnetic dipole _____ is the value assigned to two opposite magnetic poles separated by a fixed distance.

2. A _____ uses a magnetic field to bend moving charged particles into a circular orbit.

3. The relationship among the size and direction of a small segment of current, the direction to a point, and the magnetic field produced at that point is called the law of _____.

4. The _____ gives us the force that a magnetic field exerts on a current or a moving charge.

5. The _____ rule gives us the direction of the magnetic moment produced by a current loop.

6. A(n) _____ is a device for measuring currents.

7. The _____ rule for B gives us the magnetic field produced by a current or a moving charge.

8. _____ law relates the average magnetic field around a closed path to the current enclosed by that path.

9. A(n) _____ is a device for measuring voltage.

10. A(n) _____ is a current measuring device that uses the magnetic field produced by the current being measured to deflect a magnetized compass needle.

11. The magnetic effects resulting from cooperative interactions of individual ionic moments is termed _____.

12. The SI unit of magnetic field is the _____, which has dimensions of N/A·m.

13. _____ law for magnetism states that the net magnetic flux through any closed surface is zero.

14. A(n) _____ is a helical winding that may carry current.

15. The _____ law governs the behavior of paramagnetism.

16. Regions of a ferromagnetic solid in which all the ionic moments are aligned in the same direction are called magnetic _____.

17. The _____ field is produced by an electric current in analogy to the electric charge producing an electric field.

18. The magnetic _____ relates the magnetism produced in a material to the externally applied magnetic field.

19. The _____ is the magnetic analogy of the electric dipole.

Write the definitions of the following:
20. magnetic domains (19.10):
21. right-hand rule for B due to a current (19.2):
22. tesla (19.3):
23. magnetic dipole (19.1):
24. cyclotron (19.5):
25. galvanometer (19.8):
26. magnetic moment (19.2):
27. Ampere's law (19.9):
28. magnetic field (19.1):
29. law of Biot and Savart (19.6):
30. Gauss's law for magnetism (19.2):
31. ammeter (19.8):
32. voltmeter (19.8):
33. ferromagnetism (19.10):
34. solenoid (19.9):
35. right-hand rule for the magnetic moment of a current loop (19.7):
36. Curie law (19.10):
37. magnetic susceptibility (19.10):
38. right-hand rule for the force produced by a magnetic field acting on a current (192):

ANSWERS TO 1–19, IMPORTANT TERMS.

1. moment; 2. cyclotron; 3. Biot and Savart; 4. right-hand rule; 5. right-hand; 6. ammeter; 7. right-hand ; 8. Ampere's; 9. voltmeter; 10. galvanometer; 11. ferromagnetism; 12. tesla; 13. Gauss's; 14. solenoid; 15. Curie; 16. domains; 17. magnetic; 18. susceptibility; 19. magnetic dipole.

SAMPLE QUESTIONS

QUESTION 19.1: If the magnetic field of the earth is due to a circulating current below the earth's crust, does it flow from west to east or east to west?

Answer:

North poles on compasses point towards the North. That means the geomagnetic pole in the Northern hemisphere must be a South pole. Using the right hand rule, we see that the current would have to flow from East to West.

QUESTION 19.2: How can a piece of iron (such as a nail) be ferromagnetic and yet not have a net magnetic dipole moment?

Answer:

A piece of iron can be made up of many small magnetic domains, each pointing in a random direction. Thus there is no net magnetic moment.

SAMPLE SOLUTIONS

PROBLEM 19.1: Suppose a current, i, and an external magnetic field, B, lie in the plane of the paper as shown in Fig. 19.1a–c. What is the direction of the force on the current element in each case? (19.3)

(a)   (b)   (c)

Fig. 19.1:   Examples of a current and an external magnetic field, both in the plane of the paper.

Solution:

> **Strategy:** Apply the second right-hand rule. Since current and the magnetic field both lie in the plane of the paper, the force must be perpendicular to the paper. Is the force up (out of the paper) or down (into the paper)?

For case (a), if we arrange the first two fingers of the right hand along i and B as directed in Fig. 19.6a, the thumb points into the paper, so the force is in a direction into the paper.

For case (b), although we may have sore fingers, the thumb points out of the paper.

Finally, for case (c), if we point the forefinger along i and the index finger along B, the thumb is again down.

When you do these exercises do not swing your fingers back and forth with respect to each other. In all cases keep the position of the thumb and fingers fixed as shown in Fig. 19.6a. Arrange your hand by twisting it at the wrist and/or walk about the paper (or be lazy and move the paper around).

PROBLEM 19.2: A particle of electric charge $q = 1.92 \times 10^{-18}$ C is injected into a region with a magnetic field of 0.58 T with an initial velocity of 2.0×10^5 cm/s at an angle of 30° with respect to the magnetic field. What is the subsequent motion of the particle? (19.4)

Solution:

> **Strategy:** Use the second right-hand rule and determine the magnitude of the force from text Eq. 19.2.

The motion and field described are sketched in Fig. 19.2a. Coordinates have been chosen in which $x \Rightarrow y \Rightarrow z$ is right handed. It is important that this sense be preserved if the right-hand rule is to be used. The magnetic field is indicated by **B** and the particle's motion by **v**. Only one line of magnetic field **B** is shown. **B** has the same value and direction at all points.

We will take the magnetic field as being directed along the y-axis, the particle's initial velocity to be in the x-y plane, and the 30° angle to the right of **B** in the x-y plane as shown. This was an arbitrary choice since **B** and **v** define a plane. It is simplest to let this plane be one of the coordinate planes. Our choice of the x-y plane is arbitrary.

As the problem is stated, we could have chosen **v** to be 30° to the left in the x-y plane instead of to the right. Our choice was arbitrary. The left hand choice will lead to a mirror solution. For the problem, as stated, either solution is acceptable.

(a)   (b)   (c)

Fig. 19.2:   (a) The initial velocity and force. Only one line is drawn to represent the uniform B in the y direction. (b) The velocity and force at a later time. (c) The resulting motion.

Using the right-hand rule, we see that the initial force, **F**, on the particle will be exerted in the +z direction and will be of magnitude

$$F = q v B \sin \theta$$

$$= 1.92 \times 10^{-18} \text{ C} \times 2.0 \times 10^3 \text{ m/s} \times 0.58 \text{ T} \times \sin(30°) = 1.1 \times 10^{-15} \text{ N.}$$

The particle will be accelerated and hence gain velocity in the plane defined by z and the initial direction of v. The acceleration is at right angles to v (a right angle is a $90°$ angle), so initially the velocity only changes the direction of v. Once the direction of v changes, the right-hand rule gives us a force that is still perpendicular to the new v as shown in Fig. 19.2b. Thus, since the force is always perpendicular to v, the magnitude of v remains constant. This is the same arrangement that we had in Chapter 5 with centripetal force and uniform circular motion. The force is at right angles to the velocity, and hence the force changes the direction of the velocity, not its magnitude.

Thus the particle will move in a circular motion around the direction of B. Since the particle has some motion along the direction of B, the particle will spiral about the direction of B as shown in Fig. 19.2c.

PROBLEM 19.3: A long straight wire carries a current of 75 A. At what distance from the wire will the strength of the magnetic field be 10 times the earth's field. Assume the earth's field is 5.0×10^{-5} T. (19.6)

Solution:

> **Strategy:** We are looking for the distance at which the magnetic field from the wire is 10 times the earth's field. Calculate the magnetic field produced by the wire and set that magnitude equal to 10 times the magnitude of the earth's magnetic field.

The field produced by a long straight wire carrying a current i is given by text Eq. 19.5 as

$$B = k' \frac{2i}{d}$$

where B is the field and d is the perpendicular distance to the wire.

The direction of B is given by the right-hand rule. If we let B_E be the earth's magnetic field, the condition for the problem is

$$10\, B_E = 5.0 \times 10^{-4}\ \text{T} = k' \frac{2i}{d} = 10^{-7}\ N/A^2 \frac{150\ \text{A}}{d}$$

Hence,

$$d = 3.0 \times 10^{-2}\ \text{m} = 3.0\ \text{cm}$$

The field produced by the wire is equal to that of the earth's magnetic field at a distance of 3.0 cm.

PROBLEM 19.4: A rectangular coil 1 cm wide by 2 cm long is suspended vertically at the top and bottom by wire fastened to the coil at the middle of the 1 cm sides, as shown in Fig. 19.3a. The coil contains 40 turns of wire. The two long sides each have a total mass of 20 g and the short sides each have a mass of 10 g. The coil is in a uniform magnetic field B which is initially perpendicular to the plane of the coil. When a current of 3.0 A flows through the coil it is deflected so that the plane of the coil is parallel to the magnetic field as shown in Fig. 19.3b. If it takes 0.75 J of energy for this rotation, what is the magnitude of the field B? What torque is provided by the supporting wire in the final configuration? (19.7)

Fig. 19.3: (a) Initial configuration of coil and field. (b) Final configuration of coil and field. (c) Forces on the coil sides at an intermediate position.

Solution:

> **Strategy:** Use text Eq. 19.9 to calculate the magnetic moment of the loop. Then text Eq. 19.1 allows us to calculate the torque. By analogy the energy of the magnetic dipole is similar to text Eq. 17.4.

The magnitude of the force is given by text Eq. 19.2 and its direction by the right-hand rule. The force acts on the current lengths involved. We have labeled the sides α, β, γ, and δ. At all orientations between starting and stopping, the forces acting on sides β and δ tend to stretch the coil. The forces on sides α and γ try to rotate the coil as shown in Fig. 19.3c or in text Fig. 19.12. The torque is thus

$$\tau = N \, i \, A \, B \sin \theta$$

where

Quantity	Symbol	Value
number of turns	N	40
current	i	3.0 A
area of loop	A	2.0 cm^2
magnetic field strength	B	unknown
angle between magnetic field and normal to plane of loop	θ	variable

When the rotation finishes $\theta = 90°$, while at the initial position $\theta = 0°$. The loop acts as a magnetic dipole with moment,

$$\mu = i \, N \, A = 2.4 \times 10^{-2} \text{ A·m}^2 \, .$$

Since the force equation for the magnetic dipole is the same as that for the electric dipole (see text Eq. 16.4), the potential energy is given by

$$U = - \mu \, B \cos \theta.$$

The change in potential energy, ΔU, going from $\theta = 0°$ to $\theta = 90°$ is

$$\Delta U = -\mu B \cos(90°) - [-\mu B \cos(0°)] = \mu B.$$

By energy conservation

$$\Delta U = \text{work done} = 0.75 \text{ J} = 2.4 \times 10^{-2} \text{ A·m}^2 \text{ B}.$$

so

$$B = \frac{0.75 \text{ J}}{2.4 \times 10^{-2} \text{ A·m}^2} = 31 \text{ T}.$$

The torque applied by the support at maximum displacement must equal the magnetic torque since the system is in equilibrium. Thus the maximum torque provided by the wire is

$$\tau_{max} = \mu B = 0.75 \text{ N·m}.$$

PROBLEM 19.5: A cylindrical conductor of radius 0.10 cm carries a uniformly distributed current of 2.0 A. What is the magnetic field 0.050 cm from the center of the cylinder? (19.9)

Solution:

Strategy: Apply Ampere's law. Since the wire is symmetric about the z-axis, only the current interior to the 0.050 cm distance will contribute to the magnetic field.

An end view of the cylindrical conductor is shown in Fig. 19.4.

0.050 cm

0.20 cm

Fig. 19.4: A cylindrical conductor of radius 0.10 cm carrying a uniform current.

The solution of this problem will involve an application of Ampere's law. That is (see text Section 19.9)

$$\Sigma B \, \Delta l \cos \theta = \mu_0 i,$$

where the sum is taken over a loop surrounding the current i.

Since we are asked for the field at a radius of 0.05 cm from the axis without regard to the direction of this distance, we must choose the loop to be a circle of radius 0.050 cm with the circle's center at the center of the conductor.

That is, consider a circle of radius 0.050 cm centered on the z-axis and in the plane of Fig. 19.4. Let this loop be approximated by a series of very short line segments of length Δl. Since the wire is symmetric about z, B will be the same for all segments.

Then $B \cos \theta$ gives us the field tangent to the circle and $\Sigma \, \Delta l$ gives us the circumference of the circle.

If we let d be the radius of the loop (0.050 cm) then

$$2\pi \, d \times B = (2\pi \times 5.0 \times 10^{-4} \text{ m}) \, B = \mu_0 \, i$$

where i is the current interior to 0.05 cm. We do not multiply out the 2π since μ_0 also will involve π.

The current is distributed uniformly across the wire, so

$$i = \frac{\text{area of circle of radius 0.05 cm}}{\text{area of circle of radius 0.10 cm}} \times (\text{total current}) = \frac{2\pi(0.050)^2}{2\pi(0.10)^2} \times 2.0 \text{ A}$$

$$= \frac{1}{4} \times 2.0 \text{ A} = 0.50 \text{ A}.$$

Hence

$$2\pi \times 5.0 \times 10^{-4} \text{ m} \times B = 4\pi \times 10^{-7} \text{ N/A}^2 \times 0.50 \text{ A}$$

so

$$B = 2.0 \times 10^{-4} \text{ N/A·m}$$

Note that the current further than 0.050 cm from the axis does not contribute to the magnetic field at 0.050 cm.

PROBLEM 19.6: A rectangular loop of wire carrying a current of 5 A is hung on the end of a spring in a uniform magnetic field of 0.010 T strength. The loop is 10.0 cm wide and 20.0 cm long and has a mass of 500 g. The magnetic field covers only part of the loop as shown in Fig. 19.5. If the magnetic field (or the current) is turned off, the loop moves upward a distance of 10.0 cm. What is the spring constant? (19.9)

Solution:

> **Strategy:** Use the magnetic force law to calculate the forces acting on each segment of the current. For equilibrium the sum of these forces must be equal and opposite to the force produced by the spring.

Since the loop moves upward when the magnetic field is off, the force on the loop must be as shown in the figure. Only F_1 produces any movement. Since the other forces are perpendicular to the spring, we can neglect them and any edge effects. The force F_1 is given by text Eq. 19.1 as

$$F_1 = i \, l \, B \sin \theta,$$

where B is the field strength, i the current, l the length of the current segment, and θ is the angle between i and B ($\theta = 90°$ for this problem).

Thus

$$F_1 = 5.0 \text{ A} \times 0.10 \text{ m} \times 0.010 \text{ T} = 5.0 \times 10^{-3} \text{ N}.$$

This downward force must be matched by the upward force of the spring. That force would be given by $F = -kx$. Since the loop moves upwards 10 cm when the magnetic force is removed, the spring is distorted 10 cm from equilibrium. So

$$\Sigma \text{ forces } = 0 = -k \times 0.10 \text{ m} = 5.0 \times 10^{-3} \text{ N},$$

or

$$k = 5.0 \times 10^{-2} \text{ N/m}.$$

Fig. 19.5: Loop of wire hanging in a region of magnetic field. The magnetic field is perpendicular to the plane of the drawing. The magnetic forces are shown.

The mass of the loop does not enter into the solution since the position of the spring is measured from hanging equilibrium and since the gravitational force is constant in the problem.

PROBLEM 19.7: An external magnetic field of 5.0×10^{-3} T is applied to a sample of copper sulfate at a temperature of 293 K and with a volume of 1.0×10^{-6} m³. What is the magnetization produced in this sample by the external field? (19.10)

Solution:

> **Strategy:** Text Example 19.10 gives the paramagnetic susceptibility of copper sulfate at the given temperature. Use this and the definition of magnetic susceptibility to find the magnetization per unit volume and hence the total magnetization.

Text Example 19.10 gives the susceptibility per unit volume of copper sulfate at 293 K as 1.68×10^{-4}. This susceptibility is defined in text Eq. 19.13 as

$$X = \mu_0 \frac{M}{B},$$

where X is the magnetic susceptibility per unit volume, M is the magnetization per unit volume and B is the applied magnetic field.

The total magnization is then given by

$$\text{magnetization} = \frac{X\,B}{\mu_0} \times \text{volume}$$

$$= \frac{(1.68 \times 10^{-4}) \times (5.0 \times 10^{-3}\ \text{T})}{4\pi \times 10^{-7}\text{N}\cdot\text{A}^{-2}} \times 1.0 \times 10^{-6}\ \text{m}^3$$

$$= 6.7 \times 10^{-6}\frac{\text{T m}^3\ \text{A}^2}{\text{N}}.$$

These units can be further simplified by using the fundamental definitions of the magnetic field and the ampere.

GENERAL COMMENTS

We model the effects that are called "magnetic" as we did for the effects called "electric." That is; there is an effect, called the magnetic field. We represent it by the field lines. When scientists first introduced the magnetic field, the field lines were treated as real objects, rather than just representing the effects of the magnetic field.

Fig. 19.6:   (a) Identification of right hand with current, magnetic field and force.
b) Right-hand screw rule.

The magnetic force laws are very similar to the electric force laws. The major difference is that whereas electric force laws contain cosines of angles, the magnetic force laws contain sines of angles. The magnetic force laws also involve directed sources (currents) and three-dimensional relationships. For example the force on a current element in a magnetic field lies in a direction that is perpendicular to the plane defined by the current element and the magnetic field. The electric force, on the other hand, is always directed along the electric field.

To help keep the direction of magnetic forces straight, a number of right-hand rules have been introduced. We use three of them in this chapter. The first is shown in text Fig. 19.8, the second in text Fig. 19.10, and the third in text Fig. 19.20.

First right-hand rule: If the thumb of your right hand points in the direction a current flows, your curled fingers point in the direction of the magnetic field produced by that current as shown in text Fig. 19.8.

Second right-hand rule (right-hand screw rule): A current element and an external magnetic field define a plane (if they are not parallel). By the term "external" we mean do not consider the magnetic field produced by the current. The force the external field produces on the current element is perpendicular to the plane containing the current and the magnetic field and is directed the way a normal wood screw would advance if it were twisted in the direction corresponding to a rotation of the current into the magnetic field through the smallest angle. To visualize this hold your right hand in front of you with your thumb upwards, your forefinger (or index finger) to the left and your next finger somewhat towards you as shown in Fig. 19.6a. Let your index finger represent the current and your next finger the magnetic field, then your thumb points in the direction of the force. If you rotate your hand inward towards your body, making your index finger point towards the right and your next finger towards your body, then your thumb points in the direction that a normal (right-handed) screw would advance as shown in Fig. 19.6b.

Third right-hand rule: The magnetic dipole produced by a current loop can be determined by curling the fingers of your right hand in the direction of positive current in the loop. Your thumb points in the direction that the magnetic dipole points.

SUGGESTED READINGS

Becker, J. L., "Permanent Magnets." *Scientific American,* December 1970, p. 92.

Bitter, F., "Ultrastrong Magnetic Fields." *Scientific American,* July 1965, p. 64 .

Bitter, F., *Magnets, The Education of a Physicist.* Garden City, N.Y.: Doubleday-Anchor, 1959.

Bloxham, J., and D. Gubbins, "The Evolution of the Earth's Magnetic Field." *Scientific American,* December 1989, p. 68.

Cohen, D., "Magnetic Fields of the Human Body." *Physics Today,* August 1975, p. 34.

Dibner, B., *Oersted and the Discovery of Electromagnetism.* New York: Blaisdell, 1962.

Ford, K. W., "Magnetic Monopoles." *Scientific American,* December 1963, p. 122.

Franksen, O. I., *H. C. Ørsted—A Man of Two Cultures.* Birkerød, Denmark: Strandbergs Forlag, 1981.

Garrick, D., and R. Kunselman, "Magnetic Monopoles." *The Physics Teacher*, October 1971, p. 366.

Goldhaber, A. S., "Resource Letter MM-1: Magnetic Monopoles." *American Journal of Physics,* May 1990, p. 429.

Hoffman, K. A., "Ancient Magnetic Reversals: Clues to the Geodynamo." *Scientific American,* May 1988, p. 76.

Joselyn, J. A., and E. C. Whipple, "Effects of the Space Environment on Space Science." *American Scientist,* March 1990, p. 126.

Kolm, H. H., and A. J. Freeman, "Intense Magnetic Fields." *Scientific American,* April 1965, p. 66.

Lee, E. W., *Magnetism, An Introductory Survey.* New York: Dover Publications, 1970.

Rieder, W., "Circuit Breakers." *Scientific American,* January 1971, p. 76.

Runcorn, S. K., "The Moon's Ancient Magnetism." *Scientific American,* December 1987, p. 60.

Williams, L. P., "André-Marie Ampère." *Scientific American,* January 1989, p. 90.

Wolsky, A. M., R. F. Giese, and E. J. Daniels, "The New Superconductors: Prospects for Applications." *Scientific American,* February 1990, p. 60.

Chapter 20
Electromagnetic Induction

IMPORTANT TERMS

Fill in the blanks with the appropriate word or words.

1. _____ law states that the emf induced in a loop of wire is proportional to the rate of change of magnetic flux through the loop.

2. A current flowing through a coil sets up an induced current in the opposite direction. This effect is called _____ (or _____).

3. A current flowing through one coil causes a current to flow in another, nearby coil. This effect is called _____ inductance.

4. The unit of inductance in the SI system is the V·s/A, which is also called the _____.

5. A(n) _____ usually consists of two multiple-turn coils of wire wound on the same iron core.

6. Oscillating electric charges generate _____ waves.

7. The emf produced by a changing magnetic field is called a(n) _____ emf.

8. A(n) _____ is any element of a circuit that exhibits inductance.

9. A(n) _____ converts a current into rotational mechanical motion.

10. The current produced by a changing magnetic field is termed a(n) _____ current.

11. The _____ is a device that converts rotary motion into a direct voltage that can then produce a current.

12. A current which periodically reverses its direction is called a(n) _____ current.

13. _____ law tells us that the current induced by a change in an external magnetic field is such that the magnetic field produced by the current will oppose the change in the external magnetic field.

14. The set of four equations that describe the behavior of electric and magnetic fields is called _____.

15. When an electric motor shaft begins to move, the motor produces a _____ in the opposite direction to the applied emf.

Write the definitions of the following:

16. electric motor (20.3):

17. alternating current (20.3):

18. self inductance (20.5):

19. henry (20.5)

20. Maxwell's equations (20.7):

21. transformer (20.4):

22. generator (20.3):

23. (mutual) inductance (20.5):

24. back (or induced) emf (20.3):

25. Faraday's law (20.1):

26. Lenz's law (20.1):

27. electromagnetic waves (20.7):

28. induced emf (20.1):

29. inductor (20.5):

30. induced current (20.1):

ANSWERS TO 1–15, IMPORTANT TERMS.

1. Faraday's; 2. self inductance (or just inductance); 3. mutual; 4. henry; 5. transformer; 6. electromagnetic; 7. induced; 8. inductor; 9. electric motor; 10. induced; 11. generator; 12. alternating; 13. Lenz's; 14. Maxwell's equations; 15. back emf (or induced emf).

SAMPLE QUESTIONS

QUESTION 20.1: The belt on a belt-driven generator breaks while the generator is being used to charge a storage battery. You observe that the generator continues to turn. What is the explanation of this observation?

Answer:

The generator is now acting as a motor whose power is provided by the storage battery.

QUESTION 20.2: Explain why a spinning metal top slows down when a magnet is brought near.

Answer:

A current is induced in the top by the magnetic field. This current flows so as to produce a field to oppose the applied field. This will cause the top to slow.

SAMPLE SOLUTIONS

PROBLEM 20.1 A coil contains 100 turns of wire in a loop 25 cm in diameter. The loop is placed between the poles of a large electromagnet with a magnetic field strength B = 10.0 T, and with the plane of the loop perpendicular to the field. If the magnetic field is steadily reduced from 10.0 T to 1.0 T in 20 s, what is the average emf in the coil while the field is changing? (20.1)

Solution:

> **Strategy:** Faraday's law relates the induced emf to the rate of change in flux. For an average flux, we can look at the difference in flux divided by the time this change in flux takes. The flux is given by the magnetic field times the area normal to the magnetic field.

Faraday's law, given in the text as Eq. 20.2, relates the induced emf, E, and the change in flux, $\Delta\Phi$, that occurs during a change in time Δt, as

$$E = -\frac{\Delta\Phi}{\Delta t}.$$

Since the coil is perpendicular to the field, the flux at any instant through one coil is given by text Eq. 20.1 as

$$\Phi = B\,A,$$

where B is the magnetic field and A is the area of the loop. Thus the total flux through all 100 turns of wire is

$$\Phi = 100 \times B \times \frac{\pi\,(0.25\ \text{m})^2}{4} = 4.9 \times B.$$

The average induced emf is given by

$$E = -\frac{\Delta\Phi}{\Delta t} = -\frac{[4.9 \times 10.0\ \text{T}] - [4.9 \times 1.0\ \text{T}]}{20\ \text{s}} = -2.2\ \text{V}.$$

The average induced emf is 2.2 V and will oppose the change in the magnetic field.

PROBLEM 20.2: A 1.5 m long aluminum bar is held horizontally in the east-west direction and dropped from a height of 9.0 m at a place where the horizontal component of the earth's magnetic field is 3.5×10^{-5} T. What is the emf between the ends of the bar just before it strikes the ground? (20.2)

Solution:

> **Strategy**: Assume the rod maintains its horizontal orientation as it falls. The area swept out by the rod corresponds to a changing flux, so there is a motional emf as described in Section 20.2 of the text. The velocity of the rod can be found from our knowledge of the behavior of falling bodies in a constant gravitational field.

The rod sweeps out an area as it falls. Since we assume the magnetic field is constant over the region of the experiment, we can assume that the rate at which the area changes gives us the rate at which the flux changes. Hence the induced emf, E, due to the motion is given by

$$E = B l v$$

where B is the magnetic field perpendicular to the plane formed by the rod and the line in which the rod falls, l is the length of the rod, and v is the velocity at which the rod is falling.

Since the rod is falling under a constant acceleration and started from rest, we can express the velocity, v, in terms of the distance fallen, x, and the acceleration of gravity, g, as

$$v^2 = 2g\,x.$$

This gives

$$E = B\,l\,\sqrt{2g\,x} = 3.5 \times 10^{-5}\,\text{T} \times 1.5\,\text{m} \times \sqrt{19.6\,\text{m/s}^2 \times 9.0\,\text{m}} = 7.0 \times 10^{-4}\,\text{V}.$$

A voltage of 7.0×10^{-4} V is generated across the rod at the time it hits the ground.

PROBLEM 20.3: A motor which operates on 220 V has an internal resistance of 1.0 Ω. If the motor draws 5.0 A when running, what is the back emf produced by the motor? If the maximum current that the motor can draw without damage to its coils is 20 A, what resistance must be added in series to limit the current to 20 A maximum? Such a series resistor is called a starting resistor. (20.3)

Solution:

> **Strategy:** The back emf must reduce the effective emf to that produced by 5.0 A through 1.0 Ω. The starting resistor must be such that 220 V can only provide 20 A. Remember to include the voltage drop across the starting resistor when the motor is running.

When the motor is running the moving coils cut the magnetic field inside the motor and, therefore, must produce a back (or induced) emf. It is the vector sum (hence the difference) of the applied emf and the induced emf that provides the effective emf. That is, across the motor terminals

$$\text{applied emf} - \text{back emf} = \text{resistance} \times \text{current}.$$

With the starting resistor in the circuit, the line voltage is reduced by the voltage drop across the starting resistor. Call the starting resistor R_S and the current I; then we want

$$220\,\text{V} = R_S \times I + \text{back emf} + R_{\text{motor}} \times I.$$

When the motor is first started, there is no back emf. R_S is chosen to make $I = 20$ A. Thus

$$220\,\text{V} = (R_S + R_{\text{motor}}) \times 20\,\text{A} = (R_S + 1.0\,\Omega) \times 20\,\text{A},$$

and hence

$$R_S = 10\,\Omega.$$

When the motor is running $I = 5$ A, so

$$\text{back emf} = 220 \text{ V} - 10 \text{ } \Omega \times 5 \text{ A} - 1.0 \text{ } \Omega \times 5 \text{ A} = 165 \text{ V}.$$

PROBLEM 20.4: A transformer attached to a 10 kV transmission line has 20 times as many primary turns as it does secondary turns. If the output line draws 50 A, what is the current draw on the input line? (20.4)

Solution:

> **Strategy:** Assume conservation of energy; that is, the power input is the same as the power output.

If we neglect the possibility of losses in the transformer, the power input will be the same as the power output. Thus

$$\text{power}_{\text{input}} = I_{\text{in}} V_{\text{in}} = \text{power}_{\text{output}} = I_{\text{out}} V_{\text{out}}$$

where I is the current and V is the voltage. Solving for I_{in} gives

$$I_{\text{in}} = \frac{V_{\text{out}}}{V_{\text{in}}} I_{\text{out}} \, .$$

For a transformer with N_{in} turns on the input and N_{out} on the output (text Eq. 20.6),

$$\frac{V_{\text{out}}}{V_{\text{in}}} = \frac{N_{\text{out}}}{N_{\text{in}}} \, .$$

Combining these gives is

$$I_{\text{in}} = \frac{N_{\text{out}}}{N_{\text{in}}} I_{\text{out}} = 2.5 \text{ A}.$$

The primary current draw is 2.5 A.

PROBLEM 20.5: An inductor is rated as being 1.2 H. What is the induced emf across its terminals if there is a current flowing which obeys the rule $i(t) = 2.00 \text{ A} + 0.50 \text{ A/s} \times t$, where t is measured in seconds? (20.5)

Solution:

> **Strategy:** Use Faraday's law to get the magnitude and the sign of the induced emf.

Faraday's law (text Eq. 20.9) gives us the relation between the induced emf, E, and the changing current, i,

$$E = -L \frac{\Delta i}{\Delta t} = -1.2 \text{ H} \times 0.50 \text{ A/s} = -0.60 \text{ V}.$$

where L is the inductance and Δi is the current change that occurs in an elapsed time Δt.

Since the current is increasing, the induced emf must be negative to oppose the current change. This is shown in Fig. 20.1 where the sign of the applied emf is indicated by the +,– pair on either side of the words "Applied emf" and the sign of the induced emf is indicated in the same manner.

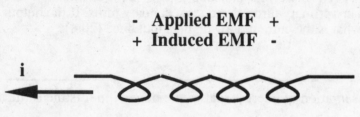

- Applied EMF +
+ Induced EMF -

Fig. 20.1: Applied and induced emf in an inductor.

PROBLEM 20.6: A "U" shaped piece of wire is held fixed in a constant magnetic field of magnitude 0.80 T directed into the plane of the wire. The separation between the two parallel arms of the wire is 0.25 m. A conducting rod is moving away from the curve of the "U", as shown in Fig. 20.2, at 3.0 m/s in contact with the wire. What current flows and what external force must be applied to the rod to maintain this velocity? Solve this problem first by assuming a total resistance for the complete circuit of 0.20 Ω and then indicate how the answer will change if the changing resistance of the circuit is included. (20.5)

Fig. 20.2: A "U" shaped piece of wire with a bar sliding along the wire with a velocity v as shown. There is a uniform magnetic field B into the plane of the paper.

Solution:

Strategy: The rod across the ends of the "U" defines a circuit. As the rod moves the flux inside the circuit increases. Use Faraday's law to calculate the induced emf and then find the current that flows. The external magnetic field produces a force on this current which opposes the motion of the rod.

We let the symbol x represent the distance of the bar from the base of the loop, v the velocity of the rod and L the distance between the two parallel sides of the "U". Then from Faraday's law as given in text Eq. 20.2 or 20.9,

$$E = -\frac{\Delta \Phi}{\Delta t} = -BL\frac{\Delta x}{\Delta t} = -BLv,$$

so

$$E = -0.80 \text{ T} \times 0.25 \text{ m} \times 3.0 \text{ m/s} = -0.60 \text{ V},$$

where we have taken the change in flux as due to the change in the area enclosed by the loop of wire and the bar.

This emf causes a current to flow around the loop of magnitude

$$i = \frac{E}{R} = \frac{0.60 \text{ V}}{0.20 \text{ }\Omega} = 3.0 \text{ A}.$$

The direction of the current can be determined in two steps. First, the magnetic field this current generates must oppose the external magnetic field and thus must point out of the paper. Then, the third right-hand rule (Chapter 19) tells us that this current must be counterclockwise.

Text Eq. 19.2 gives us the force exerted on the rod as: $F = iLB = 0.6 \text{ N}$
since the current and the rod are perpendicular. This force must oppose the motion of the rod and by the second right-hand rule (Chapter 19) is directed in the $-x$ direction. To keep the rod in motion we must apply an equal and opposite force. The applied force must be 0.6 N in the $+x$ direction.

As the rod moves, the total length of wire which comprises the circuit increases, so we expect the resistance to increase as the rod moves down the wire. This will decrease the current in the circuit and thus decrease the magnetic force on the rod.

PROBLEM 20.7: A square loop of wire with sides 10 cm long and a total mass of 4 g is held in a vertical plane as shown in Fig. 20.3a a distance h above a region with a uniform magnetic field of 0.50 T. When the loop is dropped it enters the region of magnetic field with velocity v as shown in Fig. 20.3b. The loop has a resistance of 0.20 Ω. What must the distance h be so the velocity of the loop remains constant as it enters the region with the magnetic field? (20.5)

Solution:

Strategy: Calculate the rate of flux change in the coil, the resulting current, and hence the force on each side of the loop. For constant velocity this force must equal the gravitational force.

We let B be the magnetic field strength, L the length of a side of the loop, R the resistance of the loop, and h the initial height of the loop above the region of the magnetic field. When the loop is immersed y units into the region of magnetic field the flux inside the loop is given by (see Fig. 20.3b):

$$\text{flux} = \Phi = BLy$$

where y is the distance the loop is immersed in the magnetic field region.

The change in flux is due solely to the change in y, so the emf is given by

$$\text{emf} = -\frac{\Delta \Phi}{\Delta t} = -BL\frac{\Delta y}{\Delta t} = -BLv,$$

and the current that flows has a direction given by Faraday's law and a magnitude given by

$$\text{current} = \frac{\text{emf}}{R} = \frac{BLv}{R}.$$

The external magnetic field acts on this current, producing a force given by text Eq. 19.2. For the horizontal side this force is

$$\text{force} = BL \times \text{current} = \frac{B^2L^2v}{R}$$

and in a direction so as to oppose the motion of the wire as shown in Fig. 20.3c. This figure also shows the forces on the vertical sides. Their magnitudes are equal and given by

$$\text{force on vertical sides} = BIy.$$

These forces are equal and opposite so they do not affect the motion of the loop.

Fig. 20.3:   (a) A square loop at rest a distance h above a region containing a uniform magnetic field. The magnetic field is into the plane of the paper. (b) The same loop having fallen into the magnetic field a distance y and having a velocity v. (c) The forces due to the current set up by the magnetic field. These forces act on the three sides of the loop that have entered the region of magnetic field.

The velocity when the loop first enters the region of magnetic field is the velocity when the loop has fallen a distance h. Equating the initial potential energy to the final kinetic energy gives us

$$mgh = \frac{1}{2}mv^2,$$

thus

$$v = \sqrt{2gh}.$$

For constant velocity the magnetic force is equal and opposite to the gravitational force so

$$mg = \frac{B^2L^2v}{R} = \frac{B^2L^2\sqrt{2gh}}{R}.$$

Solving this for h gives

$$h = \frac{1}{2}\frac{gm^2R^2}{(BL)^4} = 0.50 \text{ m}.$$

The loop must be dropped from a height of 50 cm.

PROBLEM 20.8: What is the equivalent inductance for the circuit shown in Fig. 20.4a? Neglect any mutual inductance and assume that inductors in series and parallel add just as resistors in series and parallel add. (20.5)

Fig 20.4:   (a) Series and parallel combination of inductors. (b) First reduction. (c) Second reduction. d) Final equivalent inductance.

Solution:

> **Strategy:** Reduce the circuit to simpler components one step at a time as was done with resistor and capacitor networks in Chapter 18.

Problems such as this with inductors or others with capacitors are best approached in steps as was done with the resistor networks of Chapter 18. First take any simple series combinations and reduce them to a single inductor, then combine simple parallel elements. Repeat these two steps until only one circuit element remains. Starting with the circuit in Fig. 20.4a this reduction is shown in Figs. 20.4b–d on the preceding page.

Since there are no simple series inductors in this circuit, the first step is to combine the three inductors (0.10 H, 0.20 H and 0.50 H) which are in parallel. Using the same rules that apply to resistors, we have

$$\frac{1}{L_p} = \frac{1}{0.10 \text{ H}} + \frac{1}{0.20 \text{ H}} + \frac{1}{0.50 \text{ H}} = \frac{17}{\text{H}},$$

so

$$L_p = 0.059 \text{ H} \qquad\qquad (\text{see Fig. 20.4b})$$

We now can combine L_p with the 0.80 H inductor to get

$$L_s = 0.059 \text{ H} + 0.80 \text{ H} = 0.86 \text{ H} \quad (\text{see Fig. 20.4c})$$

Finally,

$$\frac{1}{L_f} = \frac{1}{0.86 \text{ H}} + \frac{1}{0.30 \text{ H}} = \frac{4.50}{\text{H}}$$

so

$$L_f = 0.22 \text{ H}.$$

The equivalent inductance of this circuit is 0.22 H.

SUGGESTED READINGS

Bransky, J., "Superconductivity—A New Demonstration." *The Physics Teacher,* September 1990, p. 392.

Coltman, J. W., "The Transformer." *Scientific American,* January 1988, p. 86.

Frank, N. H., "Faraday and Field Concepts in Physics." *The Physics Teacher,* May 1968, p. 210.

Froome, K. D., and L. Essen, *The Velocity of Light and Radio Waves.* New York: Academic Press, 1969.

Mak, S. Y., and P. K. Tao, "Measurement of Self-Inductance." *The Physics Teacher,* Sept. 1988, p. 378.

Nahin, P. J., "Oliver Heaviside." *Scientific American,* June 1990, p. 122.

Reines, F., and R. Ballard, "Participatory Lecture Demonstration with an 83-Ton Bar Electromagnet." *American Journal of Physics,* April 1973, p. 566.

Spradley, J. L., "Hertz and the Discovery of Radio Waves and the Photoelectric Effect." *The Physics Teacher,* November 1988, p. 492.

Walker, J., "The Amateur Scientist: The Secret of a Microwave Oven's Cooking Action is Disclosed." *Scientific American,* February 1987, p. 134.

Chapter 21
Alternating-Current Circuits

IMPORTANT TERMS

Fill in the blanks with the appropriate word or words.

1. The time it takes the current in an inductor to build up to 1/e of its final value is called the inductive _____.

2. A(n) _____ (or a(n) _____) passes current in one direction and blocks the current in the other direction.

3. The capacitance factor analogous to resistance is called the capacitive _____.

4. The characteristic time for the current in an RC circuit to decay to 1/e of its initial value, when the emf is removed, is called the _____.

5. The process of putting some of the output of an amplifier into the input is called _____.

6. The peak ac voltage is related to the peak ac current by the _____.

7. The constant of proportionality between the current and applied voltage for an inductor is called the inductive _____.

8. A(n) _____ amplifier is one of a large class of amplifiers whose behavior can be adjusted by the appropriate choice of feedback.

9. The _____ voltage is the square root of the of the cycle average of the voltage squared.

10. A(n) _____ is a two-dimensional quantity that rotates at constant frequency. It is used to represent an oscillating current or voltage.

Write the definitions of the following:

11. inductive time constant (21.1):

12. capacitive time constant (21.2):

13. rms current and voltage (21.3):

14. operational amplifier (21.8):

15. phasor (21.5):

16. feedback (21.8):

17. capacitive reactance (21.4):

18. inductive reactance (21.4):

19. impedance (21.5):

20. rectifier (21.7):

ANSWERS TO 1–10, IMPORTANT TERMS

1. time constant; 2. diode, rectifier; 3. reactance; 4. capacitive time constant; 5. feedback; 6. impedance; 7. reactance; 8. operational; 9. rms (root-mean-square); 10. phasor.

SAMPLE QUESTIONS

QUESTION 21.1: How can an ac distribution system transmit power if the average of the current over every cycle is zero?

Answer:

The power transmitted is determined by the time average of the voltage times the current. This is the same as the product of the rms current times the rms voltage. For example, with a purely resistive load the current reverses when the voltage reverses. so the product of current times voltage is always positive.

QUESTION 21.2: Is it possible to adjust R, L, and C in some circuit so that the impedance is zero, thereby giving you free electrical energy?

Answer:

Since the impedance is given by $\sqrt{R^2 + (X_L - X_C)^2}$ you would have to have $R = 0$ and $X_L = X_C$. Since any real inductor must have some resistance and any real capicator some conductance, it would not be possible to make the resistance exactly zero. Even if inductors without resistance and capicators without losses were possible, you would not get any energy from the system. The current would be out of phase with the voltage at all times and the resulting power would be zero.

SAMPLE SOLUTIONS

PROBLEM 21.1: The switch is closed in the circuit shown in Fig 21.1 at 11:45:00.0 A.M. At what time will the current be at 90% of its maximum value? (21.1)

Fig 21.1: Series LR circuit.

Solution:

> **Strategy:** We are asked for the time dependence of the current, which is given by text Eq. 21.1. Since there is a dc voltage, the maximum current is as if the inductor were not present.

If we let t be the time since the switch was closed, V be the voltage, R the resistance and, L the inductance, then the current, i, is given by

$$i(t) = \frac{V}{R}\left(1 - e^{-\frac{R\,t}{L}} \right) = i_m\left(1 - e^{-\frac{R\,t}{L}} \right).$$

After a very long time the current will approach its maximum value i_m, given by

$$i_m = \frac{V}{R}.$$

We are asked for the time at which $i(t)/i_m = 0.90$, or the time t for which

$$0.90 = 1 - e^{-\frac{Rt}{L}}.$$

Thus

$$e^{-\frac{Rt}{L}} = 0.10.$$

The inverse of the exponential function was discussed in Chapter 14 and is denoted by ln. Most scientific calculators have both a ln and a log key; use the ln key since the log key is the inverse of 10^x and we want the inverse of e^x. Hence

$$-\frac{Rt}{L} = \ln(0.10),$$

or

$$t = -\frac{L\ln(0.10)}{R} = -\frac{3.5\text{ H} \times (-2.3)}{10\ \Omega} = 0.81\text{ s}.$$

The current is at 90% of its maximum value 0.81 s after the switch is closed. At that time the clock reads 11:45:00.8 s.

PROBLEM 21.2: Two 2000 μF capacitors are joined in parallel and charged to 100 V. The capacitors discharge through a resistance of 200 kΩ. (a) What is the time constant of the circuit? (b) How much voltage will remain across the capacitors after one hour has elapsed? (21.2)

Solution:

> **Strategy:** Two capacitors in parallel can be reduced to an equivalent capacitor. The RC circuit can then be analyzed.

In Chapter 18, Section 8 you learned how to reduce capacitors in parallel to an equivalent capacitor. Two 2000 μF capacitors in parallel are equivalent to one 4000 μF capacitor, as shown in Fig. 21.2a and b.

(a)   **(b)**   **(c)**

Fig. 21.2:   (a) Two capacitors in parallel. (b) The equivalent capacitor. (c) The equivalent capacitor discharging through a resistor.

The two capacitors are charged to 100 V and then discharged through 200 kΩ. The time constant is given by text Eq. 21.8 as

$$\text{time constant} = RC_{eqv} = (200 \times 10^3\ \Omega) \times (4000 \times 10^{-6}\ F) = 800\ s.$$

The voltage across the capacitor is given by:

$$V_C = V_0 e^{-t/RC}.$$

For these number and for t = 1 hr = 3600 s, we have t/RC = 4.5:

$$V_C = 100\ V \times e^{-4.5} = 100\ V \times 0.011 = 1.1\ V.$$

After one hour there will still be a voltage of 1.1 V across the capacitors.

PROBLEM 21.3: An inductor is attached to a voltage source whose peak amplitude can be kept constant but whose frequency can be varied. Keeping the voltage peak amplitude constant at 50 V while varying the frequency from 0.00 cycles per second (cps = Hz) to 100 cps produces the curve of reciprocal peak current versus frequency shown in Fig 21.3. The curve has a value of 0.10/A at frequency f = 0.00 and 126/A at f = 100 cps. At any frequency above 1 cps the curve appears to be a straight line. What is your best guess as to the ideal circuit elements which can approximate this real inductor in the frequency range of 0–100 cps? (20.5)

Solution:

Strategy: See if the given properties can be satisfied by a series RLC circuit. If not, it will be necessary to assume components in parallel.

The peak voltage and current are related by $V_{peak} = Z\, i_{peak}$ where, if we assume the real component can be replaced by a series combination of resistor, inductor, and capacitor, Z is given by text Eq. 20.17 as

$$Z = \sqrt{R^2 + \left(2\pi f L - \frac{1}{2\pi f C}\right)^2}.$$

So

$$\frac{1}{i_{\text{peak}}} = \frac{Z}{V_{\text{peak}}}.$$

At very low frequencies

$$Z \approx \frac{1}{2\pi fC},$$

which becomes very large when $f \ll 1/C$. Since the reciprocal of the peak current is small at $f = 0$, we can have no series capacitance. Another way to state this is to take the series capacitance as infinite. An infinite series capacitor has no effect on a circuit since there can be no voltage across it (voltage = charge divided by capacitance). Remember that the formula for a parallel plate capacitor from Chapter 17 had the capacitance proportional to the reciprocal of the plate separation. If the plate separation goes to zero the capacitance goes infinite and the capacitor becomes a short circuit.

Fig. 21.3: Reciprocal of peak current versus frequency with constant peak voltage for a real inductor.

The term with C should be left out of the expression for impedance, and hence

$$\text{at } f = 0.00, Z = R$$

and

$$\frac{1}{i_{\text{peak}}} = \frac{R}{V} = \frac{0.1}{A} = \frac{R}{50\,V},$$

so

$$R = 5\,\Omega.$$

Since we are told that the curve is essentially a straight line for frequencies above 1 cps, it will be reasonable to neglect R in the expression for Z, giving $Z = 2\pi fL$ and

$$\frac{1}{i_{peak}} = \frac{Z}{V} = 0.04\ \pi fL.$$

The slope of the curve is

$$\text{slope} = \frac{\Delta\left(\dfrac{1}{i_{peak}}\right)}{\Delta f} = 0.04\ \pi L = \frac{126/\text{A} - 0}{100\ \text{cps} - 0} = 1.26\ \frac{1}{\text{A·cps}},$$

so

$$L = \frac{\text{slope}}{0.04\pi} = 10\ \text{H}.$$

Thus the inductor is a 10 H inductor which has an effective series resistance of 5 Ω. This is not a unique analysis of the circuit since there could be, and probably is, an effective parallel capacitance as shown in Fig 21.3b; however, since the curve is essentially a straight line except near $f = 0$, any such capacitance is too small to be detected in the 0–100 cps range.

PROBLEM 21.4: Given an inductor of 0.010 H, what value of capacitance would you need in order to build a series circuit that resonates at 50,000 Hz? (21.5)

Solution:

Strategy: The circuit described is an RLC circuit but we are not given the value of the resistance. Since we are not asked to find the impedance at resonance, the resistance value is not needed.

The resonant frequency of a series RLC circuit is given by

$$f_0 = \frac{1}{2\pi\sqrt{LC}},$$

so given f_0, the resonant frequency, and L, the inductance, C, must be given by

$$C = \frac{1}{4\pi^2 f_0^2 L} = \frac{1}{39.48 \times 25 \times 10^8/\text{s}^2 \times 0.010\ \text{H}} = 1.0 \times 10^{-9}\ \text{F}.$$

A 0.001 µF capacitor will be needed to make the desired series resonance with a 0.010 H inductor.

PROBLEM 21.5: An RCL circuit consists of a 500 Ω resistor in series with a 100 mH inductor and a 20 µF capacitor. At a frequency of 60 Hz, what is the impedance of this circuit and what is the phase shift of the current relative to the voltage? (21.5)

Solution:

> **Strategy:** This is a straightforward calculation of the impedance and phase shift in an ac circuit.

If we let f be the frequency, R be the resistance, C be the capacitance, and L be the inductance, then the impedance is given by text Eq. 20.17 as

$$Z = \sqrt{R^2 + \left(2\pi fL - \frac{1}{2\pi fC}\right)^2}.$$

If we also let

$$X_C = \frac{1}{2\pi fC}$$

and

$$X_L = 2\pi fL,$$

then the phase shift is given by

$$\phi = \tan^{-1}\left(\frac{X_L - X_C}{R}\right),$$

and we can also write

$$Z = \sqrt{R^2 + (X_L - X_C)^2}.$$

A substitution of the numbers will give the requested result. We find

$$X_C = 132\ \Omega,\ X_L = 37.7\ \Omega,\ \phi = -11°\ \text{and}\ Z = 509\ \Omega.$$

PROBLEM 21.6: A series RLC circuit has the behavior shown in Fig 21.4a at low frequencies and the behavior shown in Fig 21.4b at high frequencies. In both cases the lines are nearly linear (straight). In both cases the current is the rms current and the applied rms voltage is 100 V. Neglecting the series resistance, what is the predicted resonant frequency of this series RLC circuit? (21.6)

Solution:

> **Strategy:** Analyze the low and high frequency behavior of an RLC impedance, neglecting the resistance of the circuit. At low frequencies the capacitance dominates the impedance, while at high frequencies the inductance will dominate.

The necessary analysis is similar to that of Problem 21.4. Since $V_{rms} = Z\,i_{rms}$ with Z given by text Eq. 21.20, we need to analyze Z at high and low frequencies. The terms high and low make sense by meaning capacitive reactance (low frequency) or inductive reactance (high frequency).

For low frequency

$$Z = \frac{1}{2\pi fC} \quad \text{so} \quad i_{rms} = 2\pi C V_{rms} f.$$

Thus i_{rms} versus f is a straight line, through $i_{rms} = 0$ at $f = 0$, whose slope is given by $2\pi C V_{rms}$.

Measuring the graph in Fig 21.4a gives us

$$2\pi CV = 628\ C = \text{slope} = \frac{6.28 \times 10^{-4}\ A - 0\ A}{100\ \text{cps} - 0\ \text{cps}},$$

$$C = 1.0 \times 10^{-8}\ F = 0.01\ \mu F.$$

(a)

(b)

Fig. 21.4: (a) Low frequency behavior of a series RLC circuit. (b) High frequency behavior of a series RLC circuit.

For high frequency

$$Z = 2\pi fL \quad \text{so} \quad \frac{1}{i_{rms}} = \frac{2\pi L}{V_{rms}}\ f.$$

Thus $1/i_{rms}$ versus f is a straight line whose slope is $2\pi L/V_{rms}$. Measuring the graph in Fig 21.4b gives us

$$2\pi L/V_{rms} = 6.28 \times 10^{-2}\, L = \text{slope} = \frac{62.9 - 0.1}{2.0 \times 10^4 - 1.0 \times 10^4}\ /A\cdot s.$$

so

$$L = 0.10\ H.$$

The inductance is 0.10 H and the capacitance is 0.01 µF. This gives a resonant frequency f_0 of

$$f_0 = \frac{1}{2\pi\sqrt{LC}} = 5,000\ cps.$$

PROBLEM 21.7: The transformer in a CD audio system has an rms output of 12.4 V. What is the rms voltage across the load resistor R shown in Fig 21.5a? (21.7)

Solution:

Strategy: The diode cuts off any negative voltage, so the output will be a series of peaks separated by periods of no voltage as shown in text Fig. 20.20. These periods of no voltage will affect the rms value obtained from the definition of rms. The rms input voltage will enable you to find the peak output voltage. Then apply the definition of rms again to answer the question.

| (a) | (b) |

Fig. 21.5: (a) A half wave rectifier circuit. (b) The output wave form for a sine wave input.

The circuit shown in Fig 21.5a has an output as shown in Fig 21.5b. These figures are similar to text Fig. 21.20. We let V_{rms} be the input rms voltage, V_0 the input peak voltage, V_r be the peak output voltage above zero, V_a the output rms voltage, and v the instantaneous voltage.

Text Eq. 21.10 allows us to find the size of V_0 in terms of V_{rms}; that is,

$$V_{rms} = \sqrt{\overline{v^2}} = \frac{V_0}{\sqrt{2}}$$

gives, using the first and last terms,

$$V_O = \sqrt{2}\, V_{rms}.$$

We assume that there is no loss in the diode so that the value of V_r is the same as V_O; that is,

$$V_r = V_O = \sqrt{2}\, V_{rms}.$$

However when we use text Eq. 21.10 again to get the rms output voltage V_a, the answer will change since the output voltage is zero for half a cycle. Without the diode we would have had the same contribution form the negative portion of the voltage as from the positive portion. With the diode the negative portion is zero.

Without the diode

$$\overline{v^2} = \frac{V_O^2}{2}, \qquad \text{(without diode)}$$

while with the diode we must get only half of that value, that is,

$$\overline{v^2} = \frac{V_r^2}{4}. \qquad \text{(with diode)}$$

Thus

$$V_a = \sqrt{\overline{v^2}} = \sqrt{\frac{V_r^2}{4}} = \frac{V_r}{2}.$$

For our particular problem $V_{rms} = 12.4$ V, so $V_r = 17.5$ V and hence $V_a = 8.77$ V.

The rms output voltage is 8.77 V.

PROBLEM 21.8: What is the total gain of the two-operational-amplifier combination shown in Fig 21.6 if $R_1 = 2$ kΩ, $R_2 = 10$ kΩ, $R_3 = 5$ kΩ, and $R_4 = 25$ kΩ? (21.8)

Fig. 21.6: An inverting amplifier and a noninverting amplifier connected in series. On the figure IN+ is the noninverting input, IN− is the inverting input, and OUT indicates the output of the amplifier. If wires cross without a heavy dot, there is no connection.

Solution:

> **Strategy:** Treat this problem as two amplifiers. The output of the first amplifier is the input for the second amplifier. Since the voltage output is the gain times the voltage in for each amplifier, we expect the total gain to be the product of the gains of each amplifier

Let V_t be the output voltage of the first amplifier, G_1 be the gain of the inverting stage, G_2 be the gain of the noninverting stage, V_{in} be the input voltage, and V_0 be the final output voltage. Referring to Fig 21.6, we can express the gains of each stage in terms of the resistor values as

$$G_1 = -\frac{R_2}{R_1},$$

and

$$G_2 = \frac{R_3 + R_4}{R_3}.$$

Then V_t is given by

$$V_t = G_1 V_{in}$$

and

$$V_o = G_2 V_t = G_2 G_1 V_0 = G V_{in}.$$

So the total gain, G, is given by

$$G = G_2 G_1 = -30.$$

The output voltage is inverted and 30 times the input voltage.

GENERAL COMMENTS

Our model for this and the previous chapter is to assume that we can have perfect inductors, capacitors, and resistors. For example, an inductor is just an inductor without capacitance and resistance. All real inductors will have some capacitance and resistance; all real capacitors will have some resistance and inductance; and all real resistors will have some inductance and capacitance. A properly designed circuit element will minimize these often unwanted effects. However, at high frequencies and large currents the real nature of the circuit elements must be considered when circuits are designed. Unless otherwise indicated in the problem, don't include the real complications of the circuit elements.

SUGGESTED READINGS

Amerlio, G.F., "Charge Coupled Devices." *Scientific American,* February 1974, p. 22.

Brattain, W.H., "The Genesis of the Transistor." *The Physics Teacher,* March 1968, p. 109.

Brophy, J. J., *Basic Electronics for Scientists*, 4th ed. New York: McGraw-Hill, 1983.

Chipman, R. A., "DeForest and the Triode Detector." *Scientific American,* March 1965, p. 92.

Duffy, R. J., and U. Haber-Schaim, "Establishing V = L (dI/dt) Directly from Experiment." *American Journal of Physics,* February 1977, p. 170.

Dunsheath, P., *A History of Electrical Power Engineering.* Cambridge, Mass.: MIT Press, 1962.

Fink, D. G., *The Physics of Television.* Garden City, N.Y.: Anchor Books, 1960.

Heath, F.G., "Largescale Integration in Electronics." *Scientific American,* February 1970, p. 22.

Huffsey, R. R., and C. A. Huffsey, *Descriptive Electronics.* New York: Holt, Rinehart and Winston, 1970.

Issue on Microelectronics. *Scientific American,* September 1977.

MacInnes, I., "Notes on the Condition for Maximum Power Transfer to the Load." *The Physics Teacher,* April 1985, p. 224.

Meyer, H. W., *History of Electricity and Magnetism.* Norwalk, Conn.: Brundy Library, 1972.

Shiers, G. "The Induction Coil." *Scientific American,* May 1971, p. 80.

Wilson, M., "Joseph Henry." *Scientific American,* July 1954, p. 72.

Chapter 22
Geometrical Optics

IMPORTANT TERMS

Fill in the blanks with the appropriate word or words. The laws of reflection and refraction as well as the thin lens equation are also included in these exercises.

1. The straight line used to represent the path of light is called a light _____.

2. The angle between the incoming light and the normal to a surface is called the _____.

3. The law of refraction is also known as _____ law.

4. If the angle of incidence is greater than the critical angle, _____ will occur.

5. A lens whose thickness is small compared with its focal length is called a(n) _____ lens.

6. When light rays behave as if they are diverging from a point without actually ever having been at that point we say that a(n) _____ image has been formed.

7. The _____ of a lens or mirror are the failures of that lens or mirror to give a perfect image.

8. _____ light has frequencies just below (lower than) those of visible light.

9. The plane of _____ contains the normal, the incident ray, and the reflected ray.

10. The _____ is the minimum angle of reflection at which all internal light is totally reflected.

11. A lens that bends light to separate parallel rays away from each other is called a(n) _____ lens.

12. When the light rays that form an image actually intersect at that image, we have a(n) _____ image.

13. The _____ magnification of an image is the ratio of the image size to the size of the object that produced that image.

14. The spreading of light by wavelength as the light passes through a substance is the _____ of that substance.

15. _____ light has frequencies just above (higher than) those of visible light.

16. The law of _____ gives us the relation between the angle of incidence of a light ray and the angle at which that ray is reflected.

17. The _____ for a substance is the ratio of the speed of light in a vacuum to the speed in the substance.

final reflected ray by CD. Angles α and β are used to help us determine the angle of incidence. The first angle of incidence is γ and the first reflection is at an angle of δ with respect to the normal. The light hits the second mirror at an angle λ with respect to the normal of the second mirror and is reflected at an angle η with respect to the normal. The angle we want, the angle between the ray and the vertical, is φ. The angles ψ, μ, and ρ are useful angles to help us with our calculations.

Argument to Show That δ = 90° − Tilt Angle = 45°

Since the first mirror is at 45° with respect to the horizontal, we know that α is also 45°, and by similar sides we also know that β is also 45°. In summary, tilt = 45° implies α = 45°, which implies β = 45°, which makes γ = 45°.

The law of reflection (angle of incidence = angle of reflection) gives δ = 45°.

Since γ + δ = 90°, the line BC is horizontal. We did not worry about this when the initial drawing was made. Fig. 22.1a is only a sketch to help us think about the problem. It is best not to draw figures at special angles (45°, 90°, etc.) unless you know that these are the correct angles. By avoiding these special angles you will avoid the temptation to make use of symmetries that do not exist.

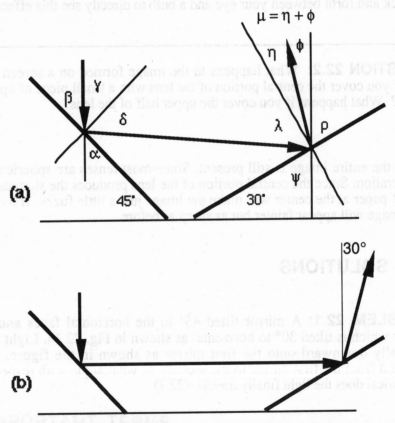

Fig. 22.1: (a) Two facing mirrors with light traveling vertically downward hitting the mirror on the left. A reflected light path is drawn which makes no assumptions about symmetry. (b) The correct light path.

Argument to Show the Angle Between Second Normal and the Horizontal, λ, Is 90° − Tilt

We now need the normal to the second mirror. Since that mirror is at 30°, the angle ψ is 60° and the angle ρ must equal ψ, and hence ρ is 60°. If ρ is 60°, then μ is 30° and the normal is tilted 60° with respect to the horizontal.

In general the normal always has a tilt λ equal to ψ. Since ψ = 60°, λ = 60°, and since λ + μ = 90°, μ = 30°.

We have already determined that the incident ray is horizontal as it hits the second mirror, so the angle of incidence is 60°. This is a result of the particular geometry of this problem. In general, the incident angle is not the tilt of the normal.

The law of reflection gives λ = η, so η = 60°.

The angle between the reflected ray and the vertical is the angle φ, and by inspection

$$\phi = \mu - \eta = -30°.$$

The minus sign indicates that φ lies on the opposite side of the vertical than we assumed in the drawing. Fig. 22.1b shows our findings.

PROBLEM 22.2: Light is incident at 60° to the normal of one side of a glass prism whose apex angle is 30°. At what angle with respect to the normal does the light exit from the other side? Assume the index of refraction of the glass is 1.5. (22.2)

Solution:

Strategy: Use Snell's law at each surface. The major problem is finding the final angle of incidence as the light passes from the glass out into the air.

Fig. 22.2 shows the geometry of the problem. The following notation is used.

Quantity	Symbol	Value
initial incident angle	ϕ_1	60°
apex angle of prism	λ	30°
first refracted angle	ϕ_2	?
second incident angle	ϕ_3	?
final refracted angle	ϕ_4	?
auxiliary angles	α, ψ	?
index of refraction of air	n_1	1.00
index of refraction of glass	n_2	1.50

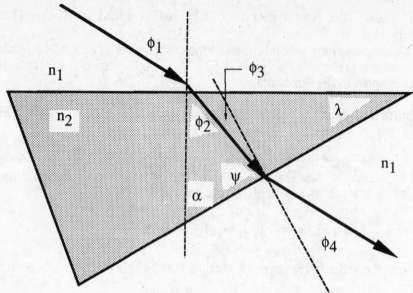

Fig. 22.2: Light passing through a prism. The normals are shown dotted and the light ray as a thick arrow.

Applying Snell's law at the first surface gives

$$n_1 \sin \phi_1 = n_2 \sin \phi_2,$$

which can be rewritten as

$$\phi_2 = \sin^{-1}\left(\frac{n_1}{n_2} \sin \phi_1\right).$$

Likewise,

$$\phi_4 = \sin^{-1}\left(\frac{n_2}{n_1} \sin \phi_3\right).$$

The relation between ϕ_2 and ϕ_3 is found by noting that $\alpha = 90° - \lambda$. This must be true since α is the third angle of a right triangle. Then

$$\psi = 180° - \phi_2 - \alpha = 90° + \lambda - \phi_2$$

and

$$\phi_3 = 90° - \psi = \phi_2 - \lambda.$$

This gives

$$\phi_4 = \sin^{-1}\left\{\frac{n_2}{n_1} \sin \left[\sin^{-1}\left(\frac{n_1}{n_2} \sin \phi_1\right) - \lambda \right]\right\}.$$

Substituting the initial numbers gives $\phi_4 = 7.9°$.

The reflected ray emerges at an angle of 7.9° with respect to the final normal.

PROBLEM 22.3: A thin lens with a focal length of 10 cm is located 25 cm to the left of another lens of focal length 5 cm. A 2 cm tall object is located 30 cm to the left of the first lens. What is the nature, location, and size of the final image? Draw the ray diagram. (22.6)

Solution:

> **Strategy:** Apply the thin lens formula two times in succession. The image formed by the first lens serves as the object for the second lens.

The lenses and spacing are shown in Fig. 22.3a. The following notation is used.

Quantity	Symbol	Value
left lens focal length	f_1	10 cm
right lens focal length	f_2	5.0 cm
separation of the lenses	d	25 cm
distance of object from 1st lens	o_1	30 cm
1st image distance	i_1	?
2nd object distance	o_2	?
final object distance	i_2	?
height of object	h	2 cm
height of final image	h_f	?
total magnification	m	?

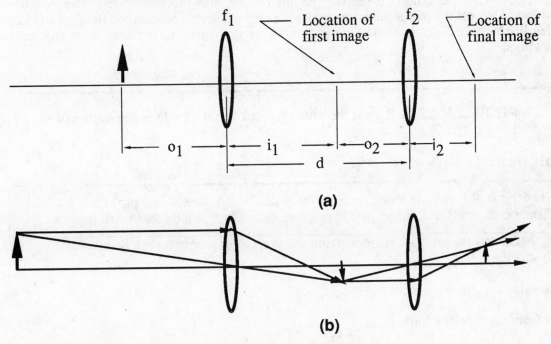

Fig. 22.3:   (a) Two thin lenses. (b) The ray diagram.

We apply the thin lens formula to the first lens:

$$\frac{1}{o_1} + \frac{1}{i_1} = \frac{1}{f_1}.$$

Our numbers give

$$i_1 = 15 \text{ cm}$$

since

$$i_1 + o_2 = d, \; o_2 = 10 \text{ cm}.$$

Then using

$$\frac{1}{o_2} + \frac{1}{i_2} = \frac{1}{f_2},$$

we find that

$$i_2 = 10 \text{ cm}.$$

Since i_2 is positive the image is to the right and hence real.

To find the image size, use the general formula for magnification, $m = -\dfrac{i}{o}$ and again note that the magnified image from the first lens is the object for the second lens, so

$$m = m_1 \times m_2.$$

For the given numbers

$$m = \left(-\frac{1}{2}\right) \times (-1) = \frac{1}{2}.$$

Thus the image is real, erect ($m > 0$), 1 cm tall, and is located 10 cm to the right of the second lens.

Fig. 22.3b is a ray diagram. To keep the picture simple, only three rays are drawn, one undefelected ray for the tail of the arrow and two rays for the tip according to the first and last rules in the text. Draw on this figure other lines according to the text rules for scale drawings.

PROBLEM 22.4: Repeat the previous problem with a lens separation of 16 cm. (22.6)

Solution:

> **Strategy:** Repeat the same process as used in the preceding problem. Positive distances indicate real images, while negative distances will indicate virtual images.

Fig. 22.3a and the table of symbols from the preceding problem may be used again. The only change is that

$$d = 16 \text{ cm}.$$

The value for i_1 is unchanged but

$$o_2 = 1.0 \text{ cm}$$

and from the thin lens formula

$$i_2 = -1.25 \text{ cm},$$

so the image is 1.25 cm to the left ($i_2 < 0$) of the second lens. Thus the image is virtual.

The value of m_1 is unchanged and

$$m_2 = -\frac{i_2}{o_2} = -\frac{-1.25 \text{ cm}}{1.0 \text{ cm}} = 1.25$$

so

$$m = m_1 \times m_2 = -0.63.$$

The image is inverted, 1.25 cm tall, virtual, and located 1.25 cm to the left of the second lens. A ray diagram is shown in Fig. 22.4.

Fig. 22.4: Ray diagram for Problem 22.4.

To keep the picture simple, only three rays are drawn, one undefelected ray for the tail of the arrow and two rays for the tip according to the first and last rules in the text.

PROBLEM 22.5: Repeat Problem 22.3 with the separation of the lenses 25 cm, as given in that problem, but with the second lens a diverging lens of focal length 5.0 cm. (22.6)

Solution:

> **Strategy:** Repeat the same process as used in the preceding problems. Positive distances indicate real images, while negative distances will indicate virtual images. A diverging lens has a negative focal length so $f_2 = -5.0$ cm.

The behavior of the light through the first lens is the same as in Problem 22.3, so the value of $o_2 = 10$ cm as before but now we must take $f_2 = -5.0$ cm. This gives

$$i_2 = -\frac{10}{3} \text{ cm,}$$

so the image is virtual and about 3.3 cm to the left of the second lens. The magnification of the second lens is 1/3, so

$$m = m_1 \times m_2 = -\frac{1}{6} = -0.17.$$

The image is inverted, 0.33 cm tall, virtual, and located 3.3 cm to the left of the second lens. A ray diagram is shown in Fig. 22.5.

Fig. 22.5: Ray diagram for Problem 22.5.

To keep the picture simple, only three rays are drawn, one undefelected ray for the tail of the arrow and two rays for the tip according to the first and last rules in the text.

PROBLEM 22.6: An object is located 2.0 m from a convex spherical mirror with radius 0.80 m. Where is the image located and what is its relative size? (22.7)

Solution:

> **Strategy:** Use the thin lens formula with focal length equal to half the radius of the mirror. A convex mirror is diverging and has a negative radius.

Let R be the radius of the mirror, f its effective focal length, o the object distance, i the image distance, h the object height, and h_i the image height. Then

$$f = -\frac{R}{2} \quad \text{and} \quad \frac{1}{o} + \frac{1}{i} = -\frac{2}{R}.$$

The image size is given by

$$\frac{h_i}{h} = -\frac{i}{o}.$$

Entering numbers gives

$$i = -\frac{1}{3} \text{ m} \quad \text{and} \quad \frac{h_i}{h} = +\frac{1}{6}.$$

The image is erect and lies 0.33 m to the left of the mirror's surface. Remember that the sign convention for mirrors measures both image and object distance positive to the same side of the mirror. This is shown in the ray diagram in Fig. 22.6. Two rays are shown. Ray 1 travels parallel to the figure axis and is chosen so as to be reflected along a line through the focal point. Ray 2 passes through the center of the sphere defining the mirror and is reflected directly back.

Fig. 22.6: Ray diagram for Problem 22.6.

SUGGESTED READINGS

Davis, J. F., and T. B. Greenslade, Jr., "Computer Modeling of Mirage Formation." *The Physics Teacher,* January 1991, p. 47.

Galili, I., F. Goldberg, and S. Bendall, "Some Reflections on Plane Mirrors and Images." *The Physics Teacher,* October 1991, p. 471.

Leinoff, S., "Ray Tracing with Virtual Objects." *The Physics Teacher*, May 1991, p. 275.

Stix, G., "Science and Business: Bug-Eyed, Gradient-Index Lenses Bend Light to Order." *Scientific American,* November 1990, p. 134.

Young, M., "The Pinhole Camera: Imaging Without Lenses or Mirrors." *The Physics Teacher,* December 1989, p. 648.

Chapter 23
Optical Instruments

IMPORTANT TERMS

Fill in the blanks with the appropriate word or words.

1. A(n) _____ lens is one whose focal length can be changed but must be re-focused at each setting.

2. A(n) _____ microscope consists of a lens close to the object under study and another lens near the observer's eye.

3. _____ is the eye's ability to focus on objects at various distances.

4. A(n) _____ glass consists of a single converging lens.

5. A(n) _____ lens is a variable focus lens that maintains its focus throughout its range.

6. The ratio of the focal length of a lens to the diameter of the lens is the _____ of the lens.

7. The lens of a microscope or telescope closest to the object is called the _____ lens.

8. The smallest distance from the unaided eye that an object can be held and still appear unblurred is the _____.

9. The _____ telescope has a mirror as its objective.

10. The lens of a microscope closest to the eye of the observer is called the _____.

11. If the human eye or other lens is distorted from its normal spherical shape, the lens is said to suffer from _____.

12. The ratio of the angular size of an object as seen through a lens to the angular size without the lens is called the _____ magnification.

13. A(n) _____ telescope has a diverging lens as its eyepiece.

14. The reciprocal of the focal length of a lens in meters is the strength of the lens in _____.

15. A(n) _____ telescope consists of two converging lenses (or two sets of lenses).

Write the definitions of the following:

16. astigmatism (23.1):

17. angular magnification (23.2):

18. objective lens (23.4):

19. refracting astronomical telescope (23.5):

20. zoom lens (23.6):

21. accommodation (23.1):

22. diopter (23.1):

23. f-number (23.3):

24. eyepiece (23.4):

25. Galilean telescope (23.5):

26. varifocal lens (23.6):

27. near point (23.1):

28. magnifying glass (23.2):

29. compound microscope (23.4):

30. ocular (23.4):

31. Newtonian telescope (23.5):

ANSWERS TO 1–15, IMPORTANT TERMS

1. varifocal; 2. compound; 3. accommodation; 4. magnifying; 5. zoom; 6. f-number; 7. objective; 8. near point; 9. Newtonian; 10. eyepiece (or ocular); 11. astigmatism; 12. angular; 13. Galilean; 14. diopters; 15. refracting (or inverting).

SAMPLE QUESTIONS

QUESTION 23.1: What is the advantage of having large-diameter objective lenses on telescopes or binoculars?

Answer:

Large diameter lenses can collect more light than small diameter optics. Since the image size for a given object depends upon the focal length of the lens, the brightness of the image will depend upon how much light is focused into the image. The larger the lens, the brighter the image is.

QUESTION 23.2: Given an additional converging lens, how can you use it to modify an inverting telescope so that the final image will be erect? You need not keep the tube length the same.

Answer:

Place the lens between the objective and the eyepiece so the image produced by the objective is the object for the new lens. The new lens must be placed so as to form a real image. This real image will be examined by the eyepiece.

SAMPLE SOLUTIONS

> **PROBLEM 23.1:** A farsighted eye has a near point of 120 cm. A converging lens is used to provide clear vision of a book placed 25 cm in front of the eye. What is the power of the lens used in diopters? (23.1)

Solution:

> **Strategy:** The idea is to use the lens to produce a virtual image of the book at the near point. Use the thin lens equation. A correct choice of units will make the problem easier.

The idea of a corrective lens for farsightness is to place a converging lens near the eye. The power of the converging lens is chosen so as to produce an image at the near point of the eye when the object is held at a convenient distance. We let o be the distance the book is held, i be the location of the image (i is chosen to be the near point), f the focal length of the lens and p the power of the lens. This is shown in Fig. 23.1.

If we neglect the distance between the eye and the lens, we have

$$\frac{1}{o} + \frac{1}{i} = \frac{1}{f}.$$

If, in addition, all distances are measured in meters, we can also write

$$\frac{1}{o} + \frac{1}{i} = p \qquad p \text{ in diopters, } i \text{ and } o \text{ in meters.}$$

Fig. 23.1: A converging lens used to allow reading at a convenient distance.

For this problem $o = 0.25$ m and $i = -1.2$ m. The minus sign for i follows from the assumption that the object and the image are on the same side of the lens.

Then, entering the numbers, we get
$$p = +3.2 \text{ diopters}.$$

This corresponds to a focal length of 0.31 m = 31 cm. The lens will probably be within 2.5 cm of the eye, so our assumption that we could neglect the distance between the eye and the lens is reasonable.

PROBLEM 23.2:   A magnifying glass held 2.0 m in front of a TV screen projects an image of the screen on a wall 3.0 m from the TV screen. What is the power of the lens? (23.2)

Solution:

> **Strategy:** The lens produces a real image of the TV screen at the given distance. Use the thin lens equation. A correct choice of units will make the problem easier.

We let o be the distance between the lens and the TV screen, i be the location of the image beyond the lens, f the focal length of the lens and p the power of the lens. This is shown in Fig. 23.2.

The thin lens formula is

$$\frac{1}{o} + \frac{1}{i} = \frac{1}{f}.$$

If, in addition, all distances are measured in meters, we can also write

$$\frac{1}{o} + \frac{1}{i} = p \qquad p \text{ in diopters, } i \text{ and } o \text{ in meters.}$$

For this problem, $o = 2.0$ m and $i = 3\,\text{m} - 2\,\text{m} = 1$ m, so

$$\frac{1}{2} + \frac{1}{1} = 1.5 = p.$$

The magnifying glass must have a power of +1.5 diopters.

Fig. 23.2:   Lens used to produce an image of a TV picture.

PROBLEM 23.3: A dime (diameter about 1.8 cm) is viewed through a 7 × magnifying glass. Approximately what angle does it appear to subtend at the eye? (23.2)

Solution:

> **Strategy:** Assume the image of the dime is formed at infinity. Knowing the magnifying power and the angular size of the dime without the magnifying glass, we can calculate the apparent size with the magnifying glass.

A 7× magnifying glass produces a ratio of angular sizes of 7:

$$M = \frac{\text{angular size with aid}}{\text{angular size without aid}} = 7 \,.$$

The angular size without aid is $\dfrac{\text{diameter}}{25 \text{ cm}} = \dfrac{1.8}{25} = 0.072 = 4.1°$,

so

$$\text{apparent angular size} = 7 \times \text{angular size without aid} = 0.50 = 29°.$$

The dime will have an apparent angular size of 0.50 radians.

PROBLEM 23.4: A camera lens has a diameter of 2.0 cm and a focal length of 6.0 cm. Using a particular film, a proper exposure of a scene can be made with the camera set to an f-number of 5.6 and an exposure time of 1/60 s. Can the camera record the same scene with the same film at a speed of 1/500 s? What is the maximum shutter speed possible to properly expose this scene? (23.3)

Solution:

> **Strategy:** Calculate the minimum f-ratio possible for this lens and compare that number with the f-number needed for the same exposure at 1/500 s.

Film responds to the total light hitting it per unit area during the exposure, so we require that

$$\frac{\text{exposure time}}{(\text{f-number})^2} = \text{constant} \,.$$

The f-number is adjusted by a diaphragm that reduces the open diameter of the lens, so the minimum f-number (which must occur with the diaphragm completely open) is given by

$$\text{f-number}_{min} = \frac{\text{focal length}}{\text{true diameter}} = 3.0 \,.$$

To change the exposure time from 1/60 s to 1/500 s we must have

$$\frac{\text{exposure time}}{(\text{f-number})^2} = \frac{1/60 \text{ s}}{(5.6)^2} = \frac{1/500 \text{ s}}{(\text{f-number})^2} \,,$$

which gives

$$\text{f-number} = 1.9 \,.$$

Since 1.9 is less than the minimum possible f-ratio, it would be impossible to take the exposure at 1/500 s.

The minimum exposure time is given by

$$\frac{\text{exposure time}}{(\text{f-number})^2} = \frac{1/60 \text{ s}}{(5.6)^2} = \frac{t_{min}}{(3.0)^2},$$

which gives

$$t_{min} = \frac{1}{209} \text{ s}.$$

Since many cameras have an exposure setting of 1/200 s, that would probably be the best minimum time and hence the maximum shutter speed.

PROBLEM 23.5: The maximum useful magnification of an ordinary visible light microscope is about 2000×. A microscope comes with a set of eyepieces, the most powerful of which is marked 25×. Assuming that the manufacturer gave you the maximum useful eyepiece, what is the focal length of the objective used to get the maximum magnification? (23.4)

Solution:

Strategy: Assume the standard spacing of 16 cm for the microscope body and use the relationship between the magnification of a microscope and the eyepiece used.

Let M be the magnification of the microscope, M_e be the magnification of the eyepiece, L the body length of the microscope, and f_0 the focal length of the objective. We can write the relation among these as

$$M = -\frac{L}{f_0} M_e.$$

Take $M = -2000\times$, $M_e = 25\times$, and $L = 16$ cm to find: $f_0 = 0.20$ cm.

The objective of the microscope has a focal length of 2.0 mm.

PROBLEM 23.6: You are watching a basketball game with a pair of 7 × 50 binoculars. If a 2.0 m tall basketball player is 50 m away from you, what is his apparent angular size in radians as seen through the glasses? (23.5)

Solution:

Strategy: Compute the true angular size of the player and use the definition of magnification.

The magnification of a pair of binoculars is defined as

$$M = \frac{\text{angular size with aid}}{\text{angular size without aid}}.$$

Since we are given $M = 7$, we need the true angular size in order to calculate the apparent angular size.

For an object whose size is small compared with its distance,

$$\text{angular size without aid} = \frac{\text{true size}}{\text{distance}} = \frac{2.0 \text{ m}}{50 \text{ m}} = 0.040 \text{ radians} .$$

The angular size as seen through the binoculars is just 7 times this, so the player seem to intercept an angular size of 0.28 radians.

PROBLEM 23.7: An astronomical telescope is used to view the moon. If the objective has a focal length of 150 cm and a diameter 10 cm and the eyepiece a focal length of 3.0 cm, what is the angular magnification of the moon? If the unmagnified moon has an angular diameter of 0.50°, what is the angular size of the lunar image? (23.5)

Solution:

> **Strategy:** Since the object is far away and since we assume the image is formed at infinity, the angular magnification is given by the ratio of the focal lengths.

The moon is far away compared to the focal length of the objective and we will assume the telescope is adjusted to form the image at infinity, so the magnifying power, M, is given by

$$M = \frac{\text{focal length of objective}}{\text{focal length of eyepiece}} = \frac{150}{3} = 50.$$

The definition of angular magnification gives

$$\text{apparent angular size} = M \times \text{true angular size} = 50 \times 0.5° = 25°.$$

The diameter of the objective is not needed to get these results.

SUGGESTED READINGS

Anicin, B. A., V. M. Babovic, and D. M. Davidovic, "Fresnel Lenses." *American Journal of Physics,* April 1989, p. 312.

Begbie, G. H., *Seeing and the Eye.* Garden City, N.Y.: The Natural History Press, 1969.

Blunk, W., and A. Burkert, "Doing Physics: A Telescope Workshop." *The Physics Teacher,* February 1987, p. 111.

Childers, R. L., "A Photographer's Rule of Thumb." *The Physics Teacher*, May 1991, p. 292.

Cox, A., *Photographic Optics: A Modern Approach to the Technique of Definition.* New York: Focal Press, 1974.

Darius, J., *Beyond Vision*. Oxford: Oxford University Press, 1984. A collection of over 100 scientific photographs selected for esthetic appeal and historical significance. Information inaccessible to the naked eye is shown over a range from submicroscopic to cosmic.

Drake, S., and C. T. Kowal, "Galileo's Sightings of Neptune." *Scientific American*, December 1980, p. 74.

Fender, D. H., "Control Mechanics of the Eye." *Scientific American*, July 1964, p. 24.

Flamsteed, S., "The View from Below." *Discover*, January 1991, p. 24.

Gibilisco, S., *Optical Illusions: Puzzles, Paradoxes, and Brain Teasers #4*. Blue Ridge Summit, Pa.: TAB Books, 1990. An entertaining look at illusions of depth, motion, color, size, and more.

Gilliam, B., "Geometrical Illusions." *Scientific American*, January 1980, p. 102.

Greenslade, T., Jr., and M. W., Green, III, "Experiments with Stereoscopic Images." *The Physics Teacher*, April 1973, p. 215.

Gregory, R. L., "Visual Illusions." *Scientific American*, November 1968, p. 66.

Gregory, R. L., *The Intelligent Eye*. New York: McGraw-Hill, 1970.

Hess, E. H., "Shadows and Depth Perception." *Scientific American*, March 1961, p. 138.

Huffman, A., "Measuring a Glasses Prescription." *American Journal of Physics*, April 1980, p. 309.

Ingals, A. G., *Amateur Telescope Making, Book One*. New York: Scientific American, 1964.

Ingals, A. G., *Amateur Telescope Making Advanced, Book Two*. New York: Scientific American, 1963.

Kennedy, D., "Inhibition in Visual Systems." *Scientific American*, July 1963, p. 122.

Koretz, J. F., and G. H. Handelman, "How the Human Eye Focuses." *Scientific American*, July 1988, p. 92.

Martin, B., J. M. Hill, and R. Angel, "The New Ground-Based Optical Telescopes." *Physics Today*, March 1991, p. 22.

Masland, R. H., "The functional architecture of the retina." *Scientific American*, December 1986, p. 102.

Minnaert, M., *The Nature of Light and Colour in the Open Air* (Reprint). New York: Dover Publications, 1954.

Ruiz, M. J., "The Physics of Visual Acuity." *The Physics Teacher*, September 1980, p. 457.

Wallington, G. G., "Present Day Photographic Lenses and their Characteristics." *The Physics Teacher*, November 1964, p. 381.

Wright, W. D., *The Rays Are Not Colored: Essays on the Science of Vision and Color*. New York: American Elsevier, 1967.

Chapter 24
Wave Optics

IMPORTANT TERMS

Fill in the blanks with the appropriate word or words.

1. A(n) _____ converts light to electrical energy and displays the intensity versus wavelength on a chart.

2. Sources with identical frequency and fixed relative phase are said to be _____.

3. The _____ surface (or _____ front) is defined as the set of points of constant phase.

4. A(n) _____ is a device to view the spectrum of a light source with the eye.

5. An array of a large number of equally spaced slits is called a(n) _____.

6. _____ shows itself in the wavelength dependence of the index of refraction.

7. The plane of _____ of light is the plane of the electric field.

8. _____ law relates the intensity of light passing through two polarizers to the angle between their planes of polarization.

9. When light is _____ by dust, the light is absorbed by the dust grains and then re-emitted in an arbitrary direction.

10. _____ are inexpensive plastic polarizers.

11. The angle of maximum polarization of reflected light is given by _____ law.

12. A(n) _____ makes a photographic record of a spectrum.

13. The rotation of the plane of polarization of incident light when the light passes through a substance is an example of _____.

14. The ability to construct a new wavefront by drawing arcs centered on the previous wavefront is called _____ principle.

15. The _____ of an optical instrument is a measure of that instrument's ability to produce a distinct image of two objects which appear very close together.

16. Scattering of light by particles smaller than the wavelength is called _____ scattering.

17. Rayleigh's _____ is to choose the minimum (best) resolution as that angle at which the first minimum of the diffraction pattern of one source falls on the maximum of the other source.

18. The _____ of light is shown by light spreading into regions which should be shadowed from the light.

19. A(n) _____ coating on a lens is made of a material with an index of refraction between that of air and that of glass.

Write the definitions of the following:

20. Malus's law (24.10):

21. wavefront (or wave surface) (24.1):

22. diffraction (24.5):

23. interference (24.3):

24. antireflection coating (20.4):

25. dispersion (24.8):

26. optical activity (24.11):

27. Rayleigh scattering (24.10):

28. coherent (24.3):

29. polarization (24.9):

30. resolving power (20.9):

31. spectrograph (24.9):

32. spectroscope (24.9):

33. scattering (24.11):

34. Rayleigh's criterion (24.7):

35. Brewster's law (24.9):

36. Huygens' principle (24.1):

37. diffraction grating (24.6):

38. spectrophotometers (24.8):

ANSWERS TO 1–19, IMPORTANT TERMS

1. spectrophotometer; 2. coherent; 3. wave; 4. spectroscope; 5. diffraction grating; 6. dispersion; 7. polarization; 8. Malus's; 9. scattered; 10. polaroids; 11. Brewster's; 12. spectrograph; 13. optical activity; 14. Huygens'; 15. resolving power; 16. Rayleigh; 17. criterion; 18. diffraction; 19. antireflection.

SAMPLE QUESTIONS

QUESTION 24.1: The source of illumination for a Young's double-slit setup is white light. Suppose you cover one slit with a filter that transmits only blue light and the other slit with a filter that transmits only red light. What would you expect to see on the screen?

Answer:

You would expect to see superimposed single slit diffraction patterns. One will be in blue light and the other in red light. The double slit interference pattern will no longer exist.

QUESTION 24.2: A card is placed between a point source and a screen, but relatively close to the screen. Do you expect the card to cast a sharp shadow? If not, what do you expect and why?

Answer:

No. There will be some diffraction of light off the edge of the card. This will produce light and dark bands for each wavelength. If the light source is white, there will be a faint fuzzy band. It may require careful inspection to see this effect.

SAMPLE SOLUTIONS

PROBLEM 24.1: A beam of light has a wavelength $\lambda = 607$ nm in air. (a) What is its wavelength in diamond? (b) What is it in ethyl alcohol? (24.2)

Solution:

> **Strategy:** Wavelength depends upon velocity (and hence index of refraction) since frequency is unchanged at boundaries. Use the information in text Table 22.1.

When light crosses boundaries, it must keep the same frequency. Since the velocity may be different the wavelength much change. Since for a wave,

$$\text{velocity} = \text{wavelength} \times \text{frequency}$$

and

$$\text{index of refraction} = \frac{\text{speed in a vacuum}}{\text{speed in material}}$$

we can write

$$\frac{\text{wavelength in material}}{\text{wavelength in air}} = \frac{\text{velocity in material}}{\text{velocity in air}} = \frac{n_{air}}{n_{material}},$$

so for diamond

$$\text{wavelength}_{diamond} = \frac{1.003}{2.417} \times 607 \text{ nm} = 252 \text{ nm},$$

while for the ethyl alcohol

$$\text{wavelength}_{alcohol} = \frac{1.003}{1.362} \times 607 \text{ nm} = 447 \text{ nm}.$$

PROBLEM 24.2: A waterproof flashlight is dropped into a pond. It lands in the muck at the bottom, sticking up at an angle of 30° from the vertical, and at a depth of 1.5 m from the surface, with the light on. The flashlight is viewed from outside the pond at a height such that it appears to be seen end on. From this position it appears that the flashlight is 4.90 m horizontally from the edge of the pond. How far away from the edge of the pond is the flashlight? (24.2)

Solution:

Strategy: Apply the law of refraction at the surface of the water. This will enable you to calculate the apparent position of the flashlight as seen by the observer.

The following notation is used.

Quantity	Symbol	Value
angle of incidence	θ	30°
index of refraction of water	n_{water}	1.33
index of refraction of air	n_{air}	1.00
angle of refraction	ϕ	unknown
depth of flashlight	h	1.5 m
horizontal distance to flashlight	d	unknown
horizontal distance from source to where the observed light leaves the water	z	unknown
apparent displacement of light	x	unknown
apparent horizontal distance to flashlight	D	4.90 m

We sketch the problem in Fig. 24.1. Since the flashlight appears to be seen end on, a light ray leaving the flashlight normally is traveling at an angle of 30° to the vertical, is refracted at the air-water interface and reaches the observer's eyes as shown. Of the unknown items listed, we are asked to find the true position of the flashlight, d. The other quantities will be needed as part of the intermediate steps. The approach will be to use Snell's law at the interface to find the angle of refraction, ϕ, which will allow us to calculate the distance $x+z$. The distance z can also be determined from the given information, so we can clearly find d.

As is shown in Fig. 24.1a, the light that is seen by the observer travels from the source to the point marked A, is refracted and then travels to the observer's eyes. To the observer it appears that the light traveled in a straight line as shown by the dotted line. Knowing that the pond is 1.5 m deep, we can see from Fig. 24.1b that

$$\tan \theta = \frac{z}{h} = \tan 30° ,$$

and from Fig. 24.1c

$$\tan \phi = \frac{x + z}{h} .$$

Hence

$$x = h \left\{ \tan \phi - \tan 30° \right\},$$

and the answer is $d = D - x = D - h \left\{ \tan \phi - \tan 30° \right\}.$

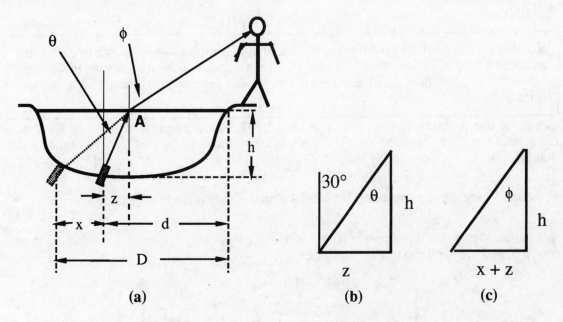

Fig. 24.1: (a) Geometry and light path for flashlight problem. (b) Geometry to determine z. (c) Geometry to determine $x + z$.

We only need to find ϕ, which can be done by using Snell's law at point A:

$$n_{\text{water}} \sin \theta = n_{\text{air}} \sin \phi$$

or

$$1.33 \times \sin 30° = 0.665 = 1.00 \times \sin \phi;$$

hence

$$\phi = \sin^{-1}(0.665) = 41.9°.$$

Since we need $\tan \phi$, it is best to do the calculation as one step on the calculator and get

$$\tan \phi = \tan \left\{ \sin^{-1} (0.665) \right\} = 0.89;$$

hence

$$d = 4.90 \text{ m} - 1.5 \text{ m} \times (\, 0.89 - 0.58 \,) = 4.4 \text{ m}.$$

The flashlight has fallen 4.4 m from the edge of the pond.

PROBLEM 24.3: A pair of narrow slits are separated 0.101 mm from each other. Light of wavelength 567 nm passes through the slits and falls on a screen 2.5 m away from the slits and parallel to the plane of the slits. What is the linear separation of the central maximum and either of the second minima? (24.3)

Solution:

> **Strategy:** The formulas for double slit maxima and minima give us the angle to the maxima and minima as seen from the slit. This angle and the distance to the screen allow us to calculate the linear separation of the central maximum and a second minimum.

We draw the problem in Fig. 24.2 and let d be the slit separation, D the slit to screen distance, x the linear distance from the central maximum to the second minima and λ be the wavelength.

From the figure we see that x and D are related to the shown angle ϕ by $\tan \phi = \dfrac{x}{D}$.

We have only shown a positive x. There is another symmetric position for negative ϕ and hence negative x with the same magnitude of ϕ.

Fig. 24.2: Light from two slits falling on a screen 2.5 m away.

If we let θ represent a general value of ϕ, the maxima for this double slit are given by (text Eq. 23.3a)

$$n\lambda = d \sin \theta \quad \text{for} \ n = 0, 1, 2, \ldots$$

and the minima by (text Eq. 23.3b)

$$\left(m + \frac{1}{2}\right)\lambda = d \sin \theta \qquad \text{for} \ m = 0, 1, 2, \ldots.$$

The central maximum is $n = 0$ or $\theta_{\text{central max}} = 0$. The second minima are given by $m = 1$, so the angle to a second minimum, ϕ, is given by

$$\left(1 + \frac{1}{2}\right)\lambda = d \sin \phi.$$

Since $\theta_{\text{central max}} = 0$, the angle denoted by ϕ is also the angle shown in Fig. 24.2. Thus

$$x = D \times \tan \phi = D \times \tan\left\{\sin^{-1}\left(\frac{3\lambda}{2d}\right)\right\}$$

$$x = 2.5 \ \text{m} \times \tan\left\{\sin^{-1}\left(\frac{3 \times 567 \times 10^{-9} \ \text{m}}{2 \times 1.01 \times 10^{-4} \ \text{m}}\right)\right\} = 0.021 \ \text{m} = 21 \ \text{mm}.$$

The second minima are 21 mm from the central maximum.

PROBLEM 24.4: A horizontal glass sheet is covered by thin water droplets containing a small amount of wetting agent. The wetting agent allows the water droplets to have a large horizontal area as shown in Fig. 24.3. How many molecules thick must a water droplet be for normally incident light of wavelength 500 nm to have a maximal reflection? Assume that the index of refraction of water is unchanged by the wetting agent (n = 1.33) and that a water molecule occupies a volume with a diameter of about 0.5 nm. (24.4)

Solution:

Strategy: Some light is reflected off the water and some passes through the water and is reflected off the glass. The optical path difference must be that given for constructive interference.

t

Fig. 24.3: Thin layer of water on a horizontal glass plate.

We assume that light is reflected off both the water and the glass as shown in Fig. 24.3. The incident ray is shown at an angle rather than the normal to show the paths taken by the light. Since the index of refraction of air is less than that of water and since the index of refraction of water is less than that for glass, constructive interference will occur for

$$m\lambda = 2nt,$$

where m is an integer 0, 1, 2, ..., and λ is the wavelength of light, n is the index of refraction of water and t is the water thickness (see text Eq. 23.5a). We assume that $m \neq 0$, since m = 0 would correspond to no water. So we take $m = 1$. Then the thickness of the water is

$$t = \frac{\lambda}{2n} = 188 \text{ nm}.$$

Since each water molecule occupies about 0.5 nm, the water layer is about

$$\frac{t}{0.5 \text{ nm}} = 4 \times 10^2 \text{ molecules thick}.$$

The water layer is about 400 molecules thick.

PROBLEM 24.5: Calculate the width of the central maximum in the single-slit diffraction pattern of yellow light of $\lambda = 589.0$ nm by a slit 0.275 mm wide viewed on a screen 3.00 m away. (24.5)

Solution:

Strategy: The central maximum will extend approximately from the first diffraction minimum on the right to the first minimum on the left.

Consider text Fig. 24.21, which shows the geometry for single slit diffraction. The minima occur at angles, θ, given by text Eq. 24.6:

$$n\lambda = b\sin\theta,$$

where n is an integer, λ is the wavelength of light and b is the slit width.

Scanning from the center ($\theta = 0$) we see light until θ satisfies the above equation with $n = \pm 1$, that is, from

$$\theta = \frac{\lambda}{b} \quad \text{to} \quad \theta = -\frac{\lambda}{b}.$$

The central width is thus given by

$$\Delta\theta = 2\,\frac{\lambda}{b} = 2\times\frac{589.0\times 10^{-9}\text{ m}}{2.75\times 10^{-4}\text{ m}} = 2\times(2.14\times 10^{-3}) = 4.28\times 10^{-3}.$$

Since this angle (given here in radians) is so small, we may convert it to a linear size by multiplying by the distance (sector of a circle S = Rθ), giving us a linear size of 1.28 cm.

The central maximum will be approximately 1.28 cm wide.

PROBLEM 24.6: A pair of narrow slits are used to generate an interference pattern. The intensities of the maxima decrease through the 3rd order and then increase for the 4th order. This suggests that the 1st diffraction minima are near the 2nd order interference minima. Assuming that the 1st diffraction minima lie on the 2nd order interference minima, what is the ratio of the slit width to the slit separation? (24.6)

Solution:

Strategy: Apply the relations giving maxima and minima for two slit interference patterns. The main difficulty with many problems such as this is to determine which minima and maxima are being discussed.

Fig. 24.4: Labeling of interference maxima and minima for Problem 24.6.

Fig. 24.4 shows the left hand half of the pattern with the numbering of the maxima and minima.

We let λ be the wavelength of light used, d the slit separation, b the slit width, and θ indicate an angle.

For an interference minimum

$$\left(m + \frac{1}{2}\right)\lambda = d \sin \theta_m \qquad m = 0, 1, 2, \ldots$$

while for a diffraction minimum

$$n\lambda = b \sin \theta_n \qquad n = 1, 2, 3, \ldots$$

For this problem $n = 1$ (1st diffraction minima) and $m = 2$ (3rd interference minima), so

$$\frac{\lambda}{b} = \sin \theta_n = \sin \theta_m = \frac{5\lambda}{2d} .$$

This gives

$$\frac{b}{d} = \frac{2}{5} = 0.4.$$

Each slit is 0.4 times as wide as the slit separation.

PROBLEM 24.7: Two identical circular apertures have been cut in a copper foil. The center to center separation of the apertures is twice the diameter of either aperture. The apertures are passing light of 500 nm wavelength and are viewed from a distance of 100 m. If they are just barely resolvable as two apertures, what is the diameter of either aperture? (24.7)

Solution:

> **Strategy:** We can express the true angular separation of the apertures in terms of the distance and the separation of the holes. Using Rayleigh's criterion, we can also relate this angular separation to the ratio of the wavelength to the diameter.

If we let D be the diameter of either aperture and L be the distance to the plate as shown in Fig. 24.5, then the angular separation of the apertures, θ, is given by

$$\theta = \frac{\text{separation}}{L} = \frac{2 \times D}{L} .$$

Use Rayleigh's criterion and equate θ to θ_m in text Eq. 23.10 to get

$$\frac{2 \times D}{L} = \theta = \theta_m = 1.22 \frac{\lambda}{D} .$$

This gives

$$D = \sqrt{0.61 \, \lambda \, L} = 5.5 \times 10^{-3} \text{ m} = 5.5 \text{ mm}.$$

The holes have a diameter of 5.5 mm.

Fig. 24.5: Two holes viewed from a distance of 100 m.

PROBLEM 24.8: Light passing through two aligned polarizers has an intensity of 3.7×10^{-5} W/m². Through what angle must one of the polarizers be turned in order to reduce the light intensity to 7.0×10^{-6} W/m²? (24.10)

Solution:

Strategy: This is a straightforward application of Malus's law.

(a)   **(b)**

Fig. 24.6: (a) Two aligned polarizers. (b) One polarizer now rotated through an angle θ.

If we let θ be the angle through which we turn one polarizer as shown in Fig. 24.6, I_m the initial maximal intensity and I the intensity after turning through θ, we have

$$I = I_m \cos^2 \theta \ .$$

Solving Malus's law for θ gives

$$\theta = \cos^{-1}\left(\sqrt{\frac{I}{I_m}} \right) = \cos^{-1}\left(\sqrt{\frac{7.0 \times 10^{-6}}{3.7 \times 10^{-5}}} \right) = 64°.$$

One of the polarizers must be turned through an angle of 64°.

GENERAL COMMENTS

Light is modeled as a transverse wave. Ripples on the surface of water provide an everyday experience of a transverse wave. Our model of light differs from that of water ripples in that there are two transverse parts to a light wave — the electric field and the magnetic field. These are at right angles to each other and each obeys the traveling wave equation of Chapter 14. If the plane of the electric field (and hence the magnetic field) is fixed in space we have a plane polarized wave. In general the plane is not fixed and obeys no simple time behavior; we call such light unpolarized. There are other types of polarization, such as elliptical and circular, that are not discussed in the text.

We call the wave picture of light a model for two reasons. First, we have not investigated the fundamental properties of the electric and magnetic fields that produce the wave and that govern the interaction of a light wave with matter. Second, as we will see in Chapter 27, there are times when light acts as a particle. The properties of light discussed in this chapter can be most easily understood by considering light as a wave. There are other properties that can be easily understood only if light is a particle. This dual property will be more fully explained in Chapters 26 and 27.

As indicated in the text, the wave property of diffraction follows from Huygens' principle. In this sense we can consider diffraction as self interference of a single wave front. For a two slit interference pattern we examine the relative phase of light from the two different slits. For a diffraction grating we examine the relative phases from the set of slits. For a diffraction pattern we examine the relative phases of the light that passed through various parts of a single slit.

SUGGESTED READINGS

Anderson, D. Z., "Optical Gyroscopes." *Scientific American,* April 1986, p. 94.

Barr, E. S., "Men and Milestones in Optics II—Thomas Young." *The Physics Teacher,* February 1967, p. 53.

Boyer, C. B., *The Rainbow: From Myth to Mathematics.* Princeton: Princeton University Press, 1987. All you ever wanted to know about the rainbow.

Burman, G. A., "Overhead Spectroscopy." *The Physics Teacher,* October 1991, p. 470.

Chiang, C., "On the Young's Interference Pattern." *American Journal of Physics,* March 1973, p. 423.

Chynoweth, A. G., "The Fiber Lightguide." *Physics Today*, May 1976, p. 28.

Davies, G. R., "Polarized Light Corridor Demonstrations." *The Physics Teacher,* October 1990, p. 464.

Feldman, A., "The Speed of Light in Different Media." *The Physics Teacher*, February 1991, p. 112.

Greenler, R. G., J. W. Hable, and P. O. Slane, "Diffraction Around a Fine Wire: How Good is the Single-Slit Approximation?" *American Journal of Physics,* April 1990, p. 330.

Hecht, E., and A. Zajac, *Optics*, 2nd ed. Reading, Mass.: Addison-Wesley, 1987.

Können, G. P., *Polarized Light in Nature.* Cambridge: Cambridge University Press, 1985. A beautifully illustrated catalog of polarization effects observed in nature.

Leung, A. F., and S. George, "Which Side of the Grating Has the Grooves?" *The Physics Teacher,* February 1990, p. 98.

Levi, B. G., "Search and Discovery: New Technology Telescope Actively Corrects for Misalignments." *Physics Today,* May 1990, p. 17.

Nussenzveig, H. M., "The Theory of the Rainbow." *Scientific American,* April 1977, p. 116.

Pepper, D. M, J. Feinberg, and N. V. Kukhtarev, "The Photorefractive Effect." *Scientific American,* October 1990, p. 62.

Rämme, G., "Colors on Soap Films—An Interference Phenomenon." *The Physics Teacher,* October 1990, p. 479.

Robinson, G. M., D. M. Perry, and R. W. Peterson, "Optical Interferometry of Surfaces." *Scientific American,* July 1991, p. 66.

Sadler, P., "Projecting Spectra for Classroom Investigations." *The Physics Teacher,* October 1991, p. 423.

Schawlow, A. L., "Measuring the Wavelength of Light with a Ruler." *American Journal of Physics,* November 1965, p. 922.

Shurcliff, W. A., and S. S. Ballard, *Polarized Light.* Princeton, N.J.: Van Nostrand-Momentum, 1964.

Walker, J., "The Amateur Scientist: Wonders With the Retroreflector, a Mirror that Removes Distortions from a Light Beam." *Scientific American*, April 1986, p. 118.

Wehner, R., "Polarized—Light Navigation by Insects." *Scientific American,* July 1976, p. 106

Wood, A., *Thomas Young, Natural Philosopher 1773–1829,* Cambridge: Cambridge University Press, 1954.

Young, A. T., "Rayleigh Scattering." *Physics Today,* January 1982, p. 42.

Chapter 25
Relativity

IMPORTANT TERMS

Fill in the blanks with the appropriate word or words.

1. The _____ theory of relativity treats accelerated reference frames and gravity.

2. The idea that being held at rest in a gravitational field is the same as uniform acceleration is called the _____.

3. A nonaccelerating frame of reference is called a(n) _____ frame.

4. The _____ theory of relativity resolved the conflicts between electromagnetic theory and mechanics for problems that do not involve gravity.

5. The time that a clock measures, as seen from the clock's rest frame, is called the clock's _____ time.

6. The measured length of a moving rod is different than its rest length. This effect is known as a(n) _____ contraction.

7. The equations relating the coordinates in one reference frame to those of another reference frame moving with constant velocity relative to the first frame are called the _____ equations.

8. The _____ mass of a body is the body's mass as measured in a frame at rest with respect to the body.

9. The relation that exists between the mass of a body and the energy contained in that body is called the _____ equation.

10. The _____ length of an object is the object's length as measured in its rest frame.

11. Newton's laws transform according to the rules of _____ relativity.

12. Two events are _____ to an observer, if in the observer's reference frame the events occur at the same time.

13. The time intervals that an observer measures with a clock moving relative to that observer are smaller than the intervals that would be measured if the clock were at rest relative to the observer. This effect is known as time _____.

14. _____ physics contains the ideas known at the end of the 19th century.

15. _____ is the idea of two events occurring at the same time.

Write the definitions for the following:

16. principle of equivalence (25.11):

17. Galilean relativity (25.1):

18. inertial (reference) frame (25.1):

19. simultaneous (25.4):

20. length contraction (25.6):

21. proper length (25.6):

22. time dilation (25.5):

23. proper time (25.5):

24. mass-energy relation (25.7):

25. general theory of relativity (25.12):

26. special theory of relativity (25.2):

27. (Lorentz) transformation equations (25.4):

28. rest mass (25.8):

ANSWERS TO 1–15, IMPORTANT TERMS

1. general; 2. principle of equivalence; 3. inertial; 4. special; 5. proper; 6. length (or Lorentz); 7. Lorentz transformation; 8. rest; 9. (Einstein) mass–energy; 10. proper; 11. Galilean; 12. simultaneous; 13. dilation; 14. classical; 15. simultaneity.

SAMPLE QUESTIONS

QUESTION 25.1: Is the mass of a red-hot piece of metal greater than its mass when it is at room temperature? Explain your answer.

Answer:

Yes. The atoms in the hot metal have a higher average velocity than those in the cool metal. The hot metal contains more energy; therefore, the mass of the system of atoms must be larger.

QUESTION 25.2: Twins separate, with one going to live on top of a very high mountain and the other living in a cave deep underground. Which twin ages faster?

Answer:

The twin living in the cave is in a lower gravitational potential relative to the twin on the high mountain. The cave dwelling twin's clock runs slower that the mountain dwelling twin's clock.

SAMPLE SOLUTIONS

PROBLEM 25.1: A train moves slowly past a station at 3.0 km/hr. A passenger on the train walks 3.0 km/hr down the aisle in the direction opposite to the train's velocity. (a) What is the passenger's speed relative to the station? (b) What would it be if he reversed his direction? (25.1)

Solution:

> **Strategy:** Since all velocities are much smaller than the speed of light this problem can be solved using the relative motion concepts of Chapter 3. The velocities are added as vectors.

Relativity is important when velocity/(speed of light) is reasonably near 1. For this problem

$$\frac{\text{velocity}}{\text{speed of light}} = \frac{3.0 \text{ km/hr}}{3.0 \times 10^5 \text{ km/s} \times 3600 \text{ s/hr}} = 2.8 \times 10^{-9} \ll 1.$$

Clearly the velocities can be added according to the rules of Chapter 3.

Referring to Fig. 25.1 we see that the velocity relative to the station for case (a) is 3.0 km/hr – 3.0 km/hr = 0.0 km/hr, while for case (b) we have 3.0 km/hr + 3.0 km/hr = 6.0 km/hr.

(a)   **(b)**

Fig. 25.1:   (a) Passenger and train moving in opposite directions. (b) Passenger and train moving in the same direction.

PROBLEM 25.2: A space ship is moving away from the Earth at a speed of 0.50 c when it launches a satellite in the same direction with a speed of 0.60 c, as seen from the ship. What is the satellite's speed as seen from Earth? (25.3)

Solution:

> **Strategy:** Use the relativistic addition of velocity formula.

322 Chapter 25

Fig. 25.2: (a) Space ship launching a satellite as seen from the earth. (b) Launch as seen from the space ship.

Fig. 25.2 shows the problem. Let u be the velocity of the satellite with respect to the earth, u' the velocity of the satellite with respect to the space ship, and v the velocity of the space ship with respect to the earth. Then a straightforward application of text Eq. 25.1 gives:

$$u = \frac{u' + v}{1 + \frac{u' v}{c^2}} = \frac{0.5c + 0.6c}{1 + \frac{(0.5c)(0.6c)}{c^2}} = \frac{1.1 \times c}{1 + 0.30} = 0.85 \ c.$$

The satellite moves away from the Earth with a speed of 0.85 c as measured by an observer on Earth — less than the result predicted by the classical addition of relative velocities.

PROBLEM 25.3: The star Alpha Centauri is 4.3 light years away. An astronaut travels to this star at 0.85 c and returns at 0.80 c, where c is the speed of light. Neglecting any problems with turning around and neglecting the time it takes to turn around, describe the elapsed time as observed from the earth and as observed by the traveling astronaut. A light year is the distance light travels in one year. (25.5)

Solution:

Strategy: Calculate the elapsed time as measured by an observer on the earth; then use the time dilation relation to find the traveler's elapsed time. Since there is no frame in which the space ship remains at rest, it is important to consider each leg of the trip separately when calculating the elapsed time to the ship.

It is important to apply the rule for time dilation separately for each leg of the trip. The time dilation rule can only be applied when considering two frames. There are three frames of reference in this problem: (1) the earth frame, (2) the outgoing rocket frame, and (3) the returning rocket frame.

We will approach this problem by using the relation between velocity and distance to calculate the elapsed time as measured from the earth. We can then use the equation for time dilation (text Eq. 25.4) to calculate the elapsed time in the astronaut's frame. When we use text Eq. 25.4, remember that Δt_0 is the elapsed time in the rest frame of the clock and Δt is the elapsed time as measured in a frame with relative velocity v.

The total elapsed time as seen from the earth is given by

$$\text{elapsed time} = \frac{\text{distance to Alpha Centauri}}{\text{speed out}} + \frac{\text{distance to Alpha Centauri}}{\text{speed back}}$$

$$= \frac{4.3 \text{ light years}}{0.85 \text{ c}} + \frac{4.3 \text{ light years}}{0.80 \text{ c}} = (5.1 + 5.4) \text{ years} = 10.5 \text{ years.}$$

We can calculate the elapsed time (Δt_0 in text Eq. 25.4) as seen by the astronaut:

$$\text{elapsed time} = (\text{time out in earth frame}) \times \sqrt{1 - \frac{v_{out}^2}{c^2}}$$

$$+ (\text{time back in earth frame}) \times \sqrt{1 - \frac{v_{back}^2}{c^2}}$$

$$= (2.7 + 3.2) \text{ years} = 5.9 \text{ years.}$$

On this trip 5.9 years will elapse to the astronaut while 10.5 years pass on earth.

PROBLEM 25.4: A meter stick ($L_0 = 1.00$ m) is moving at 0.90 c relative to us. Verify that its contracted length is less than 0.5 m. This being the case it is obvious that if we had an open box of length 0.9 m in the path of the moving stick, as shown in Fig. 25.3a, we could close the open side and have the stick entirely inside the box as shown in Fig. 25.3b for the short time before the stick hits the back side of the box. But from the frame of the stick it is the box which is Lorentz contracted, so clearly the stick is never completely inside the box. What is the resolution of this apparent paradox? (This problem requires use of the equations relating the two frames of references.) (25.6)

Fig. 25.3: (a) Meter stick approaching a 0.9 m box at a speed of 0.99 c. (b) Meter stick inside box.

Solution:

Strategy: Examine the problem in the frame of the meter stick. Use the Lorentz length contraction to calculate the length of the meter stick as it appears in the rest frame of the box. Then use the transformation equations to express what happens in the frame of the meter stick to resolve the apparent paradox.

The length of the moving meter stick can be calculated by the Lorentz contraction formula text Eq. 25.3. The measured length, L, will be

$$L = L_0 \sqrt{1 - (v^2/c^2)} = 1.00 \text{ m} \sqrt{1 - 0.90^2} = 1.00 \text{ m} \times 0.43 = 43 \text{ cm},$$

which is clearly less than 0.9 m, so it would seem that the meter stick can fit into the box. On the other hand, if we sit on the meter stick, the Lorentz contracted box is even shorter than the meter stick.

To resolve the apparent paradox we need to carefully examine the events involved. There are three events which we will call **A**, **B** and **C**.

Event **A** is when the front end of the meter stick just enters the box.
Event **B** is when the back end of the meter stick enters the box.
Event **C** is when the front end of the stick hits the back end of the box.

These three events are shown in Fig. 25.4a as measured by the observer at rest with respect to the box.

Let x and t be the space and time coordinates measured in a frame at rest with respect to the box and x' and t' be those coordinates in a frame at rest with respect to the stick. Special relativity tells us that these coordinates are related by:

$$x' = \frac{x - vt}{\sqrt{1 - (v^2/c^2)}} \text{ and } t' = \frac{t - \frac{vx}{c^2}}{\sqrt{1 - (v^2/c^2)}} .$$

Our equations look simplest if we choose $x = 0$ at the open end of the box and $t = 0$ when the meter stick enters the box. In the frame of the box the coordinates of **A**, **B**, and **C** are given by

$$x_A = 0, t_A = 0 \text{ for } \mathbf{A},$$

$$x_B = 0, t_B = \frac{L}{v} = 1.6 \times 10^{-9} \text{ s for } \mathbf{B},$$

and

$$x_C = 0.9 \text{ m}, t_C = \frac{0.9 \text{ m}}{v} = 3.3 \times 10^{-9} \text{ s for } \mathbf{C},$$

where L is the measured length of the meter stick in the frame of the box and v is the velocity of the stick.

We can calculate these events using the Lorentz transformation equations.
For event **A**:

$$x'_A = \frac{x_A - vt_A}{\sqrt{1 - (v^2/c^2)}} = 0 \text{ m}; \quad \text{and} \quad t'_A = \frac{t_A - \frac{vx_A}{c^2}}{\sqrt{1 - (v^2/c^2)}} = 0 \text{ s}.$$

For event **B**:

$$x'_B = \frac{x_B - vt_B}{\sqrt{1 - (v^2/c^2)}} = -1.0 \text{ m}; \text{ and } t'_B = \frac{t_B - \frac{vx_B}{c^2}}{\sqrt{1 - (v^2/c^2)}} = 3.7 \times 10^{-9} \text{ s}.$$

For event **C**:

$$x'_C = \frac{x_C - vt_C}{\sqrt{1 - (v^2/c^2)}} = 0.00 \text{ m}; \quad \text{and} \quad t'_C = \frac{t_C - \frac{vx_C}{c^2}}{\sqrt{1 - (v^2/c^2)}} = 1.4 \times 10^{-9} \text{ s}.$$

Fig. 25.4: (a) The three events as measured in the frame of the box. (b) The three events as measured in the frame of the meter stick.

In the frame of the meter stick, event **C** occurs before event **B**. Thus, as measured in the rest frame of the stick, the stick hits the back side of the box before the rear of the stick enters the box. There is no paradox, only the unusual, in terms of everyday experience, disagreement as to which event occurs first. The time order of the events **B** and **C** can be reversed because they are separated by a time shorter than the time it takes for light to travel between the two events. In both frames the physics makes sense, and either observer can predict what the other will measure.

PROBLEM 25.5: If 2.50 mg of matter were completely converted to energy, how long could this energy run a 60 W light bulb? (25.7)

Solution:

> **Strategy:** Use the Einstein mass-energy relation to determine the energy available and the definition of the watt to calculate the time.

The total energy contained in an amount of mass m is given by the Einstein mass energy relation $E = mc^2$, where c is the speed of light. This energy will be expressed in joules. For 2.50 mg we have

$E = m c^2 = 2.5 \times 10^{-6} \text{ kg} \times (3.00 \times 10^8 \text{ m/s})^2 = 2.25 \times 10^{11} \text{ kg·m}^2/\text{s}^2 =$

2.25×10^{11} J.

One watt = 1 J·s, so the 60 W bulb can run

$$\frac{2.25 \times 10^{11} \text{ J}}{60 \text{ W}} = 3.8 \times 10^9 \text{ seconds} = 1.0 \times 10^6 \text{ hr} \approx 1.2 \times 10^2 \text{ years}.$$

The 60 W bulb can be run for about 120 years of the energy in 2.5 mg of matter.

PROBLEM 25.6: If we use the relativistic mass given by text Eq. 25.9 and measure forces and accelerations in the frame of the observer (not in the rest frame of the particle), the equation $F = ma$ still applies. An electron moving through an electric field of 1.00×10^4 V/m (directed parallel to the electron's velocity) is observed to undergo an acceleration of 1.05×10^{15} m/s^2. What is the velocity of this electron? *Hint:* The electronic charge should be considered to be the same in all frames. (25.7)

Electric Field

Fig. 25.5:   Electron moving parallel to electric field.

Solution:

Strategy: Use $F = ma$ with F the electric force to calculate the effective mass. Use the effective mass and the rest mass to get the velocity.

The observed mass of the electron is given by

$$m = \frac{F}{a}$$

where F is the electric force and a is the measured acceleration. The electric force is given by

$$F = Ee$$

where e is the charge on the electron and E is the electric field. Thus, the measured mass, m, is given by

$$m = \frac{eE}{a} = \frac{1.60 \times 10^{-19} \text{ C} \times 1.00 \times 10^4 \text{ V/m}}{1.05 \times 10^{15} \text{ m/s}^2} = 1.52 \times 10^{-30} \text{ kg}.$$

This measured mass, m, is related to the rest mass, m_0, by text Eq. (24.8):

$$m = \frac{m_0}{\sqrt{1 - (v^2/c^2)}} .$$

Solving this equation for v gives us

$$v = c \times \sqrt{1 - \frac{m_0{}^2}{m^2}} \, ,$$

so

$$v = 3.00 \times 10^8 \text{ m/s} \times \sqrt{1 - \left(\frac{9.11 \times 10^{-31}}{1.52 \times 10^{-30}}\right)^2} = 2.4 \times 10^8 \text{ m/s} = 0.80 \text{ c}.$$

The electron is moving at 0.80 c.

PROBLEM 25.7: An electron is accelerated from rest to a velocity $v = 0.80$ c by means of an electric potential difference ΔV. Calculate ΔV, using the value 0.511 MeV for the rest energy (rest mass) of the electron. The MeV is the energy one electron gains when accelerated through a potential difference of 1,000,000 V. (25.7)

Solution:

> **Strategy:** Use the mass-energy relation to find the energy of the electron at 0.8 c. This change in energy is the work done on the electron.

The energy, E, of the electron is given by

$$E = \frac{m_0 c^2}{\sqrt{1 - \left(\frac{v}{c}\right)^2}}$$

$$= \frac{0.511 \text{ MeV}}{\sqrt{1 - \left(\frac{0.80 \text{ c}}{c}\right)^2}} = 1.7 \times 0.511 \text{ MeV} = 0.85 \text{ MeV}.$$

The increase in kinetic energy is equal to the work done on the electron by the potential, that is,

$$\text{work} = e\Delta V = E - m_0 c^2 = 0.36 \text{ MeV} = 0.36 \times 10^6 \times eV$$

where e is the electronic charge, so

$$\Delta V = 3.6 \times 10^5 \text{ V}.$$

The accelerating potential is 3.6×10^5 V.

PROBLEM 25.8: Two electrons are traveling in the same directions with energies of 1.0 MeV and 2.0 MeV as seen from a frame S. Find the velocities of each of these electrons in S and then the velocity of the most energetic one relative to the other. Use 0.511 MeV for the electron rest mass (energy). (25.8)

Solution:

> **Strategy:** Use the relativistic energy relation to find the velocities, then use the relativistic velocity addition formula.

The Einstein energy relation is

$$E = \frac{m_0 c^2}{\sqrt{1 - \left(\frac{v}{c}\right)^2}}$$

where E is the energy, m_0 is the rest mass and v is the relative velocity. Solving this for v gives:

$$\frac{v}{c} = \sqrt{1 - \left(\frac{m_0 c^2}{E}\right)^2} \; .$$

We let u_1 be the velocity of the slower electron and u_2 be the velocity of the faster particle. Choose the axis so that both velocities are positive. Entering the numbers gives

$$\frac{u_1}{c} = 0.86 , \quad \text{and} \quad \frac{u_2}{c} = 0.97.$$

The relative velocity is not the difference $u_2 - u_1$; we must use the relativistic addition formula. The text formula relates the observed velocity of body 2, denoted by u, the velocity of body 1, denoted by v, and the (relative) velocity of body 2 relative to body 1 as seen by body 2, denoted by u', as

$$u = \frac{u' + v}{1 + \frac{u' v}{c^2}} \; .$$

Solving this for u' gives

$$u' = \frac{u - v}{1 - \frac{u v}{c^2}} = 0.64 \; c.$$

The relative velocity is 0.64 c.

PROBLEM 25.9: A spaceship is approaching Alpha Centauri at 0.95 c. The maximum energy of the light from this star, in the frame of the star, is centered on a wavelength of about 5.5×10^{-7} m, which is in the center of the visible spectrum. What is the wavelength of this light as seen from the spaceship? What does this suggest would be one of the problems of space travel, if we could reach speeds near the speed of light? (25.10)

Solution:

> **Strategy:** Use the Doppler formula to calculate the observed wavelength. Then consider what the observed wavelength implies.

The problem as stated is illustrated in Fig. 25.6a. As seen from the spaceship, the star is approaching at a speed v = 0.95 c as shown in Fig. 25.6b.

Fig. 25.6: (a) Spaceship approaching star at speed v. (b) Picture as measured in frame fixed on the spaceship.

We let f be the observed frequency, f_0 be the emitted frequency and v the relative velocity. The observed wavelength is $\lambda = c/f$ and the emitted wavelength $\lambda_0 = c/f_0$. Then, since the velocity is one of approach, we can use text Eq. (24.12):

$$f = \frac{c}{\lambda} = \frac{f_0 \sqrt{1 - (v^2/c^2)}}{1 - (v/c)} = \frac{\frac{c}{\lambda_0} \sqrt{1 - (v^2/c^2)}}{1 - (v/c)} .$$

Solving this for the wavelength, λ, observed gives

$$\lambda = \frac{\lambda_0 \, (1 - (v/c))}{\sqrt{1 - (v^2/c^2)}}$$

$$= \frac{5.5 \times 10^{-7} \text{ m} \times 0.05}{0.31} = 8.9 \times 10^{-8} \text{ m}.$$

This wavelength is much shorter than visible light (visible light lies in the range 4.4 to 7.0 $\times 10^{-7}$ m). The value of λ is hard ultraviolet radiation. To protect from this harmful radiation, the spaceship will need shielding. Shielding means extra mass which will make it harder to get the spaceship to a high velocity.

GENERAL COMMENTS

Both special and general relativity provide new models of physical reality. First let us examine the differences between Newtonian theory and the special theory of relativity.

Newtonian	Special Relativity
Space and time are separate entities.	Space and time are aspects of a single entity.
Space is the same for all observers. Time is the same for all observers.	The split between space and time is made differently by different observers.

If we are dealing only with velocities that are small compared to the speed of light, the differences between classical (Newtonian) physics and special relativity become too small to measure. The predictions of either are essentially the same. This reduction is a physical requirement for any new model of the world — it must include the old model in some limit.

Special relativity leads to many surprising predictions but it does not lead to paradoxes. A **paradox** is when two mutually contradictory conclusions are reached. A surprising result is not necessarily a paradox. An example of a paradox would be that a door is observed to be both open and shut at the same time. Many problems posed in special relativity may look paradoxical until the problems are carefully examined. Several of the sample problems examined apparent paradoxes.

The time dilation and the length contraction of special relativity are not paradoxical since both ideas involve measurements made at two separate locations at the same time and hence involve simultaneity. These effects should not be called "apparent effects only" since the observed length has all the properties we expect length to have. The contraction and dilation are real to the observer. This is one of the essential "new" properties of the relativity model. Properties extending in space and time depend upon the observer. This dependence upon the observer is even more important when we consider quantum mechanics in later chapters.

Surprising results should not disconcert us as long as our ideas of causality are preserved. Let us define an event as a single position and time in one frame. In any other moving frame the event has a single position and time also. The number will, of course, depend upon the frame. A photoflash bulb firing is an example of an event. The flash occurs at a single location and time.

If event **A** "causes" event **B**, then the time order of **A** and **B** will be the same to all observers. For example, the photoflash firing causes my eyes to hurt. No moving observer should claim that my eyes hurt before the photoflash bulb was fired.

Special relativity preserves this idea of causality; however, if two events are not causally related, then it is possible that the time order will be reversed to two different observers. There is even an observer for whom the two events are simultaneous. We saw how this property relates to the measurement of lengths in the sample problems.

In general relativity the space-time of special relativity becomes a dynamic entity of the model. That is, space-time itself is modified by the presence of matter. Again, general relativity must reduce to special relativity when the gravitational fields become weak. A measure of the strength of a gravitational field for this model is the quantity

$$\frac{2Gm}{Rc^2},$$

where G is the Newtonian gravitational constant, m is the mass of the object, R is the size of the object and c is the speed of light. When this quantity is much less than 1, then special relativity holds.

In summary:

General Relativity $\xrightarrow[\frac{2\,G\,m}{Rc^2}\,\ll\,1]{}$ Special Relativity $\xrightarrow[\text{speeds}\,\ll\,c]{}$ Newtonian.

SUGGESTED READINGS

Baker, A., *Modern Physics and Antiphysics*. Reading, Mass.: Addison-Wesley, 1970.

Bondi, H., *Relativity and Common Sense*. Garden City, N.Y.: Anchor Books, 1964.

Boslough, J., "Searching for the Secrets of Gravity." *National Geographic,* May 1989, p. 563.

Bronowski, J., "The Clock Paradox." *Scientific American,* February 1963, p. 134.

Chai, A., "Some Pitfalls in Special Relativity." *American Journal of Physics,* February 1973, p. 192.

Clark, R. W., *Einstein: The Life and Times, An Illustrated Biography*. New York: Harry N. Abrams, Inc., 1984. One of the more readable books about Einstein and his work. It covers the social and political life of Einstein and many of his contemporaries. Photographs from a wide variety of sources make this book especially interesting.

Coe, L., "The Nature of Time." *American Journal of Physics,* August 1969, p. 810.

Corwin, M., and D. Wachowiak, "Lookback Time: Observing Cosmic History." *The Physics Teacher,* October 1989, p. 518.

Desloge, E. A., and R. J. Philpot, "Uniformly Accelerated Reference Frames in Special Relativity." *American Journal of Physics,* March 1987, p. 252.

DeWitt, B. S., "Quantum Gravity." *Scientific American,* December 1983, p. 112.

Dickie, R. H., "The Eötvös Experiment." *Scientific American,* December 1961, p. 84.

Einstein, A., *Relativity, The Special and the General Theory* (Trans. by R. W. Lawson). New York: Crown, 1961.

Feynman, R. P., "The Special Theory of Relativity." *The Physics Teacher,* April 1964, p.151.

French, A. P., ed., *Einstein, A Centenary Volume*. Cambridge, Mass.: Harvard University Press, 1979.

Frisch, D. H., and J. H. Smith, "Measurement of the Relativistic Time Dilation Using μ–mesons." *American Journal of Physics,* 1963, p. 342.

Goldman, T., R. J. Hughes, and M. M. Nieto, "Gravity and Antimatter." *Scientific American,* March 1988, p. 48.

Hafele, J. C., "Relativistic Time for Terrestrial Circumnavigation." *American Journal of Physics,* January 1972, p. 81.

Hawking, S. W., "The Edge of Spacetime." *American Scientist,* July 1984, p. 355.

Hawking, S. W., *A Brief History of Time: From the Big Bang to Black Holes.* New York: Bantam Books, 1988. An historical overview of cosmology from the early Greeks to modern times, including the general theory of relativity and quantum mechanics.

Hetherington, N. S., "Hubble's Cosmology." *American Scientist,* March 1990, p. 142.

Hughes, T. P., "Einstein the Inventor." *American Heritage of Invention & Technology,* Winter 1991, p. 34.

Hutchings, J. B., "Observational Evidence for Black Holes." *American Scientist*, January 1985, p. 52.

Lanczos, C., *The Einstein Decade (1905–1915).* New York: Academic Press, 1974.

Pais, A., "Knowledge and Belief: The Impact of Einstein's Relativity Theory." *American Scientist,* March-April 1988, p. 154.

Ramsey, N. F., "Precise Measurement of Time." *American Scientist,* January-February 1988, p. 42.

Rothman, M. A., "Things That Go Faster Than Light." *Scientific American,* July 1960, p. 142.

Sandin, T. R., "In Defense of Relativistic Mass." *American Journal of Physics,* November 1991, p. 1032.

Schlegel, R., "Comments on the Hafele-Keating Experiment." *American Journal of Physics,* March 1974, p. 183.

Schwinger, J., *Einstein's Legacy.* New York: Scientific American Books (W. H. Freeman), 1986. This book illustrates Einstein's place in physics and the nature of the changes in physics that he initiated.

Scott, G. D., and H. J. van Driel, "Geometrical Appearances at Relativistic Speeds." *American Journal of Physics,* August 1970, p. 971.

Shankland, R. S., "Conversations with Albert Einstein. I & II" *American Journal of Physics,* January 1963, p. 49; July 1973, p. 895.

Shankland, R. S., "The Michelson-Morley Experiment." *Scientific American* , November 1964, p. 107

Smith, R. W., "Edwin P. Hubble and the Transformation of Cosmology." *Physics Today,* April 1990, p. 52.

Staunton, L. P. and H. van Dam, "Graphical Introduction to the Special Theory of Relativity." *American Journal of Physics,* October 1980, p. 807.

Swenson, L. S., Jr., "Measuring the Immeasurable." *American Heritage of Invention & Technology,* Fall 1987, p. 42. Michelson's interferometer.

Turner, E. L., "Gravitational Lenses." *Scientific American,* July 1988, p. 54.

Chapter 26
The Discovery of Atomic Structure

IMPORTANT TERMS

Fill in the blanks with the appropriate word or words.

1. The _____ is one event per second and is the SI unit of activity.

2. _____ is the breaking up of chemical compounds by electric currents.

3. The _____ is the normal negative charge carrier.

4. Electrons (often rapidly moving) emitted by radioactive elements are called _____ particles.

5. The smallest crystal structure is called the unit _____.

6. The positive charges of an atom are located in the atom's _____.

7. _____ are the short wavelength electromagnetic waves emitted by a Crooke's tube.

8. The irregular motion of minute particles suspended in a liquid or gas is called _____ motion.

9. The time it takes one-half of a radioactive substance to decay is called the _____ of the substance.

10. _____ number is how many molecules of a substance are in one mole of the substance.

11. The _____ of a radioactive substance is the number of radioactive emissions per second.

12. Helium nuclei emitted in radioactive decay are called _____ particles.

13. The highly penetrating radiation from uranium is electromagnetic. It is called _____ rays.

14. The phenomenon of emission of particles as one element changes to another is called _____.

15. The x rays from a Crooke's tube are also called _____ rays.

16. The relation among the wavelength of a wave incident upon a crystal, the wavelength after scattering, the crystal structure size, and the direction of scattering is known as _____ law.

Write the definition for the following:

17. x rays (25.5):

18. Avogadro's number (25.2):

19. electrolysis (25.2):

20. electron (25.6):

21. alpha particles (25.7):

22. beta particles (25.7):

23. gamma rays (25.7):

24. unit cell (25.1):

25. radioactivity (25.7):

26. nucleus (25.9):

27. cathode rays (25.5):

28. Bragg's law (25.5):

29. Brownian motion (25.3):

30. activity (25.8):

31. becquerel (25.8):

32. half-life (25.9):

ANSWERS TO 1–16, IMPORTANT TERMS

1. becquerel; 2. electrolysis; 3. electron; 4. beta; 5. cell; 6. nucleus; 7. x rays (or cathode rays); 8. Brownian; 9. half-life; 10. Avogadro's; 11. activity; 12. alpha; 13. gamma; 14. radioactivity; 15. cathode; 16. Bragg's.

SAMPLE QUESTIONS

QUESTION 26.1: Assume you have a radiation detector that is sensitive to alpha, beta, and gamma radiations. Describe a simple method for determining which radiations are present in a particular radioactive source.

Answer:

First, place a thin sheet of aluminum between the source and the detector. If the signal in the detector is reduced, alpha particles were present in the emissions of the source. Place a thicker piece of aluminum between the source and the detector to eliminate most of the beta particles and see if gamma rays are present. Text Table 25.1 will help you to interpert your results.

QUESTION 26.2: Imagine that you could roll small hard balls at an obstacle on a smooth table. The obstacle is hidden from view, but you can see the directions of the outgoing balls. How much can you determine about the obstacle? Give several specific examples.

Answer:

If the obstacle is small, most of the balls you roll will not be deflected at all. A few will be deflected through a large angle. If the obstical is large, most of the balls will be deflected. You can study the energy (speed) of the scattered balls to determine if the collision is elastic. Vary the incoming speed to determine elastic limits. If the object is sticky, some balls will never come out.

A flat object will scatter balls differently than a round object. For example, the balls would come straight back if the object were flat. A curved obstacle would deflect almost all balls through an angle if they hit at all.

The preceding should give you an idea of some of the examples and answers you could give.

SAMPLE SOLUTIONS

PROBLEM 26.1: In an electrolysis experiment a current of 2.3 amperes is maintained for 20 minutes during which zinc is deposited on the negative electrode. Zinc has an atomic weight of 65.3 and its atoms are trivalent (trivalent atoms can lose up to 3 electrons). Calculate how much zinc was deposited. (26.2)

Solution:

Strategy: Since we know the current and the time we can calculate the total charge transported. Each atom carries an excess of 3 electronic charges, so if we divide the total charge by $3 \times e$, where e is the electronic charge, we will get the number of atoms that were transported. Since an Avogadro's number of zinc atoms has a mass of 65.3 g, a simple ratio will give the total mass.

First we calculate the total charge that flows:

$$\text{total charge} = \text{current} \times \text{time} = 2.3 \text{ A} \times 20 \text{ min} \times 60 \text{ s/min}$$
$$= 2760 \text{ C} = q,$$

where we have defined the symbol q to represent the total charge. Then we calculate the number of atoms that must flow if our calculated total charge flows:

$$\text{number of atoms} = n = \frac{\text{total charge}}{\text{charge per atom}}$$

$$= \frac{\text{total charge}}{3 \times \text{electronic charge}}$$

$$= \frac{q}{3e}.$$

There is less chance of error if we do not put numbers into the equations before the final step.

Finally the ratio defining Avogadro's number, N_A, gives,

$$\frac{\text{atomic weight of zinc}}{\text{mass deposited}} = \frac{\text{Avogradro's number}}{\text{number of atoms deposited}}$$

$$= \frac{N_A}{\frac{q}{3e}} = \frac{3 \times N_A \times e}{q} = \frac{3 \times F}{q},$$

where F is the faraday.

Hence, if M is the mass deposited,

$$\frac{65.3 \text{ g/mol}}{M} = \frac{3 \times 96458 \text{ C/mol}}{2760 \text{ C}}.$$

Solving for M gives

$$M = 0.62 \text{ g}.$$

A total of 0.62 g will be deposited.

PROBLEM 26.2: Silicon has a density of 2.23 g/cm^3 and an atomic mass of 28.09 g/mol. Calculate the number of atoms in a cube of silicon 2.00 centimeters on a side. (26.4)

Solution:

> **Strategy:** 1. Use the definition of density to determine the mass present. 2. Use the definition of atomic mass to determine the number of moles. 3. Convert this to the number of atoms by using Avogadro's number.

We first find the mass of silicon present by multiplying the density by the volume:

$$\text{mass} = \text{density} \times \text{volume} = 2.23 \text{ g/cm}^3 \times (2.00 \text{ cm})^3 = 17.84 \text{ g}.$$

The number of moles is the mass divided by the atomic mass, so

$$\text{number of moles} = \frac{17.84 \text{ g}}{28.09 \text{ g/mol}} = 0.635 \text{ mol}.$$

Finally, we multiply the number of moles by Avogadro's number to get the number of atoms:

$$\text{number of atoms} = 0.635 \text{ mol} \times 6.03 \times 10^{23} \text{ atoms/mol} = 3.83 \times 10^{23} \text{ atoms}.$$

There are 3.83×10^{23} atoms in a cube of silicon 2.00 cm on a side.

PROBLEM 26.3: X rays of wavelength $\lambda = 6.73 \times 10^{-11}$ m are diffracted from a crystal. The third order scattering angle is 15° as shown in Fig. 26.1. What is the spacing between the planes of the crystal? (26.5)

Fig. 26.1: X-ray scattering caused by a crystal. Note that the deviation angle is measured with respect to the incident beam and not with respect to any surfaces of the crystal.

Solution:

> **Strategy:** Eq. 26.4 relates the deviation angle, the wavelength and the crystal plane spacing.

As is indicated in Fig. 26.1, the measured deviation of the x rays is two times the tangentially incident angle, θ. This angle must be measured from the direction of the incident beam, not from any of the crystal faces. Since we are given that the scattered angle is 15°, we have $\theta = 7.5°$.

If n is the order of the scattering, λ the wavelength, d the interplanar spacing, and θ is the tangentially incident angle, then

$$n\lambda = 2\,d\,\sin\theta.$$

We are given n, λ, and have found θ, so we can solve for d as

$$d = \frac{n\,\lambda}{2\,\sin\theta} = \frac{3 \times 6.73 \times 10^{-11}\ \text{m}}{2 \times \sin 7.5°} = 7.7 \times 10^{-10}\ \text{m}$$

So the interplanar spacing is 7.7×10^{-10} m. Note that the factors of 2 and 3 do not affect the significant figures since they are exact numbers.

PROBLEM 26.4: Electrons in a cathode ray tube are accelerated through a potential of 26.2 KV. The electron beam then passes through a pair of deflection plates 1.85 cm long separated by 0.35 cm. What potential must be applied to the plates in order to have the beam deflected 2.70 cm at a screen 25.0 cm beyond the deflection plates? The physical arrangement is shown in Fig. 26.2. *Hint:* Neglect the gravitational force on the electron and the edge effects of the plates. (26.6)

Solution:

> **Strategy:** Determine the electric force acting on the electron in terms of the applied potential. Use the laws of kinematics to calculate the resulting deflection.

Fig. 26.2: Electron in a cathode ray tube. The symbols are explained in the following material.

This is a kinematics problem where the force is the electric field and the particle undergoing motion is an electron. Use the following notation.

Symbol	Physical quantity	Value
d_0	plate separation	0.35 cm
L_0	plate length	1.85 cm
m	mass of electron	9.11×10^{-31} kg
e	charge of electron	1.60×10^{-19} C
PD_0	potential used to accelerate in x direction	26.2 KV
PD	potential used to accelerate in y direction	unknown
v_0	electron velocity entering plates	$v_0 = \sqrt{2\,PD_0/m}$
L	plate to screen distance	25.0 cm
d	linear deflection at screen	2.7 cm
F	force on electron in y direction	unknown
v	y velocity of electron after leaving plates	unknown

Make the following physical arguments:
1. While the electron is between the plates a force acts in the y direction.
2. This force gives the electron a y component of velocity.
3. This y velocity produces a y deflection at the screen.
4. We cannot use energy considerations to get the y velocity since we cannot assume that the electron moves the entire distance from the top to the bottom plate.

The following steps are needed:
1. Find the force in the y direction between the plates.
2. Calculate the y acceleration between the plates.
3. Find the time spent between the plates.
4. Find the velocity in the y direction when leaving the plates.
5. Calculate the displacement at the screen distance.

We neglect the electric field produced by the plates outside of their geometric size and assume that the electric field between the plates is uniform (neglect edge effects caused by the finite sizes of the plates). We also neglect the gravitational force since it is usually much smaller than the electric force. Then, while the electron is between the plates, it is subject to a y force

$$eE = e \, \frac{PD}{d_0} = ma,$$

where a is the acceleration in the y direction.

The electron enters between the plates from the left in Fig. 26.2 and exits to the right. We know the entering x direction velocity, v_0, from energy considerations; that is,

$$PD_0 = \frac{1}{2} m v_0^2.$$

The electron is between the plates for a time t given by

$$t = \frac{L_0}{v_0} = \frac{L_0}{\sqrt{\dfrac{2e \, PD_0}{m}}}.$$

Thus, when the electron leaves the region between the plates it has a y velocity

$$v = at = \frac{e}{m} \, \frac{PD}{d_0} \, L_0 \, \sqrt{\frac{m}{2e \, PD_0}} \, .$$

During this time the y displacement becomes:

$$y_0 = \frac{1}{2} a \, t^2 = \frac{L_0^2}{4d_0} \, \frac{PD}{PD_0} \, .$$

We will assume this y displacement, y_0, is small compared to the final displacement. This assumption is reasonable provided $L \gg L_0$.

If we assume that $v \ll v_0$, then it takes a time $T = L/v_0$ to travel from the plates to the screen. During this time, the y displacement becomes

$$d = v \, T = \frac{e}{m} \, \frac{PD}{d_0} \, L_0 \, \frac{m}{2e \, PD_0} \, \frac{1}{v_0} = = \frac{PD}{PD_0} \, \frac{L_0}{d_0} \, \frac{L}{2} \, .$$

So

$$PD = 2 \, PD_0 \, \frac{d_0}{L_0} \, \frac{d}{L} \, .$$

The given numbers can now be entered and we find that the potential needed is 106 V. Note the advantage of not putting in the numbers until the last step. The values for e and m are not needed. If we put the numbers in along the way we would have to multiply and divide by e and m several times. It is left for the student to show that

$$y_0 \ll d \, .$$

PROBLEM 26.5: A radioactive sample has a half-life of 5 min. If the initial activity is 8192 disintegrations per second, what will the activity be after one hour? (26.8)

Solution:

Strategy: Apply the definition of activity.

The activity is directly proportional to the number of atoms present. If the half-life is 5 min, 1 hr will be 12 half-lives. From text Table 25.2 we conclude that

$$N = \frac{N_0}{2^n}$$

where n is the number of half lives that have occurred, N_0 is the initial number of atoms and N is the remaining number of atoms. After 12 half-lives,

$$N = \frac{N_0}{4096}$$

So, if we denote the initial activity by A_0, the activity, A, after 12 half-lives will be

$$A = \frac{A_0}{4096} = \frac{8192}{4096} = 2 \frac{\text{decays}}{\text{s}}.$$

After one hour there will be 2 disintegrations/second on the average.

PROBLEM 26.6: A radioactive isotope decays to one-eighth its original amount in 5.0 hr. What are its half-life and decay constant? How long will it take for 10% of this isotope to decay? Suppose that, at a given time, there are 1.0×10^{15} atoms of this isotope left. What is the activity of the material? (26.8)

Solution:

Strategy: Count half-lives to get the half-life and then apply the definitions of decay constant and activity.

We can refer to text Table 26.3 or note that

$$\frac{1}{2^3} = \frac{1}{8}.$$

That is, 3 half-lives have passed, so the half-life, $t_{1/2}$, is

$$t_{1/2} = \frac{5.0}{3} = 1.7 \text{ hr}.$$

The decay constant, λ, is given by

$$\lambda = \frac{0.693}{t_{1/2}} = 0.42/\text{hr}.$$

The activity A is given as

$$A = \lambda N = 0.42 \times 1.0 \times 10^{15} = 4.2 \times 10^{14} \text{ events/hr} = 1.2 \times 10^{11} \text{ Bq}.$$

PROBLEM 26.7: The ratio of radioactive to nonradioactive carbon in a sample of wood recovered from an archaeological dig is found to be one-fourth that of a piece of wood known to be 100 years old. Approximately how long ago was the first sample cut? *Hint:* Assume the natural ratio of radioactive to nonradioactive carbon for living wood has not changed in the interval. (26.8)

Solution:

Strategy: After each half-life there will be one-half of the radioactive carbon present as was present at the start of the half-life.

If the ratio has changed to one-fourth then there have been two half-lives since the carbon in the wood was deposited in the wood. Since the half-life of radioactive carbon is about 5700 years, the old piece of wood is about 1.14×10^4 years plus the 100 years. The extra 100 years is not really needed since the age is uncertain by a larger amount.

PROBLEM 26.8: What is the distance of closest approach of an alpha particle of speed $v = 1.70 \times 10^7$ m/s to a gold nucleus of $Z = 79$? (26.9)

Solution:

Strategy: Conserve energy. Since the initial speed is more than 5% the speed of light, we may have to solve this problem using relativity; however, even at 5% the speed of light, any error caused by neglecting relativity is small. We will consider both approaches.

I. Nonrelativistic Approximation
Initially all the energy is in the kinetic energy of the alpha particle. At closest approach the alpha particle and the gold nucleus are at rest and the energy is in the electrostatic energy of the alpha particle in the electric field of the gold nucleus. We neglect any effect of the electrons associated with the gold nucleus.

Let m_α be the mass of the alpha particle, v_α be the initial velocity of the alpha particle, Z be the charge on the gold nucleus, e be the electronic charge, and b be the distance of closest approach. Then

$$\frac{1}{2} m_\alpha v_\alpha^2 = k \frac{(2e)(Ze)}{b}.$$

This gives

$$b = 4k \frac{Ze^2}{m_\alpha v_\alpha^2} = 4.3 \times 10^{-14} \times \frac{79}{100} \times \left(\frac{1.8}{1.7}\right)^2 = 3.8 \times 10^{-14} \text{ m},$$

where we have made use of the results of the calculation at the end of the chapter in the text.

II. Relativistic Calculation

We let the subscript Au denote the gold nucleus. Considering relativity would require us to include the rest masses of the particles and to use the relativistic expression. If we neglect any motion of the gold nucleus, we can use a frame fixed on the gold nucleus and write:

$$E_{init} = E_{Au} + E_\alpha = m_{Au}c^2 + \frac{m_\alpha c^2}{\sqrt{1 - \frac{v_\alpha^2}{c^2}}} = E_{closest} = m_{Au}c^2 + m_\alpha c^2 + k\frac{(2e)(Ze)}{b} \, .$$

The $m_{Au}c^2$ terms cancel. To solve the remainder, expand the square root, use significant figures, and show that the equation reduces to the nonrelativistic equation of the first part of this solution.

SUGGESTED READINGS

Anderson, D.L., "Resource Letter ECAN1 on the Electronic Charge and Avogadro's Number." *American Journal of Physics,* January 1966, p. 2.

Brownell, G. L., and R.J. Shalek, "Nuclear Physics in Medicine." *Physics Today,* August 1970, p. 32.

Bushong, S. C., "Radiation Exposure in Our Daily Lives." *The Physics Teacher,* March 1977, p. 135.

Crawford, D. J., and R. W. Leggett, "Assessing the Risk of Exposure to Radioactivity." *Scientific American,* May 1981, p. 48.

Einstein, A. *Investigations on the Theory of the Brownian Movement* (edited by R. Furth, trans. by A. D. Cowper). New York: Dover Publications, 1956.

Everhart, T. E., and T. L. Hayes, "The Scanning Electron Microscope." *Scientific American,* January 1972, p. 54.

French, A. P., "Earliest Estimates of Molecular Size." *American Journal of Physics,* February 1967, p. 162.

Hauptman, H. A., "The Phase Problem of X-Ray Crystallography." *Physics Today,* November 1989, p. 24.

Hedges, F. E. M., and J. A. J. Gowlett, "Radiocarbon Dating by Accelerator Mass Spectrometry." *Scientific American,* January 1986, p. 100.

Hunton, D. E., "Shuttle Glow." *Scientific American,* November 1989, p. 92.

Jaffe, C. C., "Medical Imaging." *American Scientist,* November1982, p. 576.

Nero, A. V., Jr., "Controlling Indoor Air Pollution." *Scientific American,* May 1988, p. 42.

Wood, E.A., "Crystals." *The Physics Teacher,* January 1965, p.7.

Chapter 27
Origins of the Quantum Theory

IMPORTANT TERMS

Fill in the blanks with the appropriate word or words.

1. The lowest energy state of a system is called the _____ state.

2. The quanta of light are called _____.

3. The _____ frequency is the minimum frequency of light that causes photoelectric emission.

4. A body whose emitted radiation depends only upon its temperature and not upon its composition is called a(n) _____.

5. The relation between wavelength of maximum emission and the temperature of an ideal body is given by the _____ law.

6. _____ law tells us that the square root of the K_α x-ray frequency is proportional to the nuclear charge.

7. The set of spectral lines of hydrogen, some of which lie in the visible portion of the spectrum, is called the _____ series.

8. The _____ is the minimum energy needed to remove an electron from a surface.

9. The _____ volt is the work needed to move an electron through a potential of one volt.

10. _____ is the branch of science that studies the spectra of electromagnetic radiation absorbed and emitted by substances.

11. The _____ effect is the emission of negative charge from an object as a result of light falling on the surface.

12. The ratio of the energy of a photon to the frequency of the associated light is called _____ constant.

13. The _____ formula is an empirical relation that predicts the emitted wavelength of light from hydrogen.

14. A plot representing the allowed energies of a system is called a(n) _____.

15. The _____ of a quantity is the smallest possible unit of that quantity.

16. The _____ quantum number in the Bohr theory of the atom gives the orbital size and hence the energy of an electron bound to a nucleus.

Write the definitions of the following:

17. Wien displacement law (27.4):

18. spectroscopy (27.1):

19. energy level diagram (27.7):

20. photoelectric effect (27.5):

21. photon (27.5):

22. Planck's constant (27.4):

23. principal quantum number (27.6):

24. Rydberg formula (27.2):

25. blackbody (27.3):

26. ground state (27.6):

27. quantum (27.4):

28. Moseley's law (27.8):

29. Balmer series (27.2):

30. threshold frequency (27.5):

31. work function (27.5):

32. electron volt (27.5):

ANSWERS TO 1–16, IMPORTANT TERMS

1. ground; 2. photons; 3. threshold; 4. blackbody; 5. Wien displacement; 6. Moseley's;
7. Balmer; 8. work function; 9. electron; 10. spectroscopy; 11. photoelectric; 12. Planck's;
13. Rydberg; 14. energy level diagram; 15. quantum; 16. principal.

SAMPLE QUESTIONS

QUESTION 27.1: When the interior of an operating kiln reaches a uniform temperature, it is virtually impossible to see the objects that are being heated. Explain why this is so.

Answer:

Hot bodies at the same temperature emit the same radiation and energy per unit area. The heated bodies should look the same as the kiln walls.

QUESTION 27.2: Can x rays be emitted by hydrogen?

Answer:

The maximum emission energy for hydrogen is ionization energy. This energy is 13.6 eV, which is ultraviolet. X rays are in the keV range.

SAMPLE SOLUTIONS

PROBLEM 27.1: An incandescent lamp is designed to operate at a filament temperature of 3500 K. If the lamp is operated at a higher voltage, the temperature will be higher. By what percentage will the total wavelength of maximum emission change if the lamp temperature is increased to 3700 K? (27.3)

Solution:

> **Strategy:** Use the Wien displacement law.

The relevant physical law is the Wien displacement law:

$$\lambda_m T = 2.9 \times 10^{-3} \text{ m·K}, \qquad \text{(text 27.3)}$$

where λ_m is the wavelength at which the radiation distribution from a body at kelvin temperature T peaks. For $\lambda_m(T)$ the percentage change is

$$\frac{\lambda_m(3700) - \lambda_m(3500)}{\lambda_m(3500)} \times \frac{\frac{1}{3700} - \frac{1}{3500}}{\frac{1}{3500}} = \left(\frac{35}{37} - 1\right) \times 100 = -5\% .$$

There is about a 5% decrease in the wavelength of maximum emission.

PROBLEM 27.2: Astronomers have observed a 3.0 K blackbody radiation striking the earth uniformly from all directions. This radiation is believed to be associated with the big bang commonly held to have occurred at the beginning of the universe. This radiation is the red shifted remnant of the initial primordial fireball. What is the wavelength of peak radiation from this fireball and how much energy per unit area and per unit time is striking the earth assuming the universe is a blackbody? From thermodynamics, a slowly expanding gas without external energy being provided should obey the rule $TV^{1/3} = $ constant, where T is the kelvin temperature and V is the volume occupied by a given number of "molecules" of the gas. At what wavelength did the radiation peak when the temperature of the universe was 3000 K? What was the energy density of radiation then as compared to the current epoch? Assuming the photons that make up the blackbody radiation can be treated as a gas, what volume did a typical region of the universe the size of the solar system (radius = 6.0×10^{12} m) occupy when the temperature of the universe was 3000 K? (27.3)

Solution:

> **Strategy:** Use the given expansion law, the Wien displacement law and the Stefan-Boltzmann law (Chapter 12) to calculate the behavior.

If the temperature T is measured in kelvin, the Wien displacement law gives us

$$\lambda_m T = 2.9 \times 10^{-3} \text{ m·K}$$

where λ_m is the wavelength of the peak in the blackbody curve.

For $T = 3.0$ K, the current wavelength peak is at 9.7×10^{-4} m, while for $T = 3000$ K the peak is at 9.7×10^{-7} m.

The Stefan-Boltzmann law is

$$R = \sigma T^4$$

where R is the emission per unit area and

$$\sigma = 5.67 \times 10^{-8} \frac{J}{\text{s·m}^2 \text{ K}^4} .$$

At 3.0 K this gives

$$R_0 = 4.6 \times 10^{-6} \frac{J}{\text{s·m}^2} ,$$

while at 3000 K, $R = 10^{12} R_0$.

To answer the last question we note that the value of a sphere of radius r is given by

$$\frac{4}{3} \pi r^3.$$

Let the subscript "o" denote the present and no subscript denote the time when T = 3000 K. We can then write

$$T_0(V_0)^{1/3} = T(V)^{1/3},$$

which simplifies to

$$T_0 r_0 = Tr.$$

Then

$$T_0 r_0 = 3.0 \text{ K} \times 6.0 \times 10^{12} \text{ m} = 3000 \text{ K} \times r$$

or

$$r = 6.0 \times 10^9 \text{ m}.$$

This is about 5 times the radius of the sun, so a volume about 5 times the solar radius has expanded to the size of the solar system as the temperature of the universe dropped from 3000 K to the present 3 K. At earlier epochs the universe was even hotter.

PROBLEM 27.3: What is the threshold wavelength for photoelectric emission from gold? (27.5)

Solution:

> **Strategy:** Use the photoelectric equation and the value of the work function for gold.

From text Table 27.2, the work function of gold is 5.31 eV. At threshold the maximum kinetic energy in the Einstein photoelectric equation is zero, so we have $0 = hf_{th} - \phi$, where h is Planck's constant, f_{th} is the threshold frequency, and ϕ is the work function.

If λ_{th} is the threshold wavelength, this can be written as

$$hf_{th} = \frac{hc}{\lambda_{th}} = \phi,$$

or

$$\lambda_{th} = \frac{hc}{\phi} = \frac{(6.626 \times 10^{-34}\ \text{J·s})(3.00 \times 10^8\ \text{m/s})}{(5.31\ \text{eV})(1.60 \times 10^{-19}\ \text{J/eV})} = 234\ \text{nm}.$$

The threshold wavelength for gold lies at 234 nm, which is in the ultraviolet region of the spectrum.

PROBLEM 27.4: What is the maximum kinetic energy of the photoelectron ejected from a silver surface by ultraviolet light of wavelength 250 nm? Express this result in eV and in joules. What is the maximum wavelength that can eject photoelectrons from silver? (27.5)

Solution:

> **Strategy:** Calculate the frequency for the incident light and then use the photoelectric equation to find the kinetic energy.

The photoelectron is ejected from the silver surface as shown in Fig. 27.1. The maximum kinetic energy, KE_{max}, is given by

$$KE_{max} = hf - \phi \qquad \text{(text Eq. 27.5)}$$

where f is the frequency of the light considered as a wave, h is Planck's constant, and ϕ is the work function for silver.

Fig. 27.1: Photoelectron emitted from a silver surface.

From text Table 27.2 we find that $\phi_{Ag} = 4.74$ eV. If we let λ be the wavelength of the incident light, then

$$f = \frac{c}{\lambda}$$

and we can write

$$KE_{max} = \frac{h\,c}{\lambda} - \phi$$

$$= \frac{6.63 \times 10^{-34} \text{ J·s} \times 3.00 \times 10^8 \text{ m/s}}{250 \times 10^{-9} \text{ m}} - \phi_{Ag} = 7.96 \times 10^{-19} \text{ J} - 4.74 \text{ eV}.$$

Before we can subtract these two numbers they must be in the same units. The conversion is

$$1 \text{ eV} = 1.60 \times 10^{-19} \text{ J},$$

so

$$KE_{max} = 0.233 \text{ eV} = 3.73 \times 10^{-20} \text{ J}.$$

The work function of silver is so large that even ultraviolet light can barely liberate photoelectrons.

The maximum wavelength for photoemission will occur when $KE_{max} = 0$. This occurs when

$$\frac{h\,c}{\lambda} = \phi.$$

For silver the maximum wavelength is about 262 nm.

PROBLEM 27.5: What is the shortest wavelength of the Brackett series ($n_1 = 4$)? (27.6)

Solution:

> **Strategy:** Use the Rydberg formula or the derivation of Bohr.

If we use the Bohr model, the emitted wavelength, λ, is given by

$$\frac{1}{\lambda} = \frac{m\,e^4}{8\varepsilon_0^2 h^2 c}\left(\frac{1}{(n_1)^2} - \frac{1}{(n_2)^2}\right) \quad \text{where } n_2 > n_1. \quad \text{(text Eq. 27.12)}$$

For hydrogen

$$\frac{m\,e^4}{8\varepsilon_0^2 h^2 c} = 1.097 \times 10^5 \text{ / cm}.$$

Since n_1 is fixed, λ is smallest when

$$\left(\frac{1}{(n_1)^2} - \frac{1}{(n_2)^2}\right)$$

is largest, which is when $n_2 \to \infty$ ($1/n^2 \to 0$). So

$$\frac{1}{\lambda} = \frac{1.097 \times 10^5/\text{cm}}{4^2} = 6.856 \times 10^3/\text{cm}.$$

The shortest wavelength of the Brackett series is 1.459×10^{-4} cm.

PROBLEM 27.6: The K_α x ray from a target has a wavelength of 0.143 nm. Which element is the target? (27.8)

Solution:

Strategy: Use Moseley's law to calculate the atomic number from the observed wavelength of the K_α line.

Moseley's law is given in the text as

$$f = \frac{3}{4} f_0 \, (Z - 1)^2, \qquad \text{(text Eq. 27.15)}$$

where Z is the atomic number (number of protons), $f_0 = cR$ with R the Rydberg constant, $R = 109737/\text{cm}$, and f is the frequency of the K_α line.

Making the substitution we have

$$f = \frac{c}{\lambda} = \frac{3}{4} c R \, (Z - 1)^2,$$

and solving for Z, we get

$$Z = 1 + \frac{2}{\sqrt{3\lambda R}} \, .$$

Using our numbers (be careful: 1 nm $= 10^{-9}$ m $= 10^{-7}$ cm), we get

$$Z = 1 + 29.1 \approx 30.$$

From the periodic table (text Fig. 27.13) we see that the element is zinc.

PROBLEM 27.7: An electron bound to a hydrogen nucleus is determined to have an orbital angular momentum of 2.13×10^{-34} J·s. What is the principal quantum number and with what energy is the electron bound to the atom? (27.6)

Solution:

Strategy: Use the Bohr model. The electron's angular momentum is given uniquely by the principal quantum number. Once we know the principal quantum number, we can calculate the binding energy.

The relation between the angular momentum of an electron, L, and the principal quantum number, n, for that electron is given by:

$$L = n\frac{h}{2\pi} , \qquad \text{(text Eq. 26.5)}$$

where h is Planck's constant. Since n must be an integer, it is best to substitute the given numbers at this point and determine the proper integer:

$$n = \frac{2\pi L}{h} = \frac{2 \times 3.14159 \times 2.13 \times 10^{-34} \text{ J·s}}{6.626 \times 10^{-34} \text{ J·s}} = 2.01 .$$

Since n must be integer we take $n = 2$. This is not an unreasonable act since considering the significant figures given, the angular momentum could have been as small as 2.12×10^{-34} J·s.

Knowing n, we can calculate the binding energy from text Eq. 26.11. In electron volts this equation is

$$E = -\frac{13.6 \text{ eV}}{n^2} \qquad n + 1, 2, 3, \dots \qquad \text{(text Eq. 26.11)}$$

For $n = 2$ this relation gives

$$E = -3.4 \text{ eV}.$$

The electron is in the $n = 2$ state and is bound to the atom with an energy of -3.4 eV.

PROBLEM 27.8: A small meteoroid of size and mass approximately that of a golf ball (0.10 kg) is in circular orbit about the earth. Assume that angular momentum is quantized. Calculate the principal quantum number of the orbit and the spacing between adjacent orbits. *Hint:* Assume the mass is uncharged so that gravity is the dominant force and that the distance of the orbit from the earth's surface is small compared with the earth's radius. (27.6)

Solution:

> **Strategy:** Equate the centripetal and gravitational forces to find how the velocity depends upon the radius. Use this relation in the quantization of angular momentum to find the value for n.

Since the meteoroid is in orbit, the gravitational force must provide the necessary force for circular motion. If we let $m = 0.10$ kg be the mass of the meteoroid, M the mass of the earth, v the circular velocity and r the distance to the earth's center, we have

$$\frac{mv^2}{r} = \frac{GMm}{r^2} = gm\left(\frac{R_E}{r}\right)^2,$$

where g is the acceleration of gravity at the earth's surface, R_E is the radius of the earth, and we have made use of the definition of g,

$$g = \frac{GMm}{R_E}.$$

The condition for quantizing angular momentum is

$$mvr_n = n\frac{h}{2\pi}.$$

Let us assume that the orbits must be quantized and set the r in the first equation to some r_n given by Bohr's equation.

When we combine these two relations to eliminate the velocity we get

$$r_n = \frac{n^2h^2}{4\pi^2gm^2R_E^2}.$$

Since r_n is the same order of magnitude as R_E, we can get an approximate value for n by setting $r_n = R_E$. This gives

$$n = \sqrt{\frac{4\pi^2gm^2R_E^3}{h^2}} = 5 \times 10^{43},$$

where we have used

$$g = 9.8 \text{ m/s}^2, m = 0.10 \text{ kg}, R_E = 6.4 \times 10^6 \text{ m and } h = 6.64 \times 10^{-34} \text{ J·s}.$$

Since we have made an approximation, the results are given to only one significant figure.

To find the interval between orbits calculate the difference in the orbital radius for $n + 1$ and for n, giving

$$\Delta r = r_{n+1} - r_n = \frac{(n+1)^2h^2}{4\pi^2gm^2R_E^2} - \frac{n^2h^2}{4\pi^2gm^2R_E^2} \approx \frac{2nh^2}{4\pi^2gm^2R_E^2} = \frac{2R_E}{n},$$

where, for large n, we have taken $2n + 1$ as about the same as $2n$.

So the fractional change is

$$\frac{\Delta r}{R_E} = \frac{2}{n} = 4 \times 10^{-44}.$$

This fractional difference is completely negligible and the quantum limitations cannot be observed in planetary motions.

COMMENTS ON MODELS

We introduce two new models of the physical world in this chapter. One of them, the Bohr model of the atom, has been replaced by models that give a more accurate description of the world. The other, the photon, describes a fundamental property of light that is used in current scientific models of the world.

The Bohr model of the atom is still used in introductory physics textbooks since it provides an easy means of visualizing the structure of the atom. It gives very accurate results for the

Bohr's theory suspends the rules of electrodynamics. The electron in a Bohr orbit is accelerated (circular or elliptical motion), but it does not radiate. Bohr's theory also suspends Newton's laws since it is nonlocal information (how many wavelengths fit in an orbit) which determines the Bohr orbitals; however, the electron still remains a particle in the classical sense of being localized at a particular distance from the nucleus. Modern quantum theory modifies this model by treating the electron in an atom as a nonlocal object.

The photoelectric effect forces us into models which are even further from our everyday experience. Planck's photon, when used to explain blackbody radiation, could be considered a mathematical tool. However, when Einstein explained the photoelectric effect the particle nature of light was essential. We have a model for light in which light acts as a wave (wave optics, Chapter 24) and light also acts as a particle (photoelectric effect and absorption and emission of light from atoms). This duality (wave-particle) is an essential part of the models of modern physics. A particle acts as a wave when scientists ask questions which can be answered by wave properties and it acts as a particle when we look for particle properties. We will see in the next chapter that all particles can act as waves.

The dual model of nature seems strange since we do not experience this dual behavior in everyday life. On the other hand, we should be able to accept this model, provided

(a) the predictions match the experimental observations, and
(b) in the classical limit, waves act as waves and particles as particles.

SUGGESTED READINGS

Banet, L., "Evolution of the Balmer Series." *American Journal of Physics,* June 1966, p. 496.

Bayfield, J. E., "Rydberg Atoms." *American Scientist,* July 1983, p. 375.

Gulkis, S., P. M. Lubin, S. S. Meyer, and R. F. Silverberg, "The Cosmic Background Explorer." *Scientific American,* January 1990, p. 132.

Heilbron, J. L., *The Dilemmas of an Upright Man: Max Planck as Spokesman for German Science.* Berkeley, Calif.: University of California Press, 1986. This biography includes Planck's scientific work and his public life. It describes the conflict between his patriotism and his moral anguish over the events in Germany in the 1930s.

Chapter 28
Quantum Mechanics

IMPORTANT TERMS

Fill in the blanks with the appropriate word or words.

1. The _____ wavelength of a particle is inversely proportional to the mass of the particle.

2. The _____ principle relates the minimal lack of knowledge of the position of a particle and the minimal lack of knowledge of the momentum of a particle.

3. The _____ principle states that no two electrons may have identical quantum numbers.

4. The quantum interaction between orbital angular momentum and external magnetic fields is called _____ quantization.

5. The _____ equation is a fundamental equation of quantum mechanics describing the behavior of a particle.

6. The absolute value squared of the _____ gives the probability density in quantum mechanics.

7. The ability of a particle to pass through a region of potential where it could not classically exist is called _____.

8. The absolute square of the wave function of a particle is the _____ density of finding the particle.

9. Given the momentum of a particle there is an associated wavelength for the particle called the _____ wavelength.

10. The intrinsic angular momentum of an electron is referred to as the electron's _____.

11. The _____ is a unit used to measure the orbital angular momentum of an electron.

12. The minimal kinetic energy that a quantum system must possess is called the _____ energy.

13. A large number of waves added to have an appreciable amplitude only in a small region of space is called a(n) _____.

Write the definitions of the following:

14. de Broglie wavelength (28.3):

15. Schrödinger equation (28.4):

16. Compton wavelength (28.2):

17. spin (28.11):

18. uncertainty principle (28.5):

19. Bohr magneton (28.10):
20. Pauli exclusion principle (28.11):
21. zero point energy (28.7):
22. space quantization (28.10):
23. probability density (28.6):
24. wave function (28.4):
25. wave packet (28.5):
26. barrier penetration or tunneling (28.7):

ANSWERS TO 1–13, IMPORTANT TERMS

1. Compton; 2. uncertainty; 3. Pauli exclusion; 4. space; 5. Schrödinger; 6. wave function; 7. barrier penetration or tunneling; 8. probability; 9. de Broglie; 10. spin; 11. Bohr magneton; 12. zero point; 13. wave packet.

SAMPLE QUESTIONS

QUESTION 28.1: How would atomic shell structures be affected if electrons did not obey the Pauli exclusion principle?

Answer:

It would be possible for all the electrons to be in the $n = 1$ state. There would be no chemistry as we know it today since the electronic structure of atoms would all be identical.

QUESTION 28.2: Decide, on the basis of some numerical estimates, whether the quantized energy states of a person confined to a room can be observationally distinguished from the smooth range allowed by classical physics.

Answer:

Treat the person as a particle of mass $m = 70$ kg confined to a box of dimension 4 m. We can see, without entering the numbers, that the coefficient in text Eq. 27.12 $\{h^2/(8mL^2)\}$ is extremely small. The change in energy between the level n and the level $n+1$ will be too small to detect.

SAMPLE SOLUTIONS

PROBLEM 28.1: What is the ratio of the Compton wavelength of an electron to the Compton wavelength of a proton? If they both have the same speed, what is the ratio of the de Broglie wavelengths? (Assume all speeds are much less than the speed of light.) (28.3)

Solution:

Strategy: Use the text formulas for the Compton and the de Broglie wavelengths and equate the results.

The Compton wavelength, λ_c, of a particle of mass m is given by $\lambda_c = \dfrac{h}{mc}$,

where h is Planck's constant and c is the speed of light. If the subscript p refers to the proton and the subscript e refers to the electron,

$$\frac{(\lambda_c)_{\text{electron}}}{(\lambda_c)_{\text{proton}}} = \frac{\dfrac{h}{m_e c}}{\dfrac{h}{m_p c}} = \frac{m_p}{m_e} = 1.84 \times 10^3 \,,$$

while the de Broglie wavelength, λ, is given in terms of the momentum p by $\dfrac{h}{p}$, or for small (compared with c) velocity v by

$$\frac{h}{mv} \,,$$

so

$$\frac{\lambda_{\text{electron}}}{\lambda_{\text{proton}}}\bigg|_{\text{same speed}} = \frac{\dfrac{h}{m_e v}}{\dfrac{h}{m_p v}} = \frac{m_p}{m_e} = 1.84 \times 10^3 \,.$$

The ratios are the same.

PROBLEM 28.2: Electrons are individually accelerated through a potential difference of 1000 V. Each passes through a slit of width d, and strikes a detecting screen located a distance 0.50 m away. Over a period of time a diffraction pattern is produced. What should the slit size be in order that the first minimum in the electron diffraction pattern on the screen be 0.10 mm from the center of the pattern? (requires Chapter 24)(28.3)

Solution:

> **Strategy:** Calculate the momentum from energy conservation and use this momentum to calculate a de Broglie wavelength. Use this wavelength to calculate the first minimum in the diffraction pattern.

If V is the potential difference, e the electronic charge, m the electron's mass, and p the momentum of the electron before it passes through the slit, we have from energy conservation:

$$eV = \frac{p^2}{2m},$$

provided we assume that the velocity is nonrelativistic. You should verify this assumption with the numbers given.

So

$$p = \sqrt{2meV}.$$

The de Broglie wavelength λ is given by

$$\lambda = \frac{h}{p},$$

so

$$\lambda = \frac{h}{\sqrt{2meV}}.$$

The first minimum in the diffraction pattern for a wavelength λ, a deflection angle θ, and slit size d was given in Chapter 24 as

$$\sin \theta = \frac{\lambda}{d} = \frac{y}{D},$$

where y is the deflection of the electron and D is the distance to the screen as shown in Fig. 28.1. The last equality follows from the assumption that $y \ll D$.

The width is thus

$$d = \frac{\lambda D}{y} = \frac{hD}{y \sqrt{2meV}} = 1.9 \times 10^2 \text{ nm}.$$

This is a slit width about the size of the wavelength of visible light and only several hundred times larger than the spacing between planes of molecules in a solid.

Fig. 28.1:   Electron diffraction.

PROBLEM 28.3: An electron is located with an uncertainty in position of about 1.5×10^{-11} m. What is the maximum time after this measurement that the electron could be expected to be within 1.0×10^{-10} m of its original position (the approximate size of an atom), provided we assume motion only due to the uncertainty in the momentum? (28.5)

Solution:

Strategy: Assume that the momentum of the electron is about equal to its uncertainty. Calculate how long a time it would take an electron with this momentum to travel the given distance.

We illustrate the change in the location of the electron in Fig. 28.2.

The electron must obey the uncertainty relation,
$$\Delta x \, \Delta p_x \geq h.$$

Since the uncertainty in position is about 1.5×10^{-11} m, the uncertainty relation implies that:

$$\Delta p_x \geq \frac{h}{\Delta x} = \frac{6.63 \times 10^{-34} \text{ J·s}}{1.5 \times 10^{-11} \text{ m}} = 4.4 \times 10^{-23} \text{ kg·m/s}.$$

An electron has a mass of 9.1×10^{-31} kg, so if we assume nonrelativistic expressions,

$$\Delta v_x = \frac{\Delta p_x}{m} = \frac{\Delta p_x}{9.1 \times 10^{-31} \text{ kg}} \geq 4.8 \times 10^7 \text{ m/s}.$$

The time to travel a distance is given by

$$t = \frac{\text{distance}}{\text{velocity}} = \frac{1.0 \times 10^{-10} \text{ m}}{4.8 \times 10^7 \text{ m/s}} = 2.1 \times 10^{-18} \text{ s}.$$

Since this velocity is more than 10% of the speed of light, the relativistic expression for momentum should probably be used. It is not unreasonable in such a case to use c itself as a good approximation in the uncertainty of the electron's velocity. Thus the electron could be moving at nearly the speed of light. Its position, in that case, would be given by

$$x(t) = x_0 + \Delta v \, t \approx x_0 + ct$$

so

$$t \approx \frac{x}{c} = \frac{1.0 \times 10^{-10} \text{ m}}{3.0 \times 10^8 \text{ m/s}} = 3.3 \times 10^{-19} \text{ s}.$$

With either assumption we see that the electron rapidly moves away from its original position.

Initial uncertainty   **Uncertainty at a later time**

Motion caused by uncertainty in momentum

Position at t = 0   **Position at a later time**

Fig. 28.2: Minimal motion of a particle located at $x \pm \Delta x$ caused by the uncertainty relation.

Thus the electron will be on the order of 10 times its initial uncertainty away from its initial position in a time on the order of 10^{-18} s.

PROBLEM 28.4: Suppose an electron is confined in a very deep potential box of width $L = 0.024$ mm. What are the energy levels of the two lowest states possible? Sketch the wave functions for these two states. How does the ground state energy compare with the rest mass energy? (28.7)

Solution:

Strategy: Apply the arguments given in Section 28.7 of the text and the equation given for the energy levels. We treat this as a one-dimensional problem.

Since the box is said to be "very deep," we assume the arguments of text Section 28.7 apply and can make the sketch as shown in Fig. 28.3.

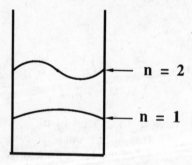

n = 2

n = 1

Fig. 28.3: Wave functions of an electron in a potential box.

The energy levels are given by text Eq. 28.12:

$$E_n = \frac{h^2 n^2}{8mL^2}$$

where n is the level number, h is Planck's constant, m is the mass of the particle, and L is the length of the box. Thus

$$E_n = \left(\frac{(6.63 \times 10^{-34} \text{ J·s})^2}{8 \times 9.1 \times 10^{-31} \text{ kg} \times (2.4 \times 10^{-5} \text{ m})^2} \right) n^2 = 1.0 \times 10^{-28} \, n^2 \text{ J}.$$

The rest mass energy of an electron is

$$E_o = m \, c^2 = 9.1 \times 10^{-31} \text{ kg} \times (3.0 \times 10^8 \text{ m/s})^2 = 8.2 \times 10^{-15} \text{ J}.$$

So the ratio of the ground state energy to the rest mass is about 10^{-14}.

PROBLEM 28.5: In the preceding problem, what must the value of L be such that the ground state energy is equal to the rest mass energy of the electron? (requires Chapter 25)(28.7)

Solution:

> **Strategy:** Find the value of L in the preceding expression for E_n that makes $E_n = E_o$ where E_o is the rest mass energy.

We have two relationships:

$$E_o = m_e c^2$$

and

$$E_n = E_{binding} = \frac{h^2 n^2}{8 m_e L^2},$$

where E_o is the rest mass energy of an electron with mass m_e, c is the speed of light, E_n is the binding energy for quantum number n, h is Planck's constant, L is the box dimension, and n is an integer from 1 to any number.

Equating these for $n = 1$ gives,

$$L = \frac{1}{2\sqrt{2}} \frac{h}{m_e c} = \frac{6.63 \times 10^{-34} \text{ J·s}}{2.83 \times 9.1 \times 10^{-31} \text{ kg} \times 3.0 \times 10^8 \text{ m/s}} = 8.6 \times 10^{-13} \text{ m}.$$

This size, 8.6×10^{-13} m, is about 20 times the typical size of a nucleus. This again shows us that electrons, as individual particles, cannot be localized in an atomic nucleus.

PROBLEM 28.6: How many hydrogenic electronic states are there corresponding to the principal quantum number $n = 4$? (28.9)

Solution:

> **Strategy:** Count the possible orbital quantum numbers, l, for $n = 4$ and count how many magnetic quantum numbers, m_l, correspond to each l. Multiply each number by two to allow for two spin states.

For each value of the principal quantum number n, the orbital quantum number l may assume integral values between 0 and $n-1$ (n values). For each value of l, the magnetic quantum number, m_l, may assume any integral value between $-l$ and $+l$ ($2l+1$ values). Finally for each value of m_l, the electron may have its spin up or down, $m_s = \pm 1/2$. The table below catalogs the possible states for n = 4.

l value	m_l values	m_s values	Number of states
3	$-3, -2, -1, 0, +1, +2, +3$	$\pm 1/2$	14
2	$-2, -1, 0, +1, +2$	$\pm 1/2$	10
1	$-1, 0, +1$	$\pm 1/2$	6
0	0	$\pm 1/2$	2
		Total number of states	32

There are 32 electronic states for $n = 4$.

PROBLEM 28.7: A magnetic storm on the sun associated with a cluster of sunspots is observed to produce a Zeeman splitting of 1.2×10^{-4} eV as shown in Fig. 28.4. What is the magnetic field that produced this splitting? (28.10)

$$\Delta E = 1.2 \times 10^{-4} \text{ eV}$$

Fig. 28.4: Zeeman splitting of hydrogen spectral line produced by the solar magnetic field associated with a sunspot cluster.

Solution:

Strategy: Calculate the energy difference in the $\pm 1/2$ spins produced by a magnetic field. Solve for the field.

The energy of each state is given by

$$E = \mu_s B = 2\mu_B m_s B,$$

so the energy difference is

$$\Delta E = E_+ - E_- = 2\mu_B \left\{ \frac{1}{2} - \left(-\frac{1}{2} \right) \right\} B$$

$$= 2 \times 9.27 \times 10^{-24} \frac{J}{T} \times \frac{1}{1.6 \times 10^{-19}} \frac{eV}{J} \times B = 1.16 \times 10^{-4} \frac{eV}{T} \times B.$$

Hence,

$$B = \frac{1.2 \times 10^{-4} \text{ eV}}{1.16 \times 10^{-4} \frac{eV}{T}} = 1.0 \text{ T}.$$

The associated magnetic field was 1.0 T.

PROBLEM 28.8: Find the energy difference between electron states of $m_s = +1/2$ and $m_s = -1/2$ in an applied magnetic field of $B = 2.0$ T. Express this energy in electron volts. What is the wavelength of a photon having this same energy? (28.11)

Solution:

> **Strategy:** Calculate the energy difference in the two states.

The energy, E, added to an electron in a magnetic field, B, is given by $E = \mu_s B$, where $\mu_s = 2 \mu_B m_s$ with μ_B the Bohr magneton and m_s the quantum number.

The change in energy from $m_s = +1/2$ to $m_s = -1/2$ is

$$\Delta E = E_+ - E_- = 2 \mu_B [+1/2 - (-1/2)]B = 2 \mu_B B = 2 \times 9.27 \times 10^{-24} \text{ J/T} \times 2.0 \text{ T}$$

$$= 3.7 \times 10^{-24} \text{ J} = 2.3 \times 10^{-4} \text{ eV}$$

The energy is 2.3×10^{-4} eV.

For a photon to have the same energy, it would have a wavelength given by $E = \dfrac{hc}{L}$.

Thus $L = \dfrac{E}{hc} = \dfrac{3.7 \times 10^{-24} \text{ J}}{6.6 \times 10^{-34} \text{J·s} \times 3.0 \times 10^8 \text{ m/s}} = 19 \text{ m}.$

The wavelength would be 19 m. That is in the radio band.

COMMENTS ON MODELS

Quantum mechanics is both an evolutionary and a revolutionary development from classical mechanics. It is evolutionary in the sense that it must, when the proper questions are asked, reduce to classical mechanics when Planck's constant is considered sufficiently small. On the other hand, quantum mechanics is revolutionary in as much as the basic structure of matter is different from what was assumed by the classical physicists.

The wave nature of matter is probabilistic but not in a classical sense. The wave function does not just represent a lack of accuracy; it is a basic physical indeterminacy. That is, when we give the wave function for the electron in the ground state of an atom, it is not possible to claim the electron is localized somewhere until a measurement of position is made. If this measurement is made with sufficient accuracy, the uncertainty relation forces us to say that we can't tell if the electron was bound to that atom. In general, quantum measurement usually changes or destroys the system, while classically the act of measurement can be done without disturbing the system under study. Quantum mechanics tells us that the probe particle must be considered as part of the system under study. Thus the physicist is no longer an uninvolved observer since the very act of observation effects the nature of the world.

SUGGESTED READINGS

Deltete, R., and R. Guy, "Einstein's Opposition to the Quantum Theory." *American Journal of Physics,* July 1990, p. 673.

Matteucci, G., "Electron Wavelike Behavior: A Historical and Experimental Introduction." *American Journal of Physics,* December 1990, p. 1143.

Chapter 29
The Nucleus

IMPORTANT TERMS

Fill in the blanks with the appropriate word or words.

1. The neutral particle in the nucleus is the _____.

2. _____ are nuclei with the same charge but different numbers of neutrons.

3. A(n) _____ has the same mass as an electron, but has opposite charge.

4. The dose _____ takes into account not only the energy deposited per unit mass by radiation but also the biological effect of the type of radiation.

5. Nuclear fissions induced by neutrons released by previous nuclear fissions are called a chain _____ if more and more fissions are induced as time passes.

6. The _____ is the measure of the ability of a particular type of radiation to cause biological damage.

7. The _____ is the energy deposited per unit mass by absorbed radiation.

8. A(n) _____ occurs when an orbital electron is captured by its nucleus.

9. The minimum energy that a particle needs in order to react with a nucleus is called the _____ energy.

10. When two nuclei combine to form another, we have the process known as _____.

11. The _____ of a nuclear decay is the difference between the initial and final mass energies.

12. The positive charges in the nucleus are _____.

13. Protons and neutrons are collectively referred to as _____.

14. The difference between the sums of the mass energies of the nucleons in a nucleus and the mass energy of that nucleus is termed the _____ of that nucleus.

15. _____ plus and _____ minus are other names for positrons and electrons, respectively.

16. The _____ is a particle with nearly zero rest mass and no charge. It has the same spin as the electron.

17. The process in which one nucleus breaks up into two or more nuclei is termed _____.

Write the definitions of the following:

18. proton (29.1):
19. chain reaction (29.10):
20. neutron (29.1):
21. nucleon (29.2):
22. isotope (29.2):
23. dose equivalent (29.8):
24. positron (29.6):

25. beta plus (β^+) and beta minus (β^-) (29.6):
26. quality factor (29.8):
27. electron capture (29.6):
28. neutrino (29.6):
29. threshold energy (29.9):
30. fission (29.10):
31. fusion (29.11):
32. dose (29.8):
33. Q value (29.4):
34. nuclear binding energy (29.3):

ANSWERS TO 1–17, IMPORTANT TERMS

1. neutron; 2. isotopes; 3. positron; 4. equivalent; 5. reaction; 6. quality factor 7. dose; 8. electron capture; 9. threshold; 10. fusion; 11. Q value; 12. protons; 13. nucleons; 14. binding energy; 15. beta (β^+ and β^-); 16. neutrino; 17. fission.

SAMPLE QUESTIONS

QUESTION 29.1: Why is the decay rate of radioactive nuclei not changed by chemical environment, heating, or pressure?

Answer:

While all of the listed effects might change the electronic structure, they cannot affect the nucleus. That is why physicists did not expect cold fusion to be possible.

QUESTION 29.2: Why is radon still found in nature? Its half-life is 1600 years and the estimated age of the universe is five billion years.

Answer:

The radon we find now has been produced by the recent decay of parent nuclei. The statement about the age of the universe tells us that no radon produced at the start of the universe is still in existance.

SAMPLE SOLUTIONS

PROBLEM 29.1: Natural potassium occurs as a mixture of three isotopes:

$$^{20}_{10}\text{K at 93.08\%,} \quad ^{21}_{10}\text{K at 0.01\% and} \quad ^{22}_{10}\text{K at 6.91\%.}$$

Calculate the atomic mass of naturally occurring potassium assuming that the atomic mass of any isotope is given by the number of nucleons in the nucleus (this assumption is good to better than 1 to 2%). (29.2)

Solution:

Strategy: Naturally occurring potassium should consist of the given mixture of isotopes.Calculate the weighted average of the masses of the three isotopes. The weighting used is the percentage of occurrence.

The general procedure to use if there are N isotopes of masses m_i and each having a percentage occurrence of P_i, i = 1 to N, is to to write

$$m_{\text{observed}} = \frac{\sum\limits_{i=1}^{N} m_i P_i}{\sum\limits_{i=1}^{N} P_i} = \frac{\sum\limits_{i=1}^{N} m_i P_i}{100}.$$

For potassium we write

$$m_{\text{observed}} = \frac{20 \times 93.08 + 21 \times 0.01 + 22 \times 6.91}{100} = 20.1 \text{ u}.$$

the atomic mass of naturally occurring potassium is nearly the same as that of the most abundant isotope.

PROBLEM 29.2: Calculate the binding energy of $^{9}_{4}\text{Be}$ using the data in Table 29.2. (29.3)

Solution:

Strategy: The binding energy is the difference between the mass of the nucleus and the sum of its components.

In text Table 29.2 we find that $^{9}_{4}\text{Be}$ has a mass of 9.012182 atomic mass units, which is 8394.796 MeV. This nucleus is composed of 4 protons and 9 − 4 = 5 neutrons. In text

Table 29.1 we find the mass of a proton is 938.272 MeV and that of a neutron is 939.566 MeV.

This gives a binding energy, ΔE, of

$$\Delta E = 8394.796 \text{ MeV} - 4 \times 938.272 \text{ MeV} - 5 \times 939.566 \text{ MeV} = -56.122 \text{ MeV}.$$

The binding energy is 56.122 MeV, with the minus sign indicating that to break up ^9_4Be we will need to provide energy.

PROBLEM 29.3: One of the decays of an isotope of polonium is:

$$^{208}_{84}\text{Po} \rightarrow {}^{204}_{82}\text{Pb} + {}^4_2\alpha \, ,$$

where the lead (plumbium) isotope has an atomic mass of 204.0363 u. The emitted α particle has a kinetic energy of 5.108 MeV. What is the atomic mass of the polonium isotope, assuming no other particles are emitted? What is the Q value of this decay? (29.4)

Solution:

Strategy: Add the rest energy of the α particle to its kinetic energy to get the total energy in the α particle. This energy plus the mass energy in the Pb nucleus must be the total initial energy in the Po nucleus. Use conservation of momentum and energy to determine how this energy is distributed.

We can used text Table 29.2 to help us answer this question. The initial mass of the Po must be the sum of the final Pb mass plus the final kinetic energy of the Pb nucleus plus the mass of the α particle plus the kinetic energy of the α particle. That is,

mass of Po = mass of Pb + mass equivalent of Pb kinetic energy

+ mass of α + mass equivalent of α particle kinetic energy.

Text Table 29.2 gives the α particle mass as 3728.431 MeV, so the total energy associated with the α particle is

$$(3728.431 + 5.108) \text{ MeV} = 3733.539 \text{ MeV}.$$

In order to compare masses we must be able to convert mass in u to mass in Mev and conversely. This conversion factor can be calculated by comparing the quoted value of the mass of the proton to the quoted value in MeV. This ratio is

$$\frac{\text{mass of p in u}}{\text{mass of p in MeV}} = \frac{1.007276}{938.279} \frac{\text{u}}{\text{MeV}} = 1.0735 \times 10^{-3} \frac{\text{u}}{\text{MeV}}.$$

The emitted alpha particle has an equivalent mass of

$$3733.539 \text{ MeV} \times 1.0735 \times 10^{-3} \frac{u}{\text{MeV}} = 4.0081 \text{ u.}$$

This gives two of the four terms we need (mass of α + mass equivalent of α particle kinetic energy) To calculate the kinetic energy term we will need to consider the recoil of the Pb nucleus and conserve momentum and energy. The decay occurs as shown in Fig. 29.1, where v_α is the velocity if the alpha particle, v_{Pb} is the velocity of the Pb nucleus, m_{Pb} is the mass of the Pb nucleus, and m_α is the mass of the α particle.

Conservation of momentum gives

$$v_{Pb}\, m_{Pb} + v_\alpha\, m_\alpha = 0,$$

Fig. 29.1: Momentum conservation for the decay of Po to Pb by α particle emission.

Since the kinetic energy of the α particle is small compared with the rest energies, all velocities will be small compared with the speed of light, so we can use the nonrelativistic kinetic energy expression:

$$(\text{kinetic energy})_{Pb} = \frac{1}{2} m_{Pb} v_{Pb}{}^2 = \frac{1}{2} m_{Pb} \left(\frac{m_\alpha}{m_{Pb}} v_\alpha \right)^2$$

where the last equality results from substituting in the expression for conservation of momentum. Thus

$$(\text{kinetic energy})_{Pb} = \frac{1}{2} m_\alpha v_\alpha{}^2 \frac{m_\alpha}{m_{Pb}} = \frac{m_\alpha}{m_{Pb}} KE_\alpha$$

$$= 0.01961 \times 5.108 \text{ MeV} = 0.100 \text{ MeV.}$$

The mass equivalent of this energy is

$$0.100 \text{ MeV} \times 1.0735 \times 10^{-3} \frac{u}{\text{MeV}} = 10^{-4} \text{ u.}$$

Therefore,

$$\text{mass of Po} = 204.0363 \text{ u} + 0.0001 \text{ u} + 4.0081 \text{ u} = 208.0445 \text{ u.}$$

So $^{208}_{84}$Po has a mass of 208.0445 u.

The Q value is the sum of the two kinetic energy terms, so

$$Q = (5.108 + 0.100) \text{ MeV} = 5.208 \text{ MeV.}$$

PROBLEM 29.4: Show that an α particle with a kinetic energy of 8.00 MeV need not be treated as a relativistic particle. (29.6)

Solution:

> **Strategy:** A particle needs to be treated as a relativistic particle if its kinetic energy is a reasonable fraction of its rest energy.

Text Table 29.2 gives us the rest energy of an α particle $\left(^4_2\text{He}\right)$ as 3728.402 MeV. Clearly 8.00 MeV is very small compared to the rest energy. We can verify that the speed of an 8.00 MeV alpha particle is much less than that of light by using the expression for relativistic kinetic energy from Chapter 25:

$$\text{kinetic energy} = m_0c^2\left(\frac{1}{\sqrt{1 - \dfrac{v^2}{c^2}}} - 1\right).$$

We know the kinetic energy and the rest energy, so

$$1 + \frac{\text{Kinetic Energy}}{m_0c^2} = \frac{1}{\sqrt{1 - \dfrac{v^2}{c^2}}} = 1 + \frac{8.00 \text{ Mev}}{3728.402 \text{ MeV}} = 1.002.$$

Solving for v/c gives

$$\frac{v}{c} = 0.07.$$

The 8 MeV alpha particle is traveling at only 7% the speed of light.

PROBLEM 29.5: A small radioactive source is monitored with a detector that subtends a solid angle of 0.080 steradians. If the average count rate in the detector is 7396 counts per minute, what is the activity of the source in Bq? (29.7)

Solution:

> **Strategy:** Assume the radiation is emitted spherically. Calculate the fraction of a sphere that the detector intercepts. Multiply the counts/s by the inverse of this fraction to get the activity.

The geometry of this problem is shown in Fig. 29.2.

Fig. 29.2: Detector with a 0.080 steradian intercept angle.

We must assume the radiation is emitted isotropically in all directions. That is, the radiation is emitted uniformly over the entire sphere which is 4π steradians. This means

$$\frac{\text{total counts/min}}{4\pi \text{ steradians}} = \frac{\text{observed counts/min}}{0.080 \text{ steradians}}.$$

So

$$\text{Total counts/min} = \frac{4\pi}{0.080} \times 7396 = 1161761 = 1.16 \times 10^6 \text{ counts per minute.}$$

The activity in Bq is the counts per second so

$$\text{activity} = \frac{1.16 \times 10^5 \text{counts per minute}}{60 \text{ seconds per minute}} = 1.9 \times 10^4 \text{ Bq.}$$

The activity is about 20,000 Bq.

PROBLEM 29.6: A proton with a kinetic energy of 5.00 MeV elasticity scatters head-on off a deuteron $\left(\begin{smallmatrix}2\\1\end{smallmatrix}H\right)$ which was initially at rest. What is the kinetic energy of the deuteron in MeV after the collision? Use the mass for the deuteron atom given in text Table 29.2. (29.9)

Solution:

> **Strategy:** Since the kinetic energy of the proton is small compared with its rest energy, this is a nonrelativistic problem. Conserve energy and momentum.

Fig. 29.3: (a) Proton and deuteron before collision. (b) Particles after collision.

This is a collision problem identical to those done in the mechanics portion of this book. The main difference is that energies are expressed in MeV instead of J. The term "head-on" implies the collision can be treated as a one dimensional problem. Let the subscript "p" denote the proton and the subscript "d" denote the deuteron. Let V be the initial velocity of the proton and v and m denote velocities and masses; then, conserving momentum and energy as shown in Fig. 29.3, we have:

$$m_p V = m_p v_p + m_d v_d \qquad \text{(momentum conservation)}$$

$$\frac{1}{2} m_p v^2 = \frac{1}{2} m_p v_p^2 + \frac{1}{2} m_d v_d^2 \qquad \text{(energy conservation)}.$$

Solve the momentum conservation equation for v_p and eliminate v_p from the energy equation. This gives

$$2V = \left(1 + \frac{m_d}{m_p}\right)v_d \quad \text{or} \quad v_d = \frac{2V}{1 + \left(\frac{m_d}{m_p}\right)}.$$

We can check our calculations to this point by noting that if $m_d = m_p$ then $v_d = V$ and $v_p = 0$ as expected.

Finally

$$KE_d = \frac{1}{2}m_d v_d^2 = \frac{1}{2}m_d \times \left(\frac{2V}{1 + \left(\frac{m_d}{m_p}\right)}\right)^2$$

$$= 4\frac{m_d m_p}{(m_p + m_d)^2}KE_p$$

where KE_p is the initial kinetic energy of the incoming proton. Thus,

$$KE_d = 0.889 \times KE_p = 4.44 \text{ MeV}.$$

The scattered deuteron has a kinetic energy of 4.44 MeV.

PROBLEM 29.7: Determine the kinetic energy available to the α decay of

$$^{235}_{92}U \rightarrow {}^{231}_{92}Th + {}^{4}_{2}\alpha \text{ if the observed masses are}$$

$$m\left({}^{235}_{92}U\right) = 235.0494 \text{ u, and } m\left({}^{231}_{92}Th\right) = 231.03631 \text{ u.}$$

(29.10)

Solution:

> **Strategy:** The decay must conserve energy.

We assume the decay products are nonrelativistic and write

mass of U = mass of Th + kinetic energy of Th + mass of alpha + kinetic energy of α.

Thus

kinetic energy of decay products = kinetic energy of Th + kinetic energy of α

$$= \text{mass of U} - \text{mass of Th} - \text{mass of } \alpha$$

$$= 235.0494 \text{ u} - 231.03631 \text{ u} - 3728.431 \text{ Mev} \times 1.0735 \times 10^{-3} \frac{u}{\text{MeV}}$$

$$= 0.0106 \text{ u} = 0.0106 \text{ u} \times \frac{1}{1.0735 \times 10^{-3} \frac{u}{\text{MeV}}} = 9.87 \text{ MeV}.$$

This kinetic energy will be divided between the Th and the α. The division is determined by conservation of momentum. See Problem 29.6.

PROBLEM 29.8: Determine the identity of the nucleus X in the reaction:

$$_{6}^{13}C + n \rightarrow X + \gamma. \qquad (29.11)$$

Solution:

> **Strategy:** Reactions must conserve charge and mass numbers.

The reaction may become more clear if we write the mass and charge numbers for the neutron and the γ ray:

$$_{6}^{13}C + _{0}^{1}n \rightarrow X + _{0}^{0}\gamma.$$

Then we see that the nucleus X has a mass number of $13 + 1 - 0 = 14$ and a charge number of $6 - 0 - 0 = 6$. Thus the decay product is carbon 14.

COMMENTS ON MODELS

The major portion of this chapter deals with the models used to describe nuclear decay. Three different but related models are discussed. These give a description of α, β, and γ decay.

The model for α decay involves the classical and quantum ideas of a potential well. That is, we assume the α particle is able to tunnel out of the nucleus. The remaining nucleons in the nucleus are used to define this potential well that retains the α particle. The β decay model invokes energy conservation to force us to accept the idea of the neutrino. The neutrino is given the properties needed to conserve energy during β decay reactions. Until recently one of the properties of the neutrino was that it had no mass. Recent experiments hint that the neutrino may have a small mass. Remember that the terms mass, spin and charge are only properties that we assign to our models. The γ decay model assumes that the nucleus has energy levels analogous to those in the electronic structure of atoms. In this model the entire nucleus is assumed to have these energy levels. Some nuclear physicists do spectrometry of these nuclear energy levels and attempt to deduce the nuclear structure in the same fashion that atomic physicists do spectrometry on atomic levels to deduce the electronic structure of the atom. All three of these models are used to make predictions about the behavior of nuclei.

The α and γ decay models are similar since they both can be considered as potential well problems, but they also have their fundamental differences. In the α-decay case all the nucleons together define the potential well. At this point in our studies we are where the atomic spectroscopists were before the Bohr theory was devised. Chapter 32 will help you to understand some of these ideas.

SUGGESTED READINGS

Badash, L., "The Age-of-the-Earth Debate." *Scientific American,* August 1989, p. 90.

Bahcall, J. N., "The Solar-Neutrino Problem." *Scientific American,* May 1990, p. 54.

Barth, R. F., A. H. Soloway, and R. G. Fairchild, "Boron Neutron Capture Therapy for Cancer." *Scientific American,* October 1990, p. 100.

Bethe, H., and G. Brown, "How a Supernova Explodes." *Scientific American ,* May 1985, p. 60.

Bettis, C., and C. Throckmorton, "What Teachers Should Know about Radon." *The Physics Teacher*, September 1991, p. 338.

Cobb, C. E., and K. Kasmauski, "Living With Radiation." *National Geographic,* April 1989, p. 403.

Cohen, B. L., *The Nuclear Energy Option: An Alternative for the 90s*. New York: Plenum, 1990. Analysis of energy needs and the risks of nuclear energy as well as the risks of other sources of energy.

Emmett, J. L., J. Nuckolls, and L. Wood, "Fusion Power by Laser Implosion." *Scientific American,* June 1974, p. 24.

Golay, M. W., and N. E. Todreas, "Advanced Light-Water Reactors." *Scientific American,* April 1990, p. 82.

Inglis, D. R., *Nuclear Energy: Its Physics and Social Challenge*. Reading, Mass.: Addison Wesley, 1973.

Lewis, H., *Technological Risk*. New York: Norton, 1990.

Lewis, K. N., "The Prompt and Delayed Effects of Nuclear War." *Scientific American,* July 1979, p. 35.

Lieberman, J. A., "Ionizing-Radiation Standards for Population Exposure." *Physics Today,* November 1971, p. 32.

Mathews, G. J., "The Cosmic Synthesis of Lithium, Beryllium and Boron." *Scientific American,* May 1987, p. 38.

NCRP Report 93, *Ionizing Radiation Exposure of the Population of the United States*. Bethesda, Md.: National Council on Radiation Protection and Measurement, September 1, 1987.

Spradley, J. L., "Women and the Elements." *The Physics Teacher,* December 1989, p. 656.

Upton, A. C., "Health Effects of Low-Level Ionizing Radiation." *Physics Today*, August 1991, p. 34.

Weart, S. R., *Nuclear Fear: A History of Images*. Cambridge, Mass.: Harvard University Press, 1988. An exploration of our culture's interaction with nuclear science and technology.

Chapter 30
Lasers, Holography, and Color

IMPORTANT TERMS

Fill in the blanks with the appropriate word or words. The various types of holograms are included for identification.

1. The emission of light by excited atoms triggered by the presence of other light of the proper frequency is termed _____ emission.

2. When more ions are in an excited state than are in the lower state we have a(n) _____.

3. The measure of an optical instrument's ability to produce distinct images of objects which appear close together is called the _____ of the instrument.

4. A(n) _____ medium contains atoms or molecules which can be made to emit coherent light.

5. A(n) optical _____ contains mirrors that reflect light back and forth many times.

6. A(n) _____ hologram is viewed by observing its image with nearly monochromatic light which is passed through the hologram.

7. The inverse of the number of lines per unit length that can be resolved by a holographic film is termed the _____.

8. The coherence of light extending in space is called _____ coherence.

9. A(n) _____ hologram can be viewed with a broad wavelength source provided the source is nearly a point source.

10. A photographic recording of an interference pattern is called a(n) _____.

11. The making and study of the photographic recording of interference patterns is called _____.

12. The light emitted by the random de-excitation of atoms is known as _____ emission.

13. The _____ is an optical example of stimulated emission.

14. A nearly monochromatic light beam is said to have a high _____ coherence.

15. _____ holograms are made by stamping the interference pattern into plastic.

16. _____ is the term for the process of exciting the laser medium.

17. _____ holograms are produced by having the reference and object beams strike the opposite sides of the film.

18. The _____ hologram is a transmission hologram which may be viewed with white light. The color of the holographic images varies as the angle of viewing is changed.

19. It is possible to create and maintain inverted populations in an active medium because there exists a(n) _____ state with a longer lifetime than might be expected.

20. The _____ primary colors of light are red, green, and blue.

21. The _____ primary colors are those which combined with one of red, green, or blue produce white light.

Write the definitions of the following:

22. holography (30.5):

23. laser (30.1):

24. active medium (30.2):

25. inverted population (30.2):

26. spatial coherence (30.4):

27. additive primary colors (30.7):

28. spontaneous emission (30.1):

29. pumping (30.2):

30. metastable state (30.2):

31. hologram (30.5):

32. subtractive primary colors (30.7):

33. stimulated emission (30.1):

34. optical resonator (30.2):

35. temporal coherence (30.4):

ANSWERS TO 1–21, IMPORTANT TERMS

1. stimulated; 2. inverted population; 3. resolving power; 4. active; 5. resonator; 6. transmission; 7. spatial frequency; 8. spatial; 9. white light; 10. hologram; 11. holography; 12. spontaneous; 13. laser; 14. temporal; 15. embossed; 16. pumping; 17. reflection; 18. rainbow; 19. metastable; 20. additive; 21. subtractive.

SAMPLE SOLUTIONS

PROBLEM 30.1: Calculate the separation between modes in a helium-neon laser whose cavity length is 37.5 cm. (30.2)

Solution:

Strategy: The spacing between nodes in a laser is given by text Eq. 29.2.

The mode separation, Δf, for a laser of length L is given by

$$\Delta f = \frac{c}{2L} = \frac{3.0 \times 10^{10} \text{ cm/s}}{2 \times 37.5 \text{ cm}} = 4.0 \times 10^8 \text{ Hz.}$$

PROBLEM 30.2: The interference pattern produced by light from a laser falling on two slits has a fringe visibility of 0.60. What is the ratio of the maximum to the minimum intensity? Assuming that one of the slits is illuminated two times as strongly as the other, what is the coherence? (30.4)

Solution:

Strategy: Use the definitions of fringe visibility and coherence to calculate the requested quantities.

If I_{max} is the maximum intensity and I_{min} is the minimum intensity, then the visibility V is defined in the text as

$$V = \frac{I_{max} - I_{min}}{I_{max} + I_{min}} \; .$$

This can be rearranged to give

$$\frac{I_{max}}{I_{min}} = \frac{1 + V}{1 - V} = \frac{1 + 0.60}{1 - 0.60} = 4.0$$

The coherence is defined as

$$\gamma = \frac{I_1 + I_2}{2\sqrt{I_1 I_2}} V \; ,$$

where I_1 and I_2 are the illuminations at slits 1 and 2, respectively, and V is the visibility. This gives

$$\gamma = \frac{3I_1}{2\sqrt{2I_1^2}} V = \frac{3}{2\sqrt{2}} \, 4.0 = 4.2.$$

The intensity ratio is 0.40 and the coherence is 4.2.

PROBLEM 30.3: Two beams of parallel laser light with a wavelength, λ, of 632.8 nm are brought together on a photographic plate to form a sinusoidal holographic grating. One beam strikes the plate at normal incidence. The other beam makes an angle of 32.0° with the normal. (See Fig. 30.1.) The processed hologram is illuminated with light of wavelength 488.0 nm. If this light strikes the hologram at normal incidence, how many beams emerge and at what angles? (30.5)

Solution:

Strategy: Calculate the spacing of the grating. This can be combined with the formula for a diffraction pattern from Chapter 24 to get the maximum order permitted by the grating.

If one laser beam (the reference) strikes the film normally and the other strikes at an angle θ_0 with respect to the normal as shown in Fig. 30.1, then the interference pattern formed has an effective grating spacing of d given by

$$d = \frac{\lambda_1}{\sin \theta_0} \, ,$$

where λ_1 is the wavelength of the laser light.

If this diffraction grating is now illuminated normally with laser light of wavelength λ_2, diffraction maxima will occur at (see Chapter 24) angles θ_n given by

$$\sin \theta_n = \frac{n \, \lambda_2}{d} = n \, \frac{\lambda_2}{\lambda_1} \sin \theta_0,$$

for $n = 1, 2, 3, \ldots, n_{max}$, where the largest n possible, n_{max}, is that for which $n_{max} + 1$ would make the right hand side greater than 1. Entering our numbers gives

$$\sin \theta_n = n \, \frac{488 \text{ nm}}{632.8 \text{ nm}} \sin 32° = 0.409 \, n.$$

Since n = 3 makes the right hand side greater than 1, there are two diffraction orders whose angles can be found by taking the inverse sine; that is,

$$\theta_n = \sin^{-1} 0.409 \, n \, .$$

The diffraction maxima occur at 24.1° and at 54.8°.

Fig. 30.1: Two laser beams combining to form a sinusoidal holographic grating.

PROBLEM 30.4: The temperature of the star Sirius is about 10,000 K. At what wavelength is the most light emitted? What would be the color of this star? (30.6)

Solution:

Strategy: Use the Wien displacement law to find the wavelength of maximum emission and compare this with the visible color wavelengths.

The Wien displacement law of Chapter 27 gives us the wavelength of maximum emission, λ_{max}, in terms of the kelvin temperature, T, as

$$\lambda_{max} = \frac{0.00290 \text{ K·m}}{T} = 2.90 \times 10^{-7} \text{ m} = 290 \text{ nm.}$$

This is shorter than visible light (400 nm < $\lambda_{visible}$ < 700 nm), so we would expect Sirius to look blue. Since the human eye's response drops off towards the blue, Sirius appears blue-white.

PROBLEM 30.5: A photon has a measured energy of 1.77 eV. What color sensation would this photon produce? (30.6)

Solution:

> **Strategy:** Compute the wavelength corresponding to this energy to decide the color sensation.

To convert energy in eV to energy in joules, multiply by the electron charge, e (magnitude, not the negative number). Let E_0 denote the energy in eV and E the energy in joules; then
$$E = E_0 e.$$

The frequency, f, of this light is given by
$$f = \frac{E}{h},$$

and the wavelength, λ, by
$$\lambda = \frac{c}{f},$$

where c is the speed of light.

Combining gives
$$\lambda = \frac{ch}{eE_0} = 7.01 \times 10^{-7} \text{ m} = 701 \text{ nm.}$$

This is at the long wavelength end of the visible spectrum, so the sensation would be deep red.

PROBLEM 30.6: White light passes through a magenta filter, which is, in turn, followed by a cyan filter as shown in Fig. 30.2. What color emerges? (30.7)

Solution:

> **Strategy:** Determine what colors are passed by the first filter and then which of these will be passed by the second filter.

Fig. 30.2: White light passing through a magenta filter followed by a cyan filter

Both of these filters are subtractive primary color filters. The magenta filter passes the colors complementary to green and the cyan filter passes the colors complementary to red.

The magenta filter removes the green color. Thus white light minus green enters the cyan filter. The cyan removes the red color. The two filters together have removed both the red and the green. Only the blue is left.

The light will appear blue.

SUGGESTED READINGS

Berns, M. W., "Laser Surgery." *Scientific American,* June 1991, p. 84.

Brecher, K., "Do Atoms Really 'Emit' Absorption Lines?" *The Physics Teacher,* October 1991, p. 454.

Deutsch, T. F., "Medical Applications of Lasers." *Physics Today,* October 1988, p. 56.

La Brecque, M., "Retinex: Physics and the Theory of Color Vision." *Computers in Physics,* November 1988, p. 16.

Leff, H. S., "Illuminating Physics with Light Bulbs." *The Physics Teacher,* January 1990, p. 30.

Nathans, J., "The Genes for Color Vision." *Scientific American,* February 1989, p. 42.

Rennie, R., "Science and Business: Move Over, Mr. Spock—Sound-Sensitive Crystals Bring Computerized Holograms to Life." *Scientific American,* July 1990, p.89.

Romer, G. B., and J. Delamoir, "The First Color Photographs." *Scientific American,* December 1989, p.. 88.

Tonomura, A., "Electron Holography: A New View of the Microscopic." *Physics Today,* April 1990, p. 22.

Walker, J., "The Amateur Scientist: How to Stop Worrying About Vibration and Make Holograms Viewable in White Light." *Scientific American,* May 1989, p. 134.

Weissberger, A., "A Chemist's View of Color Photography." *American Scientist,* November 1970, p. 648.

Chapter 31
Condensed Matter

IMPORTANT TERMS

Fill in the blanks with the appropriate word or words.

1. A(n) _____ crystal has a regular periodic arrangement in only one or two dimensions.

2. The energies of electrons in a crystal are only permitted to be in regions called _____.

3. The constant of proportionality between the applied electric field and the resulting current density is called the electrical _____.

4. A pure material with a conductivity value between that of a good metal and a good insulator is called a(n) _____ semiconductor.

5. A solid forming a three-dimensional array or structure is said to be _____.

6. The energy of the most energetic electron in a substance at 0 K is called the _____ energy.

7. Semiconductors whose properties depend upon the addition of a small number of other atoms are called _____ semiconductors.

8. A solid made up of many individual crystalline subunits is called a(n) _____ solid.

9. A(n) _____ semiconductor is made by adding impurities which donate negative carriers to the conduction band.

10. The _____ are those values of energies which electrons in a solid cannot have.

11. A solid without any long range order is called a(n) _____ solid.

12. The _____ is a measure of the drift velocity given electrons in a solid per unit applied electric field.

13. An insulator with a small energy gap between the last filled band and the empty conduction band becomes a(n) _____.

14. If a magnetic field and an electric field at right angles to the electric field are both applied to a conductor, the _____ voltage is generated at right angles to both the magnetic field and the electric field.

15. A(n) _____ semiconductor is made by adding impurities which trap electrons out of the conduction band.

Write the definitions of the following:

16. amorphous solid (31.1):

17. electrical conductivity (31.2):

18. forbidden zones (31.4):

19. Hall effect (31.6):

20. p-type semiconductor (31.7):

21. crystalline solid (31.1):

22. liquid crystal (31.1):

23. mobility (31.3):

24. semiconductor (31.4):

25. impurity (extrinsic) semiconductors (31.7):

26. polycrystalline solid (31.1):

27. Fermi energy (31.2):

28. allowed bands (31.4):

29. pure (intrinsic) semiconductors (31.5):

30. n-type semiconductor (31.7):

ANSWERS TO 1–15, IMPORTANT TERMS

1. liquid; 2. allowed bands; 3. conductivity; 4. pure (intrinsic); 5. crystalline; 6. Fermi;
7. impurity (extrinsic); 8. polycrystalline; 9. n-type; 10. forbidden zones; 11. amorphous;
12. mobility; 13. semiconductor; 14 Hall; 15. p-type

SAMPLE SOLUTIONS

PROBLEM 31.1: Palladium (Pd) has an atomic mass of 106.7 u and a mass density of 1.22×10^4 kg/m^3. Assume that solid palladium has 2 free electrons per atom. Calculate the Fermi energy of palladium. (31.2)

Solution:

> **Strategy:** Use the provided numbers to calculate the number of atoms per unit volume and then the free electron density.

We will let ρ represent the mass density of palladium, m_{Pd}, and n_e the number of free electrons per atom. The the free electron density, n, is given by

$$n = n_e \times \frac{\rho \times N_A}{m_{Pd}},$$

where N_A is Avogadro's number. Entering numbers, and remembering to convert atomic mass units to kg/mol, we find that

$$n = 2 \times \frac{1.22 \times 10^4 \text{ kg/m}^3 \times 6.02 \times 10^{23} \text{ atoms/mol}}{1.067 \times 10^{-1} \text{ kg/mol}} = 1.38 \times 10^{29} \text{ /m}^3.$$

Using this value of n, we find the Fermi energy is

$$E_F = \frac{h^2}{2m}\left(\frac{3n}{8\pi}\right)^{2/3} = 1.56 \times 10^{-18}\,\text{J} = 9.76\,\text{eV}$$

The Fermi energy of palladium is 9.76 eV.

PROBLEM 31.2: The electrical resistivity of Palladium is 10.2×10^{-8} $\Omega\cdot\text{m}$. What is the electronic mobility and the collision time for the conduction electrons? (31.3)

Solution:

> **Strategy:** Use the free electron density from Problem 31.1 and the electrical conductivity to calculate the collision time for the carriers. From this find the mobility.

The resistivity is defined as the reciprocal of the conductivity. The conductivity, σ, is given in terms of the free electron density, n, the electronic charge, e, the electron mobility, μ, and the electron mass, m, as

$$\sigma = ne\mu,$$

which can be solved for

$$\mu = \frac{\sigma}{ne}$$

$$= \frac{\dfrac{1}{1.02 \times 10^{-7}\,\Omega\cdot\text{m}}}{1.38 \times 10^{29}\,/\text{m}^3 \times 1.60 \times 10^{-19}\,\text{C}}$$

$$= 4.44 \times 10^{-4}\,\frac{\text{m}^2}{\text{V}\cdot\text{s}}.$$

The collision time is given by

$$\tau = \frac{\mu m}{e}$$

$$= \frac{4.44 \times 10^{-4}\,\dfrac{\text{m}^2}{\text{V}\cdot\text{s}} \times 9.1 \times 10^{-31}\,\text{kg}}{1.60 \times 10^{-19}\,\text{C}}$$

$$= 2.53 \times 10^{-15}\,\text{s}$$

The electron mobility in palladium is $4.44 \times 10^{-4}\,\dfrac{\text{m}^2}{\text{V}\cdot\text{s}}$ and an electron will suffer a collision about every 2.53×10^{-15} seconds.

PROBLEM 31.3: A 100 foot extension cord is made of copper wire. It is carrying sufficient current that a 2.0 V drop occurs along each direction of the cord. Given the electrical resistivity of copper as 1.7×10^{-8} $\Omega\cdot\text{m}$ and that copper has 8.45×10^{28} free electrons per cubic meter, how long will it take on the average for an electron to travel one way on the wire? Assume that this is a dc current and a dc potential difference. (31.3)

Solution:

> **Strategy:** The resistivity allows us to calculate the collision time. The time of travel for the electron is determined by the electric field and the average velocity.

The average drift velocity v is given in terms of the electric field E and the collision time τ as

$$v = \frac{eE}{m}\tau.$$

Since the electric field is

$$E = \frac{PD}{L},$$

where PD is the potential difference and L is the length of the wire, and since $\tau = \frac{\sigma m}{ne^2}$,

where σ is the conductivity, m is the electron mass, e is the electron charge, and n is the free electron density, we can write

$$v = \frac{\sigma\, PD}{neL}.$$

The time of travel t will be given by

$$t = \frac{L}{v} = \frac{neL^2}{\sigma\, PD}$$

$$= \frac{8.45 \times 10^{28} \times 1.6 \times 10^{-19}\ \text{C} \times (30.5\ \text{m})^2}{\dfrac{1}{1.7 \times 10^{-8}\ \Omega\cdot\text{m}} \times 2.0\ \text{V}} = 1.1 \times 10^5\ \text{s}.$$

This result may seem surprising. The current shows up almost immediately in the conductor; however, it will take an individual electron about 30 hours to travel the length of the extension cord. Clearly we have a cooperative effect where the electrons we use are not the ones we put into the cord from the power supply.

The preceding calculation ignores the heating effect on the conductivity, which would result in a smaller wire having an even smaller drift velocity. With an ac power supply the drift velocity is not changed by much in magnitude (provided $\tau \ll 1/\text{frequency}$), and the reversal of the voltage means that the electrons from the power lines never get to whatever we have attached to the cord. You are buying power, not electrons, from the power company.

PROBLEM 31.4: A ribbon of p-type semiconductor has a z dimension of 1.5 mm and a y dimension of 4.0 mm. The ribbon is suspended in a magnetic field of strength 1.5 T in the z direction. The material has a free carrier density of 3.2×10^{22} carriers/m^3. What is the Hall voltage if 200 mA is passed in the x direction through the sample? (31.6)

Solution:

Strategy: Use the relation defining the Hall field given in the text.

Let l_z represent the z dimension of the ribbon, l_y the y dimension, B_z the magnetic field, i the current in the x direction and n the density of carriers as shown in Fig. 31.1.

Figure 31.1: Semiconductor ribbon in an external magnetic field.

The Hall field, E_y is given by

$$E_y = \frac{j_x B_z}{en}$$

$$= \frac{i B_z}{l_z l_y en},$$

where j_x is the current density and e is the charge on the carriers. Since this is a p-type semiconductor, $e > 0$, so E_y points in the $+y$ direction.

The Hall voltage is then

$$V_y = \frac{i B_z}{l_z l_y en} \times l_y = \frac{i B_z}{l_z en}$$

$$= \frac{2.0 \times 10^{-1} \text{ A} \times 1.5 \text{ T}}{1.5 \times 10^{-3} \text{m} \times 1.6 \times 10^{-19} \text{ C} \times 3.2 \times 10^{22} \text{ m}^{-3}}$$

$$= 0.039 \text{ V}.$$

The Hall voltage will be 39 mV in the $+y$ direction.

PROBLEM 31.5: A diode has a saturation current of 0.20 μA. What current would it carry at a forward voltage of 0.30 V if it exactly obeyed the rectifier equation? (31.8)

Solution:

Strategy: The forward current depends upon three quantities, two of which are given in the statement of the problem. We will assume the diode is at room temperature.

The forward current, I, for a diode is given by the rectifier equation, text Eq. 31.11:

$$I = I_0\left(e^{eV/kT} - 1\right),$$

where I_0 is the saturation current and V is the applied voltage.

We assume a room temperature $T = 27°C = 300$ K, $V = 0.30$ V and $I_0 = 0.20$ μA:

$$I = 0.20 \, \mu A \times \left(e^{\left(\frac{0.30 \text{ eV}}{8.6 \times 10^{-5} \text{ eV/K} \times 300 \text{ K}} \right)} - 1 \right)$$

$$= 0.20 \, \mu A \times (e^{11.6} - 1) = 0.20 \, \mu A \times 1.1 \times 10^5 = 0.022 \text{ A}.$$

The forward current will be 22 mA.

PROBLEM 31.6: What size area of solar cells are needed to produce 50 kW if the incident solar flux is 1.0 kW·m^{-2} and the solar cells are 20% efficient? (31.10)

Solution:

Strategy: The power produced is proportional to the area and the incident energy and the efficiency of the cell.

If we let P be the desired power, A the area of the solar cell array, E the efficiency, and I_0 the incident power per unit area, then

$$P = AEI_0.$$

The required area is

$$A = \frac{P}{EI_0}$$

$$= \frac{50 \text{ kW}}{0.20 \times 1.0 \text{ kW·m}^{-2}}$$

$$= 250 \text{ m}^2.$$

You would need an array about 16 m on a side.

SUGGESTED READINGS

Bardeen, J., "Superconductivity and Other Macroscopic Quantum Phenomena." *Physics Today,* December 1990, p. 25.

Issue on Solid State and Properties of Materials. *Scientific American,* September 1967.

Shimada, I, and J. F. Merkel, "Copper-Alloy Metallurgy in Ancient Peru." *Scientific American,* July 1991, p. 80.

Chapter 32
Elementary Particle Physics

IMPORTANT TERMS

Fill in the blanks with the appropriate word or words.

1. Particles just like the electron except for their charge are called _____.

2. The _____ has a mass of about 170 times that of the electron and has the same spin.

3. Particles which interact through the strong nuclear force are called _____.

4. _____ theories attempt to explain the basic forces of nature in terms of a single force.

5. The fundamental building blocks of atoms are called _____ particles.

6. The _____ is a particle with no spin and a mass larger than the muon. It was first predicted by Yukawa.

7. The _____ are intermediate mass particles which may be produced singly in nuclear reactions.

8. Protons and neutrons are believed to be composed of three _____.

9. _____ are about a thousand times as massive as electrons. When they decay, they produce mesons and other light particles but never baryons.

10. The elementary particles consist of hadrons, which experience the strong nuclear force, and photons and _____, which do not. The electron is an example of the latter.

11. The _____ theory has the origin of the universe 10 to 20×10^9 years ago.

12. _____ are about a thousand times as massive as an electron and always include a baryon in their decay products.

13. _____ are the heavy particles which feel the strong nuclear force. They can be produced in pairs only.

14. A particle which has the same mass as another particle but a different charge is termed a(n) _____. Such pairs will annihilate when they meet.

Write the definitions of the following:

15. muon (32.2):

16. kaon (32.3):

17. hadron (32.4):

18. lepton (32.4):

19. unified theories (32.6):

20. big bang (32.7):

21. positron (32.1):

22. antiparticle (32.1):

23. pion (32.2):

24. hyperon (32.3):

25. meson (32.4):

26. baryon (32.4):

27. quark (32.5):

28. elementary (or fundamental) particle (32.0):

ANSWERS TO 1–14, IMPORTANT TERMS

1. positrons; 2. muon; 3. hadrons; 4. unified; 5. fundamental (or elementary); 6. pion; 7. mesons; 8. quarks; 9. kaons; 10. leptons; 11. big bang; 12. hyperons; 13. baryons; 14. antiparticle.

COMMENTS ON MODELS

The quark model described in the text applies to the hadrons. It provides us a picture of how the hadrons are interrelated. The quark model appears to bring us back to the point particles or classical physics. This is not a completely correct view. The interaction of these particles is basically nonclassical.

SAMPLE SOLUTIONS

Handy conversion factors:
1 MeV corresponds to a mass of 1.7827×10^{-29} kg

$$1 \text{ MeV} = 1.60 \times 10^{-13} \text{ J}$$

PROBLEM 32.1: A proton and an antiproton collide directly. Each is moving towards the other at a speed of 0.010 times that of light. Assuming that these velocities are nonrelativistic, what is the total energy released in MeV when these two particles meet and annihilate? In what form (number of photons) would you expect the radiation to appear? (32.1)

Solution:

Strategy: Calculate the total energy (rest + kinetic) before the collision. The initial energy must all be converted to light. The produced photons must satisfy the symmetry of the problem.

proton   **antiproton**

(a)   **(b)**

Fig. 32.1: (a) Colliding proton and antiproton. (b) The reaction products.

Since we are told to assume nonrelativistic motion for the particles we should be able to write:

$$E_p = m_0 c^2 + KE_p$$

and

$$E_{\bar{p}} = m_0 c^2 + KE_{\bar{p}},$$

where p denotes the proton and $\bar{p}$ the antiproton. We assume these two have the same rest mass energy and that we can use the value of 938.279 Mev given in text Table 28.2 for the rest mass term.

Since the particles have the same rest mass, conservation of momentum implies that the two kinetic energies are equal; so we can write

$$KE_{total} = m_0 v^2 = m_0 c^2 \frac{v^2}{c^2} = 938.279 \text{ MeV} \times \left(\frac{0.010 \text{ c}}{c}\right)^2$$

$$= 938.279 \text{ MeV} \times 10^{-4} = 0.094 \text{ MeV}.$$

The total energy is then

$$E_{total} = 2 m_0 c^2 + KE_{total} = 1.876652 \times 10^3 \text{ MeV}.$$

The reaction products must be one or more photons. One is not possible since the initial system of particles had no net momentum before collision. The minimal possible solution must consist of two photons of equal energy (and hence frequency) traveling in opposite directions. Although we have not covered all the material needed for the complete solution, it should be pointed out that the two photon result is only possible provided the spins of the two particles are opposed. That is, to end up with two photons, it is necessary to assume that the two particles form a singlet s state in analogy to the s state of atomic spectra.

PROBLEM 32.2: A π^0 (neutral pion) at rest in the laboratory decays into two photons. What are the frequencies of the photons? Assume the speed of light is exactly 3.000×10^8 m/s. (32.2)

Solution:

Strategy: Assume energy conservation and symmetry. The two photons have the same frequency and their total energy must be the rest energy of the pion.

The reaction can be written as

$$\pi^0 \rightarrow \gamma_1 + \gamma_2.$$

Since there is no momentum before the decay, we expect the photons will have equal energy and be directed oppositely to each other as shown in Fig. 32.2.

π^0

f_1 f_2

(a)   **(b)**

Fig. 32.2:   (a) Neutral pion at rest. (b) Final decay photons.

The energy of a photon is given by hf where h is Planck's constant and f is the frequency of the light. Thus, conserving energy and letting m_0 be the pion rest mass:

$$m_0 c^2 = 2\,hf \quad \text{or} \quad f = \frac{m_0 c^2}{2h}.$$

We can get the mass of the pion from text Table 28.1; however, it is in MeV. 1 MeV corresponds to 1.7827×10^{-29} kg, so the mass of the pion is

$$m_0 = 135.0 \times 1.7827 \times 10^{-29} \text{ kg} = 2.407 \times 10^{-27} \text{ kg}.$$

Then the frequency is

$$f = \frac{2.407 \times 10^{-27} \text{kg} \times (3.000 \times 10^8 \text{ m/s})^2}{2 \times 6.63 \times 10^{-34} \text{J} \cdot \text{s}} = 1.64 \times 10^{23} \text{ Hz}.$$

This frequency corresponds to very hard gamma rays.

PROBLEM 32.3: What is the de Broglie wavelength of a 20 GeV electron? (32.3)

Solution:

Strategy: Use the relation defining the de Broglie wavelength.

The de Broglie wavelength was given in text Chapter 28, Eq. (28.2) as

$$\lambda = \frac{h}{p}$$

where h is Planck's constant and p is the momentum of the particle.

We are given the kinetic energy of the electron, 20 GeV = 20×10^3 MeV. Since this is much larger than the energy associated with the rest mass (rest mass energy = 0.511 MeV) we can neglect the rest mass. The relativistic relation between mass and energy (text Eq. 24.12) then becomes

$$p \approx \frac{E}{c}$$

so

$$\lambda = \frac{hc}{E} .$$

We must put the energy in joules (1 MeV $= 1.60 \times 10^{-13}$ J), so

$$\lambda = \frac{6.63 \times 10^{-34} \text{J·s} \times 3.00 \times 10^8 \text{m/s}}{2.0 \times 10^4 \times 1.60 \times 10^{-13} \text{J}} = 6.22 \times 10^{-17} \text{ m} .$$

The wavelength is a very short 6.22×10^{-17} m .

PROBLEM 32.4: What is the charge of a particle whose quark composition is d$\bar{\text{s}}$? Is this a baryon or a meson? (32.6)

Solution:

> **Strategy:** Add the individual quark charges.

The d quark has a charge of $-1/3$, while the $\bar{\text{s}}$ has a charge of $+1/3$ so this particle has a charge of zero. Since it is composed of only two quarks, the particle must be a meson.

PROBLEM 32.5: Can the reaction p + p $\rightarrow$ p + n + π^+ occur? Explain why or why not. (32.6)

Solution:

> **Strategy:** Reactions must satisfy charge conservation, energy conservation, spin conservation, and baryon number conservation.

To determine if a decay can occur we must check to see if it satisfies conservation laws. In addition to the three conservation laws from previous chapters (energy, charge, spin), this chapter adds a fourth conservation law (baryon number).

Fig. 32.3: The reaction p + p $\rightarrow$ p + n + π^+.

Charge conservation: The right hand side has a total charge of +2e, as does the left-hand side. Charge conservation is satisfied.

Energy conservation: Energy conservation is not relevant in this case since we have two incoming particles which we are allowed to give any relative velocity. This allows the energy on the right-hand side to be as large as needed to satisfy energy conservation. Energy conservation is very useful if there is only a single particle at the start.

Spin conservation: The proton and the neutron are spin 1/2 particles while the pion is a spin 0 particle. The spins on the right-hand side can add to 1 or 0 as can the spins on the left-hand side, so spin conservation can be satisfied.

Baryon number conservation: The proton and neutron each have a baryon number of +1 and the π^+ has a baryon number of 0. The right-hand side has a baryon number of +2 as does the left-hand side. Baryon number conservation is satisfied.

The reaction can occur.

SELECTED READINGS

Barrow, J. D., and J. Silk, *The Left Hand of Creation: The Origin and Evolution of the Expanding Universe*, New York: Basic Books, 1983.

Drees, W. B., *Beyond the Big Bang: Quantum Cosmologies and God.* La Salle, Ill.: Open Court, 1990. The book assumes no background in cosmology or theology. It addresses the constructive nature of human knowledge.

Glashow, S. L., *The Charm of Physics*. New York: American Institute of Physics, 1991. Essays on physics, physicists, and how science affects our lives.

Hegstrom, R. A., and D. K. Kondepudi, "The Handedness of the Universe." *Scientific American,* January 1990, p. 108.

Hovis, R. C., and H. Kragh, "Resource Letter HEPP-1: History of Elementary Particle Physics." *American Journal of Physics,* September 1991, p. 779.

Langacker, P., and A. K. Mann, "The Unification of Electromagnetism with the Weak Force." *Physics Today,* December 1989, p. 22.

Moe, M. K., and S. P. Rosen, "Double-Beta Decay." *Scientific American,* November 1989, p. 48.

Renteln, P., "Quantum Gravity." *American Scientist*, November 1991, p. 508.

Silk, J., *The Big Bang: The Creation and Evolution of the Universe.* San Francisco: W. H. Freeman and Company, 1988

Weinberg, S., *The First Three Minutes: A Modern View of the Origin of the Universe.* New York: Basic Books, 1977.

Zee, A., *Fearful Symmetry: The Search for Beauty in Modern Physics*. New York: MacMillan, 1986. The story of how the search for basic symmetries has brought a clearer picture of nature.